AF598132

METHODS IN MOLECULAR BIOLOGY

Series Editor
John M. Walker
School of Life Sciences
University of Hertfordshire
Hatfield, Hertfordshire, AL10 9AB, UK

For further volumes:
http://www.springer.com/series/7651

Biomedical Literature Mining

Edited by

Vinod D. Kumar

Computational Biology, GlaxoSmithKline R&D, King of Prussia, PA, USA

Hannah Jane Tipney

GlaxoSmithKline, Hitchin, Hertfordshire, United Kingdom

Editors
Vinod D. Kumar
Computational Biology
GlaxoSmithKline R&D
King of Prussia, PA, USA

Hannah Jane Tipney
GlaxoSmithKline
Hitchin, Hertfordshire, United Kingdom

Additional material to this book can be downloaded from http://extras.springer.com

ISSN 1064-3745 ISSN 1940-6029 (electronic)
ISBN 978-1-4939-0708-3 ISBN 978-1-4939-0709-0 (eBook)
DOI 10.1007/978-1-4939-0709-0
Springer New York Heidelberg Dordrecht London

Library of Congress Control Number: 2014937662

Printed on acid-free paper

Humana Press is a brand of Springer
Springer is part of Springer Science+Business Media (www.springer.com)

Preface

The genomic era of biomedicine has been defined by unprecedented growth of data sampling capacity and increasing publication rates discussing it. While such technical advances have heralded a period of intensive scientific discovery, the associated deluge of biomedical literature has reached a volume exceeding the capacity of any researcher to process and assume, critically limiting the ability to realize the full benefit of these findings.

The need to rapidly survey the published literature, synthesize, and discover the embedded knowledge without compromising the integrity of published data is critical if researchers are to conduct "informed" work, avoid repetition, and generate new hypotheses. It is therefore unsurprising that within the scientific community a great deal of interest and effort is focused on the development of techniques that can identify, extract, and exploit this knowledge in a meaningful manner. To do so in an efficient way requires methods that can reduce complexity without compromising the integrity of published data. Consequently, over the last two decades one has seen a surge of publications related to biomedical text mining with the primary intent of aiding scientific researchers cope with the information overload.

This volume of Methods in Molecular Biology discusses the multiple facets of modern biomedical literature mining and its many applications in genomics and systems biology. The volume has been designed as a useful bioinformatics resource in biomedical literature text mining for both those long experienced in and entirely new to the field. As such, this book serves two purposes: (a) to provide a timely and comprehensive overview of the current status of this field, including a survey of present challenges; (b) to empower researchers to decide how and when to integrate text-mining tools to facilitate their own research. It comprises 15 chapters including an introductory chapter giving the fundamental definitions and some important research challenges. The 15 chapters are organized in three sections encompassing information retrieval, integrated text-mining approaches, and domain-specific mining methods.

Saffer and Burnett introduce the volume by providing a current perspective on the role of text mining in biomedical research and health care. While addressing the importance of text mining in drug discovery the authors also outline the continuing challenges relating to improved search methodologies, discovering hidden information, and improved rate of discovery.

The first section of the book reviews information retrieval methods:

- *Khare et al.* describe the current state of practice of biomedical literature access and state-of-the-art information retrieval systems in areas related to text and data mining, text similarity search, and semantic search. The authors discuss emerging trends in improving biomedical literature access using portable devices and the adoption of open access policy systems.
- One of the first steps towards making full use of the information encoded in biomedical text is the task of recognizing biological terms, such as gene and protein names. *Bada* provides a detailed survey of the various lexical terminological resources currently available and how best to utilize them to improve entity recognition tasks in biomedical text-mining applications.

- Generating useful Pharmacokinetics (PK), Pharmacodynamics (PD) models to understand Drug-Drug Interaction (DDI) is a critical step during drug development process. However, an appropriate PK ontology and a well-annotated PK corpus which provide the background knowledge for determining DDI have been lacking. To overcome this information gap, Wu et al. developed comprehensive pharmacokinetics ontology capable of encompassing in vitro and in vivo pharmacokinetics studies.
- Once biological entities have been identified within the text fragments, the next step consists of identifying the potential relationships among them. *Pavlopoulos et al.* describe how relationships between bioentities are detected by co-occurrence analysis of single sentences and/or entire abstracts.

The second section outlines how, through the integration of text-mining efforts, hidden or implicit functional information leading to new biological hypotheses generation can be discovered. Key examples are described.

- *Verspoor* describes how the application of novel biomedical text-mining strategies is being utilized for novel protein function prediction, a problem at the forefront of modern biology.
- The advent of high-throughput "omics" approaches to generate data has outstripped the ability to interpret and assign biological relevance. *Heinzel et al.* outline a method for interlinking omic data and biomedical literature towards identifying markers as representatives for a specific disease-relevant pathophysiological (mechanistic) process.
- *Czarnecki and Shepherd* present a practical guideline for constructing a text-mining pipeline from existing code and software components capable of extracting protein–protein interaction networks from full text articles. Their approach demonstrates how literature mining can be used to identify functionally coherent gene groups to facilitate the reconstruction of protein interaction networks in the formulation of novel biological processes.
- The combination of scientific knowledge and experience is the key success for biomedical research. *Jonnalagadda et al.* outline some of the strategies used to identify key scientific opinion leaders in order to support increased collaborative biomedical research.
- *Petric et al.* demonstrate the use of creative literature-mining methods to advance valuable new discoveries from existing literature and provide application examples from their research findings.

The third section of the book focuses on the utility of specialized text-mining applications that are suited to address particular domains or purposes related to drug discovery. The use of literature-mining approaches to extract novel but not yet recognized associations between concepts such as genes, diseases, drugs, and cellular processes can aid the discovery of novel drug targets and increase insight into the mode of action of a drug or find novel applications for known drugs.

- The ability to identify accelerating areas of science for a given disease area highlights scientific advancements in aspects of biology, and offers opportunities for both near and long-term strategy development for innovative medicines with early translational possibilities. *Rajpal et al.* present a literature-mining methodology that evaluates trends, and points to gene-disease associations that can be employed in making various important scientific and strategic decisions during drug development.

- Discovering novel disease genes is a key step in the drug discovery pipeline and requires not only the identification, prioritization, and selection of reliable druggable targets. *Wu et al.* review recent advances in literature- and data-mining approaches for gene prioritization, and describe a computational approach to identify and rank candidate genes by finding associations between known disease genes and disease relevant pathways.
- Drug toxicity remains a major reason why new drug candidates which enter clinical trials fail to ever reach the market. However, there are vast amounts of information in the public domain concerned with pharmacological interactions, biomedical literature, consumer posts in social media, and narrative electronic medical records (EMRs); all of which can be relevant and informative for predicting the safety of novel drugs. *Liu et al.* describe the use of text-mining techniques from these diverse document resources to uncover hidden knowledge and help predict their toxicity profiles.
- Systematically seeking novel associations between existing drugs and new indications has recently emerged as an alternative to the limited productivity issues associated with traditional drug discovery. *Tari and Patel* describe various strategies including application examples that use biomedical literature as a source for systematic drug repositioning.
- In the concluding chapter of this section, *Chen and Sarkar* present a knowledge discovery framework for mining the electronic health records (EHR) to gather phenotypic descriptions of patients from medical records in a systematic manner to identify comorbidities occurring in patients more often than expected. Currently available resources and their caveats are also discussed.

We are very grateful to the authors for contributing to this volume. The editors would like to thank Professor John Walker (Series Editor) for suggesting this project and guiding us through till the end so that we can produce this important scientific work. We hope that the reader will share our excitement to present this volume on “Biomedical Literature Mining” and will find it useful.

King of Prussia, PA — *Vinod D. Kumar*
Hitchin, UK — *Hannah Jane Tipney*

Contents

Part III From Electronic Biology to Drug Discovery and Development

Contributors

MICHAEL BADA • *Computational Bioscience program, School of Medicine, University of Colorado, Denver, CO, USA*
VICKI L. BURNETT • *Quertle LLC, Henderson, NV, USA*
BOJAN CESTNIK • *Temida, d.o.o., Ljubljana, Slovenia*
ELIZABETH S. CHEN • *Center for Clinical and Translational Science, University of Vermont, Burlington, VT, USA*
CHIEN-WEI CHIANG • *Center for Computational Biology and Bioinformatics, School of Informatics, Indiana University, Indianapolis, IN, USA*
JAN CZARNECKI • *Department of Biological Sciences, Institute of Structural and Molecular Biology, Birkbeck, University of London, London, UK*
RAUL FECHETE • *emergentec biodevelopment GmbH, Vienna, Austria*
JOHANNES M. FREUDENBERG • *Computational Biology, GlaxoSmithKline R&D, Research Triangle Park, NC, USA*
ANDREAS HEINZEL • *emergentec biodevelopment GmbH, Vienna, Austria*
YONG HU • *Institute of Business Intelligence, Guangdong University of Foreign Studies, Sun Yat-sen University, Guangzhou, People's Republic of China*
IOANNIS ILIOPOULOS • *Division of Basic Sciences, University of Crete Medical School, Heraklion, Greece*
ANIL G. JEGGA • *Division of Biomedical Informatics, Cincinnati Children's Hospital Medical Center, Cincinnati, OH, USA*
SIDDHARTHA R. JONNALAGADDA • *Department of Preventive Medicine, Health and Biomedical Informatics, Chicago, IL, USA*
RITU KHARE • *National Center for Biotechnology Information, U.S. National Library of Medicine, NIH, Bethesda, MD, USA*
VINOD D. KUMAR • *Computational Biology, GlaxoSmithKline R&D, King of Prussia, PA, USA*
ROBERT LEAMAN • *National Center for Biotechnology Information, U.S. National Library of Medicine, NIH, Bethesda, MD, USA*
LANG LI • *Center for Computational Biology and Bioinformatics, School of Informatics, Indiana University, Indianapolis, IN, USA*
MEI LIU • *Department of Computer Science, New Jersey Institute of Technology, Newark, NJ, USA*
ZHIYONG LU • *National Center for Biotechnology Information, U.S. National Library of Medicine, NIH, Bethesda, MD, USA*
BERND MAYER • *emergentec biodevelopment GmbH, Vienna, Austria*
IRMGARD MÜHLBERGER • *emergentec biodevelopment GmbH, Vienna, Austria*
CHRISTOS A. OUZOUNIS • *Donnelly Centre for Cellular & Biomolecular Research, University of Toronto, Toronto, ON, Canada*

JAGRUTI H. PATEL • *Scientific Knowledge Discovery, Merck Research Laboratories, Boston, MA, USA*
GEORGIOS A. PAVLOPOULOS • *Division of Basic Sciences, University of Crete Medical School, Heraklion, Greece*
RYAN G. PEELER • *Lnx research LLC, Orange, CA, USA*
PAUL PERCO • *emergentec biodevelopment GmbH, Vienna, Austria*
INGRID PETRIČ • *University of Nova Gorica, Nova Gorica, Slovenia*
VASILIS J. PROMPONAS • *Bioinformatics Research Laboratory, Department of Biological Sciences, University of Cyprus, Nicosia, Cyprus*
XIAOYAN A. QU • *Computational Biology, GlaxoSmithKline R&D, Research Triangle Park, NC, USA*
DEEPAK K. RAJPAL • *Computational Biology, GlaxoSmithKline R&D, Research Triangle Park, NC, USA*
JEFFREY D. SAFFER • *Quertle LLC, Henderson, NV, USA*
INDRA NEIL SARKAR • *Center for Clinical and Translational Science, University of Vermont, Burlington, VT, USA*
ADRIAN J. SHEPHERD • *Department of Biological Sciences, Institute of Structural and Molecular Biology, Birkbeck, University of London, London, UK*
EDWARD J. SILVERMAN • *Lnx research LLC, Orange, CA, USA*
BUZHOU TANG • *Department of Computer Science and Technology, Harbin Institute of Technology Shenzhen Graduate School, Shenzhen, People's Republic of China*
LUIS B. TARI • *Knowledge Discovery Lab, Software Science and Analytics, GE Global Research, Niskayuna, NY, USA*
PHILIP S. TOPHAM • *Lnx research LLC, Orange, CA, USA*
KARIN M. VERSPOOR • *Victoria Research Lab, National ICT Australia, Melbourne, VIC, Australia; Department of Computing and Information Systems, The University of Melbourne, Melbourne, VIC, Australia*
CHAO WU • *Division of Biomedical Informatics, Cincinnati Children's Hospital Medical Center, Cincinnati, OH, USA*
HENG-YI WU • *Center for Computational Biology and Bioinformatics, School of Informatics, Indiana University, Indianapolis, IN, USA*
CHENG ZHU • *Division of Biomedical Informatics, Cincinnati Children's Hospital Medical Center, Cincinnati, OH, USA*

Chapter 1

Introduction to Biomedical Literature Text Mining: Context and Objectives

Jeffrey D. Saffer and Vicki L. Burnett

Abstract

Information: If you are reading this, you know how important it is and almost certainly look to the biomedical literature for a large part of the information you need. We work hard to find more and more biomedical literature, seeking new content from multiple sources. But, can there be too much of a good thing?

Most science is reductionist by nature. It is difficult enough finding the relevant nuggets of information from 1,000 documents. It is at least ten times harder to do so from 10,000 documents. And, with 25 million biomedical journal articles and many times that of other textual information sources, we are faced with significant challenges.

In this introduction, we identify some of those challenges to prepare you for the remaining chapters.

Key words Biomedical, Text mining, Text analytics search and retrieval

1 Defining Text Mining

There are a variety of methods that have been described as "text analytics," "text exploration," and "text mining." These terms are often used interchangeably and describe a full spectrum of approaches, including analytical methods such as bibliometrics (e.g., who are the most active authors or which terms are used and at what frequency), exploratory approaches such as clustering documents together based on related content and co-occurrence analysis, and techniques for discovery of facts that are hidden in or even just implied by text documents. Common to all approaches is an attempt to understand and utilize the information within.

Vinod D. Kumar and Hannah Jane Tipney (eds.), *Biomedical Literature Mining*, Methods in Molecular Biology, vol. 1159, DOI 10.1007/978-1-4939-0709-0_1, © Springer Science+Business Media New York 2014

2 The Objectives of Text Mining

There are many goals for text mining. In the biomedical field, the importance of text mining is reflected in the sixfold increase in journal articles on text mining over the last decade. Pharmaceutical companies consider text mining a "basic necessity" [1] using it to find proteins, drugs, biological processes, and complex interactions among them all. Some of the most important goals are to improve search, reveal prior results that inform research projects and treatment approaches, discover hidden information on biological pathways or adverse effects, and improve the rate of discovery. A good example of successful text mining is the work of LePendu et al. [2], who were able to demonstrate pharmacovigilance from mining electronic health records. Additional examples will be discussed in this book.

2.1 Search and Retrieval

Search and retrieval are fundamental for effective use of the biomedical literature and other text. There are two main approaches for searching: (a) find anything that might possibly be relevant (maximize recall) and, in some but unfortunately not all cases, then use a ranking method to help the user find what is important and (b) find only the documents that are truly relevant (maximize precision) and present the user with the best documents to start with. A balance between precision and recall is required.

Currently, most search engines use the first approach and are based on Boolean search methods that have existed since the beginning of time (at least computationally speaking). These search engines, which include the most well-known sites, find an article if it contains all of your search terms—without regard to how those terms may be used. Many bells and whistles have been added to these search engines, including the use of ontologies to expand the number of potentially relevant documents. And, tremendous effort has been put into ranking algorithms. But what is important to one researcher is not necessarily relevant to another researcher's goals. The end result is that the user can spend an enormous amount of time going through the search results and can miss critical documents.

The second approach is where text mining plays a central role. Computational linguistics can be applied to determine if the author of a document made an assertion, or a relationship, that ties all of the search terms together in a meaningful way—thus providing the user with immediate access to what is important for their search. For instance, if you were looking for "aging and diseases," you would want to find direct relationships among your terms, such as "*Aging*-related defects in mitochondrial energetics have been proposed to be causally involved in *sarcopenia*" [3] but not statements with simple co-occurrences such as "However, although there are numerous experiments that have been carried out with respect to the beneficial effects of fruits and vegetables in *cardiovascular disease* and *ischemia*, until recently their putative positive effects on CNS *aging* and behavior have not been examined" [4], where no connection between aging and disease is implied. An example of a search engine using this type of relationship-based searching is Quertle (www.quertle.info), developed by the authors to address the need for identifying the most relevant documents [5, 6]. Additional text mining methods can then be applied to the search results themselves, so that more effective exploration of those results is possible. For instance, Quertle automatically extracts the key concepts to enable rapid filtering.

2.2 Finding What Is Already Known

With over one million new biomedical journal articles published every year, just keeping up with the findings requires text mining. The days of "I know everybody in my field" are over; discoveries in one area can be relevant to diverse fields, but those discoveries are easily lost in the sheer volume of articles. As such, text mining, as a means for uncovering these relevant facts, becomes essential for revealing current results that should inform ongoing research projects and treatment approaches.

2.3 Finding What Was Not Obvious

With it being a challenge to keep up with the clearly stated facts in the literature, it is an even bigger problem to uncover the less obvious information implied by the volume of text. This "hidden" information, particularly when coupled with the analysis of experimental data, can be used to predict function and uncover novel biological pathways. For example, two documents talking about different proteins in a similar biological context might suggest a functional relationship between those proteins. Teasing out these connections, piecing together biological networks, or uncovering potential adverse effects in this way can have a huge impact on research decisions and health care.

Text mining can also be used to uncover a different type of "hidden" information: trends and emerging concepts. Identification of trends was traditionally based on expert analysis, but this has shifted to text mining methods due to the volume of information that needs to be assimilated. Emerging concepts are also critical to follow and are themselves the beginnings of a new trend. There are, of course, cases where an emerging concept would be obvious (such as a new paper on the discovery of antigravity). But, in general, it is only in a historical perspective that many emerging concepts are recognized—that is, until the application of text mining methods.

2.4 Research Networks

Although there is considerable focus on underlying biomedical information, the literature can also provide insights into the underpinnings of how knowledge progresses. There are two major types of efforts in this area. Citation networks show the dependence of one body of work on another and provide windows into the evolution of knowledge. Collaboration or co-authorship networks document the professional and, to some degree, the social connections among the authors.

Collectively, both types of networks identify "who is doing what and where" and how those people are themselves connected. Both networks also reveal the individual thought leaders as well as the organizations where the key work is carried out. This information can provide important insights, such as potential collaborations, and can help drive forward research and discoveries.

3 Text Mining Methods

The methods for both analytics and mining can take advantage of natural language processing (also known as computational linguistics) and statistical methods. These methods are often supported by semantic analysis and ontologies.

3.1 Natural Language Processing

Natural language processing (NLP) or computational linguistics dates back to the early days of computers with the notions identified in Turing's classic 1950 paper on Computing Machinery and Intelligence [7]. The goal of NLP is to use rules (either predefined or learned through supervised or unsupervised methods) to process text for specific purposes (translation, extraction of assertion, summarization, and much, much more).

3.2 Statistical Methods

There are many aspects of text mining that rely more on statistical methods. These include document clustering, document classification through probabilistic models, and measurement of document similarity. Statistical methods can be used alone or in conjunction with NLP methods.

3.3 Semantics

Both NLP and statistical methods are dependent on entity recognition and sense disambiguation. Hence, semantics becomes critical.

Semantics refers generally to the meaning of language. As you encounter "semantics" within this volume or in other contexts, though, its actual meaning is, well, a matter of semantics. Often, when "semantics" is used to describe text searching or mining, it refers to the use of an ontology to expand a term into its constituent members (such as "mammal" would include "human," "dog," and "walrus") or to assign a term to its class (such as "human" is a member of "mammal").

It is critical to remember, however, that the meaning of a term in text is also highly dependent on the context. Polysemy (multiple meanings for a word) can be a major problem when mining.

For example, "Can a bear bear to bear?" (or, more understandably: "Can an ursine mammal tolerate giving birth?"). Such issues are exacerbated in the biomedical field where gene names are mostly short strings of characters which often match regular text (e.g., "snail" could be protein or gastropod) and abbreviations (e.g., "AMD" could be the gene for adenosyl methionine decarboxylase or the disease age-related macular degeneration). The potential confusion is not even limited to terminology across entity types: "Mad" is an alias both for the human adenosine monophosphate deaminase 1 gene and for human MAX dimerization gene. In Drosophila, "Mad" is the gene "mothers against dpp" (mothers against decapentaplegic). Therefore, search and mining methods need to deal with such issues explicitly, or the value of the resulting data is somewhat tempered.

The contextual issue also applies to the meaning of any statement. For example, a statement such as "The risk of diseases, including diabetes and leukemia, was investigated." does not imply a connection between "diabetes" and "leukemia." Hence, the conclusions that can be derived from mining techniques such as co-occurrence need to be tempered, unless true relationship identification methods are used.

3.4 Ontologies

As noted above, ontologies are often used to expand search and mining methods so that all related members of a class are found. In attempting to uncover a relationship between certain proteins and "cancer," one very well might want the text mining method to consider not only the term "cancer" itself but also "carcinoma," "melanoma," "lentigo maligna," and so forth. However, the expansion of potential findings that result from the use of ontologies requires concomitantly greater precision (relevancy) to be effective; otherwise, all that is accomplished is an increasingly greater list of results without necessarily greater value to the text miner.

In considering the use of ontologies for text mining, another critical point to consider is that most ontologies have been created for the purpose of classification and organization. They are not necessarily well tuned for search/retrieval or data mining. An important lesson we have learned over years of text mining research is that, like all problems of matching the correct tool to the purpose at hand, to be effective, ontologies must be designed for the specific goal.

4 Data Sources

For any discussion of text mining, there is also the question of what to mine. As with ontologies, the specific tools (sources of text content) that are relevant vary with the questions being asked.

The biomedical field has been aided immensely by the US National Library of Medicine effort to make the bibliographic information, including the abstract, for most journal articles available publically. There are few, if any, biomedical text miners who are not dependent on PubMed and its related resources.

Despite that resource, text mining has been hampered by the lack of availability of full-text articles. Although the most salient conclusions from the work described in a research article will be described in the abstract, the details, methods, and background information remain important sources for text mining. Open access has gained some momentum with approximately 20 % of biomedical journal articles available in 2009 [8] with new open-access journals being added all the time. Nonetheless, it is clear that open access is a challenge for most publishers and not all articles will be available.

In addition to journal articles, the relevant information is truly "big data" in that it encompasses patent literature, books, medical records, social network surveillance, corporate internal documents, and much more. And, like other big data problems, as much as we feel the answers must be there, we are limited in our ability to ask the right questions in the right way, unless we use text mining to extract the relevant results.

5 Text Mining Challenges

The issues with big data and text mining noted above are high-level problems. There are, in addition, the challenges that must be solved at the level of applying individual methods. As well as the semantic issues previously discussed, biological variations can be a complicating factor. For example, what happens in one species does not necessarily happen in another species. Furthermore, what happens in one species may not happen in the same species under different conditions (gender, genetics, natural history, or treatment differences). As such, the conclusions that can be derived are often obfuscated by what appears to be conflicting answers to text mining algorithms. To the extent that details of the biological system being written about can be discerned, the text mining can be improved. For some applications, such as defining biological pathways, the "answers" are driven by the preponderance of evidence (same assertion made several times), but it may be the exceptions that contain the most valuable nuggets of information for some applications.

6 Article Access

Finally, we remind you of an issue you know well. Just because it is written does not mean that it is correct. Information gleaned by text mining cannot be acted upon until there is a critical reading of the underlying documents. As with the data mining itself, access to the articles for this step is again hampered by the difficulties in easily reaching the original full-text articles. This is only partially solved through library subscriptions; thus, document delivery solutions need to be part of any comprehensive solution.

7 Conclusion: What Is Needed

Text mining is now critical for all aspects of biomedical science and health care. Integrating these approaches is no longer an option and should include:

- A comprehensive collection of information (especially full-text articles).
- An intelligent system that reads your mind and finds the information for you (or, short of that, intelligent search engines).
- Integrated text mining tools to identify concepts, define pathways, extract relationships, perform bibliometrics, and more.
- A means to access the documents for critical assessment.

The following chapters address many of the issues mentioned and demonstrate successes in current projects.

References

1. Van Noorden R (2012) Trouble at the text mine. Nature 483:134–135
2. LePendu P, Iyer SV, Bauer-Mehren A, Harpaz R, Mortensen JM, Podchiyska T, Ferris TA, Shah NH (2013) Pharmacovigilance using clinical notes. Clin Pharmacol Ther 93(6):547–555
3. Gouspillou G, Bourdel-Marchasson I, Rouland R, Calmettes G, Biran M, Deschodt-Arsac V, Miraux S, Thiaudiere E, Padois P, Detaille D, Franconi JM, Babot M, Trézéguet V, Arsac L, Diolez P (2013) Mitochondrial energetics is impaired in vivo in aged skeletal muscle. Aging Cell 13:39–48. doi:10.1111/acel.12147
4. Cantuti-Castelvetri I, Shukitt-Hale B, Joseph JA (2000) Neurobehavioral aspects of antioxidants in aging. Int J Dev Neurosci 18: 367–381
5. Coppernoll-Blach P (2011) Quertle: the conceptual relationships alternative search engine for PubMed. J Med Libr Assoc 99:176–177
6. Giglia E (2011) Quertle and KNALIJ: searching PubMed has never been so easy and effective. Eur J Phys Rehabil Med 47:687–690
7. Turing AM (1950) Computing machinery and intelligence. Mind 59:433–460
8. Björk B-C, Welling P, Laakso M, Majlender P, Hedlund T, Guðnason G (2010) Open access to the scientific journal literature: situation 2009. PLoS One 5:e11273. doi:10.1371/journal.pone.0011273

Part I

Principles of Text Mining: Getting Started

Chapter 2

Accessing Biomedical Literature in the Current Information Landscape

Ritu Khare, Robert Leaman, and Zhiyong Lu

Abstract

Biomedical and life sciences literature is unique because of its exponentially increasing volume and interdisciplinary nature. Biomedical literature access is essential for several types of users including biomedical researchers, clinicians, database curators, and bibliometricians. In the past few decades, several online search tools and literature archives, generic as well as biomedicine specific, have been developed. We present this chapter in the light of three consecutive steps of literature access: searching for citations, retrieving full text, and viewing the article. The first section presents the current state of practice of biomedical literature access, including an analysis of the search tools most frequently used by the users, including PubMed, Google Scholar, Web of Science, Scopus, and Embase, and a study on biomedical literature archives such as PubMed Central. The next section describes current research and the state-of-the-art systems motivated by the challenges a user faces during query formulation and interpretation of search results. The research solutions are classified into five key areas related to text and data mining, text similarity search, semantic search, query support, relevance ranking, and clustering results. Finally, the last section describes some predicted future trends for improving biomedical literature access, such as searching and reading articles on portable devices, and adoption of the open access policy.

Key words Biomedical literature search, Text mining, Information retrieval, Bioinformatics, Open access, Relevance ranking, Semantic search, Text similarity search

1 Introduction

Literature search is the task of finding relevant information from the literature, e.g., finding the most influential articles on a topic, finding the answer to a specific question, or finding other (bibliographic or non-bibliographic) information on citations. Literature search is a fundamental step for every biomedical researcher in their scientific discovery process. Its roles range from reviewing past works at the beginning of a scientific study to the final step of result interpretation and discussion. Literature search is also important for clinicians seeking established and new findings for making important clinical decisions. Furthermore, since current biomedical research is heavily dependent on access to various kinds of online

Vinod D. Kumar and Hannah Jane Tipney (eds.), *Biomedical Literature Mining*, Methods in Molecular Biology, vol. 1159, DOI 10.1007/978-1-4939-0709-0_2,

biological databases, literature search is also a key component of transforming knowledge encoded in nature language data, such as journal publications, into structured database records by dedicated database curators. In addition, literature search has other uses such as biomedical citation analysis for academic needs and data collection for biomedical text mining research.

To meet the diverse needs of literature access by the scientific community worldwide, a number of Web-based search tools, e.g., PubMed [1] and Google Scholar [2], and online bibliographic archives, e.g., PubMed Central [3], have been developed over the last decades. As a result, the literature access process typically includes the following consecutive steps: searching for citations on a search tool, retrieving full text on a bibliographic archive, and reading the article. Despite advances in information technologies, the ease of searching the biomedical literature has not kept pace for two main reasons. First, the size of the biomedical literature is large (dozens of millions) and it continues to grow rapidly (over a million per year), thus making the selection of proper search keywords and reviewing results a daunting task [4, 5]. Second, biomedical research is becoming increasingly multidisciplinary. As a result, the information most relevant to an individual researcher may appear in journals that are not usually considered relevant to his or her own research. For example, a 2006 study [6] found that half of the renal information is published in non-renal journals.

In response to the aforementioned challenges, there has been a recent surge in improving the literature access through the use of advanced information technologies in information retrieval (IR), data mining, and natural language processing (NLP). For instance, recent IR research includes relevance-ranking algorithms aimed at improving retrieval effectiveness. Data mining algorithms can group similar results into clusters, thus providing users with a quick overview of the search results before focusing on individual papers. Text mining and NLP techniques can be used to automatically recognize named entities (e.g., genes) and their relations (e.g., protein–protein interaction) in the biomedical text, thus enabling novel entity-specific semantic searches as opposed to the traditional keyword-based searches.

A number of literature search assistants using aforementioned information techniques have been developed over the years, some of which have been shown to be effective in real-world uses. For instance, by comparing words from the title and abstract of each citation, and the indexed MeSH terms using a weighted IR algorithm, related papers can be grouped together into clusters [7]. When used in most search tools, such a technique is known as "related articles" where users can easily find all papers relevant to a search result through a simple mouse click. The "related articles" application has been frequently used [8] since its appearance in PubMed. Because of its success in PubMed, this feature has been adopted by many journal websites as well as commercial search tools.

Retrieving full text of bibliographic archives poses another challenge for literature access. While most article abstracts are freely accessible, their full texts are still locked by the publishers: in order to read the full text, one would need either an institutional subscription or pay-per-view. Such an access model is inconvenient to the researchers and the global scholarly community [9]. In recognition of such a problem, a number of initiatives began to promote open access to the scientific literature. For instance, the Budapest Open Access Initiative reaffirmed in its tenth anniversary in 2012 that its goal is to make open access the default method for distributing new peer-reviewed research in every field and country. Agreed with such initiatives, a number of publishers and journals are adopting the open access paradigm for publishing articles. For instance, two major open access publishers include the BioMed Central (BMC) and Public Library of Science (PLoS). To accelerate open access, the US National Library of Medicine started PubMed Central (PMC), a free digital repository of full-text articles in biomedical and life sciences in early 2000. With a little over 10 years' development, PMC currently contains approximately three million items and continues to grow at least 7 % per year [10] despite some criticisms from professional societies and commercial publishers [11].

This chapter describes all of the abovementioned issues in more depth. It first introduces some existing literature search tools and bibliographic archives, that are commonly used to access the biomedical literature, in three consecutive steps: searching, retrieving, and reading articles. Next, it presents a selection of five key categories of text mining and IR applications that address challenges in searching literature. Finally, there is a discussion on the future trends of biomedical literature access, with a focus on the open access activities in the biomedical domain and recent transition to reading articles on portable devices.

2 Current Access to Biomedical Literature

Open access availability of biomedical literature has led an increasing number of users to resort to online methods of literature access. Journals and online databases are currently the most frequently accessed resources among biomedical information seekers, followed by books, proceedings, newsletters, technical reports, author web pages, etc. [12, 13]. Given the rising quality, volume, and diversity of biomedical literature [14], the information seeking trend has advanced to multiple layers of information access. Current framework comprises a *search tool* that provides unified access to multiple *literature archives*; these archives store the full text of articles and offer multiple *viewing media* to read those articles. Current practice begins with the user crafting a keyword or a faceted (structured) query and submitting on the search tool.

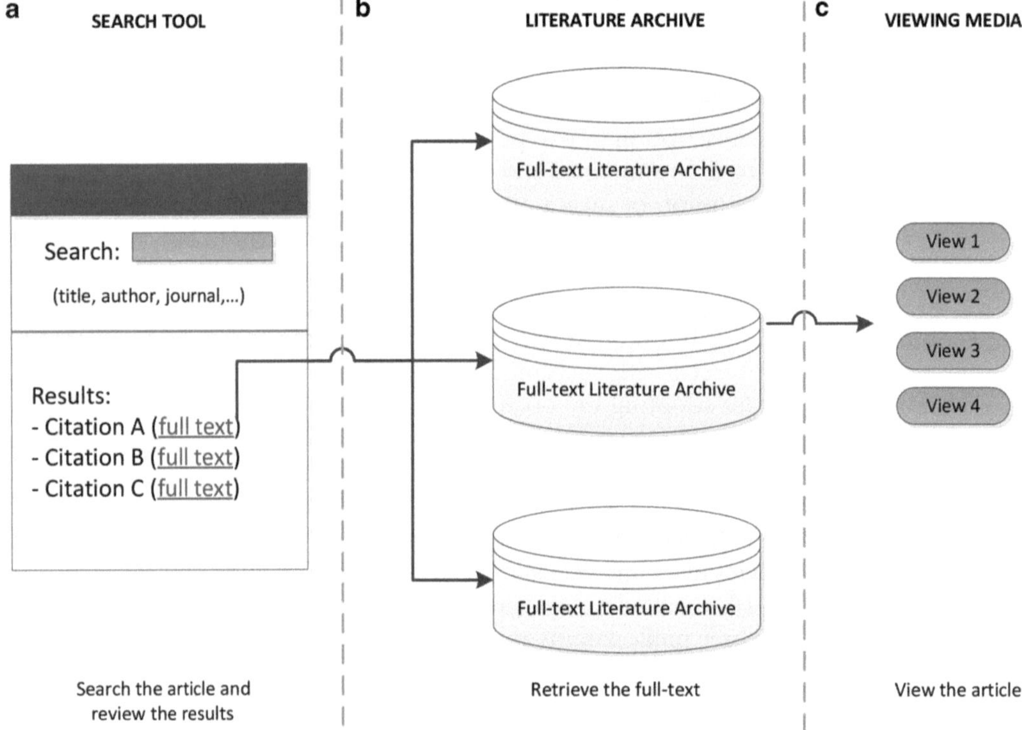

Fig. 1 The three steps of biomedical literature access: (**a**) Searching the literature and reviewing results using a search tool (e.g., PubMed), (**b**) retrieving the full text on a literature archive (e.g., PubMed Central), (**c**) consuming the article on a viewing media (e.g., PubTator)

In response, the tool presents a ranked list of citations relevant to the user query. The user has the option to go to a specific citation, access the full text on the linked literature archive, and view the article using a particular medium. Figure 1 demonstrates the three-step process of literature access.

2.1 Literature Search Tools

A search tool provides a single access point to multiple literature archives. At the core, the tool contains a citation database developed by indexing articles (abstract or full text) from different sources. The tool interface serves two purposes: (1) provides search functionality supporting queries ranging from the standard keyword search to the comprehensive faceted search (e.g., search by author, journal, title, etc.) and (2) presents ranked list of citations relevant to the query, with several options to filter and re-sort the results; in addition to bibliographic information, each citation contains a link to retrieve the full text of the article on a literature archive.

PubMed [1] is the most widely used search tool dedicated to biomedical and life sciences literature. Launched in 1996, PubMed is a publicly available citation database developed and maintained by the US National Library of Medicine. To date, PubMed contains more than 22.9 million citations for biomedical literature belonging to MEDLINE indexed journals, manuscripts deposited in PMC, and the NCBI Bookshelf. PubMed articles are indexed by the controlled vocabulary thesaurus, Medical Subject Headings (MeSH®). The search algorithm is based on PubMed's automatic term mapping algorithm [15]. The PubMed citation database is updated daily. PubMed citations date back to the early 1950s, and approximately half a million also date back to 1809. The PubMed interface offers the keyword search and allows the advanced queries by various fields such as author name, publication date, PubMed entering date, editor, grant number, and status of MeSH indexing for MEDLINE citations. A noteworthy feature of PubMed is the related articles algorithm [8] based on document similarity.

Embase [16] is a subscription-based biomedical citation database developed by Elsevier in 2000. This search service was developed primarily for biomedical and clinical practice with particular focus on drug discovery and development, drug safety, and pharmacovigilance research. Embase contains 25 million indexed records and indexes full-text articles from 8,306 journals, out of which 7,203 publish English language articles. Embase is often compared with MEDLINE, contains five million records, and covers 2,000 journals not included by MEDLINE. The Embase database is updated every day, and nearly one million records are added per year. Embase has digitally scanned the articles from 1947 to 1973. While the official reported temporal coverage of Embase dates backs to 1947, some articles also date back to 1880s. The records are indexed by Emtree thesaurus for drug and chemical information. This allows for deep indexing of articles and flexible keyword searching using term mapping [17]. The search capability is enhanced using auto complete and synonym suggestion features. The results can be filtered by drug and disease mentions in the article.

While several other state-of-the-art biomedical specific search tools [14, 18–20] have been designed since the inception of PubMed, these are not widely used as yet. Instead, other than PubMed, biomedical information seekers prefer rather generic tools that index articles from several disciplines in addition to biomedical and life sciences. Based on the popularity and discussion in previous studies [21–23], we describe one publicly available (Google Scholar [2]) and two subscription-based (Web of Science and Scopus [24, 25]) tools and describe their unique features.

Google Scholar [2], launched in 2004, is a Web search engine owned by Google Inc. Google Scholar indexes full-text articles from multiple disciplines from most peer-reviewed online journals of European and American publishers, scholarly books, and other non-peer-reviewed journals. The size and coverage of biomedical articles in Google Scholar are not revealed; theoretically, it consists of all biomedical articles available electronically. In addition to the keyword search, the tool offers searching by various fields such as author, publication date, journal, and words occurring in title and body, with different methods of term matching. The results are sorted by relevance as determined by full text of each article, author, journal, and number of citations received.

Web of Science [24], developed by Thompson Reuters in 2004, is a citation database that covers over 12,000 top-tier international and regional journals, as per their selection process [26], in every area of the natural sciences, social sciences, and arts and humanities. The science citation database of Web of Science, which is likely to contain biomedical specific articles, covers more than 8,500 notable journals from 150 disciplines and is updated weekly. The temporal coverage is dates back to 1900. The total number of biomedical citations cannot be approximated. The citations can be searched by various bibliographic fields, and the results display the total number of citations, comprehensive backward and forward citation maps, and additional keyword suggestions to improve the query. The result is ranked based on the overlap between the search terms and the terms in the articles. Also, the results can be filtered by Web of Science subject areas that are preassigned to journals.

Scopus [25], launched in 2004 by Elsevier, is a citation database for peer-reviewed literature from life sciences, health sciences, physical sciences, social sciences, and humanities. Scopus, as of November 2012, includes citations from 19,500 peer-reviewed journals, 400 trade publications, and 360 book series and is updated one to two times weekly. Temporally, citations date back to 1823. Scopus contains more than 18,300 citations from the life, health, and physical science subject areas. The faceted search is comprehensive and includes fields such as publication date, document type, subject area, author, title, keywords, and affiliation. For each result citation, the number of incoming citations, Emtree drug terms, and Emtree medical terms are displayed. For a given citation, Scopus also displays the related articles computed based on shared references. The relevance rank of results is calculated based on relative frequency and location of the search terms in the article.

Table 1 summarizes various search tools based on some key features. The biomedical coverage and size of the generic search tools, Google Scholar, Web of Science, and Scopus, could not be accurately computed, as they do not provide a breakdown for biomedical and life science-specific journals or articles. To get some insight into the coverage, we conducted a small experiment and submitted the query "type 2 diabetes mellitus" on various search tools. The results are shown in Table 2. Google Scholar returns the highest number of results. This is expected given the crawling nature of the search engine and the liberal inclusion criteria. Embase returns more citations than PubMed. While Web of Science returns the least number of results, it is discussed in higher number (10,356) of PubMed articles as compared to Scopus and

Table 1
Summary of various popular biomedical literature search tools

	PubMed	Google Scholar	Web of Science	Scopus	Embase
Developer	US National Library of Medicine	Google Inc.	Thompson Reuters	Elsevier	Elsevier
Launch year	1996	2004	2004	2004	2000
Fee based	No	No	Yes	Yes	Yes
Temporal coverage	1809 to present	Unknown	1900 to present	1823 to present	1880 to present
Total biomedical citations (approx.)	22.9 million	Unknown	Unknown	18,300	25 million
Covered journals	MEDLINE indexed journals. Manuscripts from PubMed Central, NCBI Bookshelf	Peer-reviewed journals of the USA and Europe, scholarly books, non-peer-reviewed journals	8,500 strictly selected science journals	Peer-reviewed journals, trade publications, book series	7,600 biomedical and pharmacological journals from 90 countries and 2,500+ conferences
Update frequency	Daily	Unknown	Weekly	1–2 times weekly	Daily
Relevance ranking features	Not applicable	Full text, author, journal, number of incoming citations	Overlap of search terms with the terms in the article	Frequency and location of search terms in the article	Unknown
Non-bibliographic information	MeSH keywords, related articles	Incoming citations	Keyword recommendations for query refinement, incoming citations, backward and forward citation maps	Emtree drug and medical terms, related articles, incoming citations	Emtree drug, disease and other terms, PubMed link, incoming citations (linked to Scopus)

Table 2
Comparison of search results for "type 2 diabetes mellitus" on July 15, 2013

	PubMed	Google Scholar	Web of Science	Scopus	EMBASE
Number of results	83,025	1,380,000 (approx.)	52,351	117,875	207,444
Publication year of the oldest article	1967	1853	1951	1947	1909
PubMed ID for the most recent article	23847327	23846835[a]	23504683[a]	22968324	23668792

[a]There exist other more recent articles not found in PubMed

Google Scholar which are discussed in 3,231 and 1,621 articles, respectively. The most recent and the oldest articles differ for each tool. With PubMed as reference point, Google Scholar shows the most up-to-date result. Also, the number of incoming citations for a 2001 article [27] is 7,092, 3,655, and 4,722, on Google Scholar, Web of Science, and Scopus (and Embase), respectively. This highlights the differences in coverage of various tools.

Out of the abovementioned tools, PubMed and Embase stand out in that they are the foremost developments, biomedical specific, and the most frequently updated search tools. In addition, their inbuilt search algorithms utilize controlled vocabularies. The other three generic tools differ from PubMed and Embase in that they perform citation analysis and provide indications of scholarly impact of articles; Google Scholar and Scopus provide the number of incoming citations for each article, and Web of Science offers thorough analysis including visual summaries of citation distributions. Also, all tools but PubMed employ a ranking algorithm that computes the relevance score of a given article with respect to search terms, incoming citations, journal, etc.

PubMed, Embase, and Scopus are similar in terms of their use of controlled vocabularies such as MeSH and Emtree in curating the articles. PubMed and Scopus are similar in their employment of the related articles algorithm, though internally quite different from each other. Web of Science is unique in that it has the keyword recommendation feature and a strict criterion for journal selection. The selling point of Embase is that it covers significant number of biomedical articles and journals that are not covered by PubMed. Google Scholar is unique in the comprehensiveness of its ranking algorithm. Another advantage of Google Scholar is that it links to free full-text articles more than the other search tools that might point to a locked journal [22]. Google Scholar, however, unlike others, does not support bibliography management, such as integration with bibtex, RefWorks, EndNote, and EndNote Web.

In sum, currently, there is no one-stop shop available for biomedical literature search as each tool has its own strengths and weaknesses. The choice of tool would thus depend on the subject matter, publication year, and usage context, and a wise search strategy would use multiple tools instead of relying on one [21].

2.2 Full-Text Literature Archives and Viewing Media

A literature search tool is integrated with multiple literature archives where full-text articles can be retrieved for further consumption. As of June 18, 2013, out of the 22.9 million citations in PubMed, 4 million citations are linked to their free full-text archives. Out of the citations linked to free full-text archives, 2.3 million are archived in the PMC [3] literature archive, and the remaining contain direct links to either journal's website (e.g., Journal of Cell Biology, Oncotarget, Anticancer Research, BMJ Journals) or comprehensive literature archives developed by major publishing companies.

PMC [3], launched in 2000, is a free digital archive of full-text biomedical and life science articles maintained by the US National Library of Medicine. Currently, the PMC archives approximately 2.7 million articles provided by about 3,700 journals including full participation, NIH portfolio, and selective deposit journals. PMC also contains supplemental items optionally accompanying each article. Another domain-specific archive, EBSCO's Cumulative Index to Nursing and Allied Health Literature (CINAHL) Plus with full text [28], is a subscription-based full-text literature archive designed for nurses, allied health professionals, researchers, nurse educators, and students. The content dates back to 1937 and includes full text from 768 journals and 275 books from nursing and allied health disciplines. CINAHL is also a widely used search tool among nursing professionals.

Springer's SpringerLink [29] was launched in 1996 and archives full-text content available from 1996. SpringerLink covers approximately 7.7 million full-text articles from electronic books and journals from all disciplines, out of which 6.4 million could be classified under the categories of biomedical, chemical, life, public health, and medical sciences. Supplementary material is also archived with each article.

ScienceDirect [30] is a subscription-based literature archive launched by Elsevier in 2000. ScienceDirect contains more than 11 million peer-reviewed journal articles and book chapters from more than 2,500 peer-reviewed journals and more than 11,000 books, including 8,077 journals and book chapters from life and health sciences. ScienceDirect's coverage goes back to 1823. Elsevier, which is also the host of Scopus search tool, has digitalized most of the pre-1996 content. Some additional content such as audio, video, datasets, and supplemental items are also archived. Since its launch, more than 700 million articles have been downloaded from the ScienceDirect website [31]. Recently,

Table 3
Comparison of biomedical full-text literature archives

Literature archive (provider, year)	Temporal coverage	Full-text biomedical articles and archive coverage (approx.)	Viewing media
PubMed central (US National Library of Medicine, 2000)	1950 to present	2.7 million from 3,700 journals, including full participation, NIH portfolio, selective deposit	Classic, PDF, EPUB, PubReader
CINAHL Plus with Full Text (EBSCO, 2010)	1937 to present	768 journals and magazines, 275 books and monographs from nursing and allied health disciplines	PDF
SpringerLink (Springer, 1996)	1860 to present	6.4 million from biomedical, chemical, life, public health, and medical sciences	Classic, PDF, EPUB
ScienceDirect (Elsevier, 2000)	1823 to present	8,077 life and health science journals and book chapters	PDF
Wiley Online Library (Wiley-Blackwell, 2010)	Unknown	Journals, online books, and reference works (biomedical coverage unknown)	Classic, PDF

Elsevier has integrated its search tool, Scopus, and the literature archive, ScienceDirect, into a new platform, SciVerse.

Wiley online library is a subscription-based full-text archive, developed in 2010 by Wiley-Blackwell publishing company. Wiley online library contains multidisciplinary collection of 4 million full-text articles from 1,500 journals, over 13,000 online books, and hundreds of reference works. The subject areas include chemistry, life sciences, medicine, nursing, dentistry and healthcare, veterinary medicine, physical sciences, and non-biomedical subjects [32]. The coverage of biomedical subject areas is not known.

Table 3 summarizes the above-discussed literature archives by their provider, launch year, temporal coverage, content coverage, and supported viewing media. Similar to the generic search tools, the total number of biomedical articles in the generic archives such as SpringerLink and ScienceDirect could not be precisely computed. SpringerLink does provide a subject-wise breakdown and archives the most number of full-text articles from biomedical and related areas. In PubMed, CINAHL is discussed in the highest number of articles (8,595), followed by ScienceDirect (298), SpringerLink (116), and Wiley Online Library (38). These articles are related to information seeking and retrieval studies focused on biomedical articles. It should be noted that these numbers might not give a complete picture on the coverage of various archives, as there might be other studies published in journals not indexed by PubMed.

Each literature archive offers one or more media or formats where the retrieved literature can be consumed (read) by the user. Currently, the aforementioned literature archives offer at least four types of viewing media. The first view is the classic view wherein the article can be viewed on the archive website itself. This view does not have any page breaks and needs to be read by scrolling vertically through a single long page. This is the default HTML format view offered by most literature sources for quick reference. The second viewing media is the PDF format (.pdf extension) wherein the article can be downloaded onto a device. All literature sources archive full text in the PDF format that can be used to read on laptops, desktops, and Kindle and can be printed into a hard copy. PDF format appears exactly as it would appear on a piece of paper; it allows paging, zooming, annotation, and commenting. The third viewing media is the open e-book standard EPUB (.epub extension) offered by PMC and ScienceDirect. This format offers a downloadable file that can be displayed on several devices and readers such as Calibre, iBooks, Google Books, and Mobipocket, on various platforms such as Android, Windows, Mac OS X, iOS, Web, and Google Chrome Extensions. Finally, PMC offers a new view, PubReader [33], a user-friendly modification to the classic view that emulates the ease of reading the printed version of an article. PubReader was launched by the National Center for Biotechnology Information (NCBI) in 2012 and is coded in CSS and JavaScript. The PubReader display allows an article to be read on a Web browser through laptops, desktops, and tablet computers. PubReader offers ease of navigation and readability by organizing the article into columns and pages to fit into the target screen. In addition, ScienceDirect also offers mobile applications to be used on iPad, phones, and tablet computers.

3 Text Mining Solutions to Address Search Challenges

Given the exponential growth and increasing diversity of biomedical literature, the default querying mechanism (keyword or faceted search) would no longer be enough to meet the user needs. There is a need to provide alternative methods of writing queries and interactive support in query formulation [34–36]. Existing biomedical search tools have made a few efforts in this direction, such as keyword recommendation feature by Web of Science and flexible keyword searching by Embase. Even when the user finds the right query to input, identifying the few most relevant articles among thousands of citations is not getting any easier. While most search tools employ a ranking algorithm to compute the relevance score of a given article, relevance remains an important topic in IR research [37, 38]. Existing tools also provide filters to narrow down the results by different fields. Their ability to present the results in a summarized manner remains largely unexplored, however.

In response to the shortcomings of the existing tools, the literature describes many alternative or experimental search interfaces. In this section, we discuss advanced NLP and IR techniques, primarily by discussing alternative interfaces implementing methods not yet available in the major literature search tools. We categorize these techniques into five sections: text similarity search, semantic search, query support, relevance ranking, and clustering; the former three primarily address the search challenges, and latter two address the result presentation issues.

3.1 Text Similarity Search

It can be difficult for users to make their exact information need explicit and then translate it into a query. Several alternative search interfaces have implemented another type of search where the query consists of one or more documents known to be relevant. The relevance of the documents to be retrieved is then calculated based on their similarity to the relevant documents.

eTBLAST is a tool for searching the literature for documents similar to a given passage of text, such as an abstract [39]. The tool extracts a set of keywords from the text and uses these to gather a subset of the literature. A final similarity score is computed for each document in the set by aligning the sentences in the input passage with the document retrieved. MedlineRanker allows the user to input a set of documents and then finds the set of words most discriminative of the documents within the set [40]. These are then used as features in a classifier (Naïve Bayes), which is applied to unlabeled documents to return the most relevant results. While effective, this approach requires a sufficiently large training set, between 100 and 1,000 abstracts.

Recent work by Ortuno et al. [41] partially alleviates the need for a large training set by allowing the user to enter a single abstract as query. The articles cited by the input article are then used to enrich the input set. This approach significantly improves the quality of the results over using only the input article and also typically returns significantly better results than pseudo relevance feedback. Tbahriti et al. [42] significantly improved the ability to determine whether two articles were related by classifying each sentence according to its purpose in the argumentative structure of the abstract (Purpose, Methods, Results, Conclusion). They found that the best results were obtained by increasing the weight of Purpose and Conclusion sentences relative to sentences classified as Methods or Results.

MScanner is similar to the other textual similarity tools in that it learns a classifier (Naïve Bayes) from a set of relevant documents input by the user [43]. In the case of MScanner, however, the only features are the set of MeSH terms associated with each article and the name of the journal where the article appears, resulting in a very-high-speed retrieval system. Other systems have experimented with using inputs other than text. Caipirini, for example, allows the user to specify a set of genes that are of interest and a set of genes that are not of interest [44]. The system locates abstracts mentioning the genes specified, and the system extracts keywords that appear more frequently than chance in these abstracts. These keywords are then used as features for a classifier (SVM), which provides a score representing the similarity of a text to the abstracts that mention the genes of interest versus the background set. This classifier is then applied to all of Medline, and the top results are returned.

3.2 Semantic Search

A large part of the meaning of biomedical texts is captured by the entities they mention and the relationships discussed. This observation can be exploited to support semantic search in both the queries and in the way the results are displayed to the user. Systems supporting semantic search also differ in the types of entities and relationships extracted and the methodology employed.

MedEvi is a semantic search tool intended for finding evidence of specific relationships [45]. The tool recognizes ten keywords representing entity types, such as "(gene)" and "(disease)." In addition, the tool orders results by preferring results containing the terms in the same order as they appear in the query and within close proximity. Kleio supports keyword searches of multiple prespecified fields [46], including both semantic types (including protein, metabolite, disease, organ, acronyms, and natural phenomenon) and article metadata (e.g., author). A specialized tool for specifically querying authors is Authority [47]. Authority uses a clustering approach over article metadata to determine whether ambiguous author names represent the same person or not. When the users query for an author, the system displays the matching author clusters.

MEDIE allows queries to specify any combination of subject, verb, and object [48]. For example, the query representing "What causes cancer?" would be *verb* = "cause" and *object* = "cancer." This query returns a list of text fragments where the verb matches "cause" and its object is "cancer" or any of its hyponyms, such as "leukemia." The results highlight genes and diseases in different colors. PubNet extracts entities and relationships from the articles returned by a standard PubMed query and then visualizes the results as a graph [49]. Entities supported include genes and proteins, MeSH terms, and authors.

The EBIMed service uses keyword queries as input and provides a listing of the entities most common in documents matching the query [50]. EBIMed supports a fixed set of entity types (protein, cellular component, biological process, molecular function, drug, and species) and locates both entities and relationships between two entities. All results are linked to the biological database that defines the entity. Quertle locates articles that describe relationships between the entities provided in a query, and results are grouped by the relationship described [51, 52].

Users may also switch to a keyword search with a single click. Quertle also supports a list of predefined query keywords that refer to entity types of varying granularity.

A Web-based text mining application named PubTator [53, 54] was recently developed to support manual biocuration [55–58]. Because finding articles relevant to specific biological entities (such as gene/protein) is often the first step in biocuration, PubTator supports entity-specific semantic searches based on the use of several competition-winning named entity recognition tools [59–64].

3.3 Query Support

An important aspect of improving the relevance of query results is to help the users translate their information need into a query. While text similarity, as discussed in Subheading 3.1, is a useful method for reducing this barrier, another method is to directly support the creation and revision of both keyword and faceted queries [34].

The iPubMed tool allows searching MEDLINE records to be more interactive through the *search-as-you-type* paradigm. Query results are dynamically updated after every keypress. iPubMed also supports approximate search, allowing users to dynamically correct spelling errors.

PubMed Assistant is a stand-alone system which includes a visual tool for creating Boolean queries and a query refinement tool that gathers useful keywords from results marked relevant by the user [65]. PubMed Assistant also supports integration with a citation manager.

Schardt et al. [66] demonstrated that search interfaces supporting the PICO system for focusing clinical queries improved the precision of the results. PICO is a framework for supporting evidence-based medicine and is an acronym for *P*atient problem, *I*ntervention, *C*omparison, and *O*utcome [67, 68]. SLIM is a tool emphasizing clinical queries which uses slider bars to quickly customize query results. Modifiable parameters include the age of the article, the journal subset, and both the age group and the study design of the clinical trial reported. askMEDLINE is a system which accepts clinical queries in the form of natural language questions. The system is particularly designed to support users who are not medical experts.

3.4 Relevance Ranking

Ranking results in order of their relevance to the query is a well-supported technique for reducing the workload of the user and is supported in most existing tools for searching the literature except PubMed itself. While straightforward measurements such as TF-IDF are known to work well [37], there are still aspects that can be improved.

A common ranking technique in web search is to incorporate a measurement of the importance of the document into the score. The scientific record contains many types of bibliometric information that can be used to infer the quality or the importance of an article. The PubFocus system, for example, ranks relevant documents according to an importance score that includes the impact factor of the journal and the volume of citations [69].

Bernstam et al. [70] demonstrated that algorithms that use citation data to determine document importance—including both simple citation counts and PageRank—significantly improve over algorithms that do not use citation data. Unfortunately, however, citation data suffers from "citation lag"—the period of time between when an article is published and when it is cited by another article. Tanaka et al. [71] partially overcome this limitation by using the data available at publication to learn which articles are likely to eventually be highly cited.

Lin [72] takes a different approach and instead uses the PubMed related articles tool to create a graph by linking similar document pairs. PageRank is then applied, producing a score for each document where higher scores imply that the document contains more of the content from its neighbors in the graph. Scores are thus independent of any query, but documents with higher scores will naturally be relevant for a wider range of queries.

Yeganova et al. [73] examined PubMed query logs and found that users frequently enter phrases such as "sudden death syndrome" without the quotes to indicate that the query contains a phrase. While PubMed interprets such queries as the conjunction of the individual terms, the authors demonstrate a qualitative difference between results that contain all terms and results that contain the terms as a phrase. They conclude that it would be beneficial to attempt to interpret such queries as containing a phrase and in particular suggest that documents containing the terms in close proximity are more relevant than results that merely contain all terms.

The RefMed system employs relevance feedback to explicitly model the relevance of query results [74]. In relevance feedback, the system returns an initial set of results, allows the user to indicate whether each result is useful, and then uses the input as feedback to improve the next round of results. While relevance has traditionally been considered to be binary, RefMed uses a learning to rank algorithm (rankSVM) to allow the user to specify varying degrees of relevance along a scale.

The MiSearch tool uses an implicit form of relevance feedback to model the relevance of articles to the user [75]. The system automatically collects relevant documents by recording which documents are opened while browsing. This data is used to create a model of the likelihood that the user will open a document that can be used to rank the results of any query. Features for the model include authors, journal, and PubMed indexing information.

3.5 Clustering Results

Clustering the results of the user query into topics helps in several ways. First, clustering the results helps to differentiate between the different meanings of ambiguous query terms. Second, in large sets of search results it can help the users focus on the subset of documents that interest them. Third, the clusters themselves can serve as an overview of the topic. This method has been considered in several PubMed derivatives that vary in their method of determining the clustering methodology. Popular variations include MeSH terms, other semantic content (such as UMLS concepts and GO terms), keywords, and document metadata (such as journal, authors, and date).

Anne O'Tate provides additional structure and a summary of the query results by clustering the content of the documents retrieved and also by extracting important words, publication date, authors, and their institutions [76]. Users are allowed to extend the query by any of the summarized information simply by clicking on it, and any query returning less than 50 results can be expanded to include the articles most closely related.

The McSyBi tool clusters query results both hierarchically and non-hierarchically [77]. Whereas the non-hierarchical clustering is primarily useful for focusing the query on particular subsets, the hierarchical clustering provides a brief summary of the query results. McSyBi also provides the ability for the user to adjust or reformulate the clustering by introducing a MeSH term, which is interpreted as a new binary feature for each document, depending on whether the document has been assigned the specified MeSH term. Users can also introduce a UMLS Semantic Type, which is considered present if the document is assigned at least one MeSH term with the specified type.

GoPubMed originally used the Gene Ontology (GO) [78] to organize the search results [79]. It currently groups search results according to categories "what" (biomedical concepts), "where" (affiliations and journals), "who" (author names), and "when" (date of publication). The "what" category is further subdivided into concepts from Gene Ontology, MeSH, and UniProt. GO terms are located in the abstracts retrieved, even if they do not appear directly, and are highlighted when the abstract is displayed.

XplorMed is a tool for multifactored analysis of query results [80, 81]. Results are displayed grouped both by coarse MeSH categories and important words that are shown both in summary and in context. Users may then explore the important words in more depth or display results ranked by inclusion of the important words.

Boyack et al. [7] sought to determine which clustering approach would produce the most coherent clusters over a large subset of MEDLINE. The analysis considered five analytical techniques: a vector space approach with TF-IDF vectors and cosine similarity, latent semantic analysis, topic modeling, the Poisson-based language model BM25, and PubMed related articles.

The analysis also considered two data sources, MeSH subject headings and words from titles and abstracts. The article concluded that PubMed related articles created the most coherent clusters, closely followed by BM25, and also concluded that the clusters based on titles and abstracts are significantly better than those based only on MeSH headings.

SEACOIN (Search Explore Analyze COnnect INspire) is a system that merges important word analysis with clustering and a graphical visualization to achieve a simple interface suitable for novice users [82]. The SEACOIN visualization combines a word cloud that allows the user to add additional terms to the query, a multi-level treelike graphic that allows users to see the relative number of documents containing different terms and term combinations, and a table listing the documents returned.

SimMed presents users with clusters of documents ranked by their degree of relevance to the query [83]. The interface emphasizes the clusters found to provide a summary of the query topic, thereby explicitly supporting exploratory searches. The clusters used are computed off-line, allowing high retrieval performance.

4 Future Trends to Improve Biomedical Literature Access

In terms of the search tools, in addition to the research directions highlighted in Subheading 3, we can expect to see more reader-friendly and smart applications based on advanced IR and NLP techniques in order to help readers find and digest articles more effectively and efficiently. Furthermore, with the use of social media such as blogging and tweeting, new ways of sharing and recommending papers will gain more importance in the future, in addition to the traditional search-based mechanism. For instance, using social media makes it easier to make and share comments on papers, thus providing alternative views with respect to the impact of individual papers. In the future, biomedical literature search could also be personalized. That is, search results are tailored towards the interests of individual researchers based on their own work and/or past searches. In other words, the same query by two different users may return different search results. This is desirable in certain cases. For instance, a bench scientist and medical doctor are likely to search for different information (biological vs. clinical, respectively) even though they both search for the same drug and disease pair. In the general web search domain, personalized search has been shown to be useful [84]. Therefore, such a feature could also be helpful to users when it comes to the biomedical literature search.

With regard to open-access papers, we believe that its size will continue to grow rapidly over the next 5–10 years. This is evidenced by the increasing number of publishers and journals interested in adopting the open access policy as well as by the ever-growing interests from the scientific community. That is, more and more authors are considering open-access journals as their preferred choice for publishing their work. As this happens and together with Web and computer technology advances, we can imagine free access to most research articles anywhere, anytime on any device.

Finally, with increased use of portable devices such as smartphones and computer tablets to access the Internet, there are growing needs and interests in searching and reading literature on those devices. Portable devices provide a great deal of benefits such as convenience but also present new challenges. First, it is less likely to print out the papers with portable devices—a common way for reading papers. As a result, reading directly on those devices becomes necessary. But compared to desktop or laptop computers, the screen size of portable devices is usually much smaller. As such, readability becomes a real issue on those small-screen devices, especially when it comes to reading papers. This is because unlike reading e-mails or news articles, people do not generally read straight through an article. Instead, they often need to go back and forth when reading a journal article in order to understand and digest its content. Scrolling up and down on a modern computer screen is hard but still viable; this kind of operation becomes almost impossible on small-screen devices. As mentioned in Subheading 2.2, there has already been work on supporting convenient reading on small-screen devices from new reading apps to reader-friendly Web interfaces. Although these tools already provide better readability than the traditional Web browsers, further improvement is needed in order to make users to read and digest articles comfortably on those devices. We also expect advances in Web technology to help facilitate such a transition.

Acknowledgments

This research was supported by the Intramural Research Program at the National Institutes of Health, National Library of Medicine.

References

1. PubMed. US National Library of Medicine, National Institutes of Health. http://www.ncbi.nlm.nih.gov/pubmed
2. Google Scholar. Google. http://scholar.google.com/
3. PubMed Central. US National Library of Medicine, National Institutes of Health. http://www.ncbi.nlm.nih.gov/pmc/
4. Hunter L, Cohen KB (2006) Biomedical language processing: what's beyond PubMed? Mol Cell 21(5):589–594. doi:10.1016/j.molcel.2006.02.012
5. Islamaj Dogan R, Murray GC, Neveol A et al (2009) Understanding PubMed user search behavior through log analysis. Database 2009:bap018. doi:10.1093/database/bap018
6. Garg AX, Iansavichus AV, Kastner M et al (2006) Lost in publication: half of all renal practice evidence is published in non-renal journals. Kidney Int 70(11):1995–2005. doi:10.1038/sj.ki.5001896
7. Boyack KW, Newman D, Duhon RJ et al (2011) Clustering more than two million biomedical publications: comparing the accuracies

of nine text-based similarity approaches. PloS One 6(3):e18029. doi:10.1371/journal.pone.0018029

8. Lin J, Wilbur WJ (2007) PubMed related articles: a probabilistic topic-based model for content similarity. BMC Bioinformatics 8:423. doi:10.1186/1471-2105-8-423
9. Yiotis K (2005) The open access initiative: a New paradigm for scholarly communications. Inform Tech Libr 24(4):157–162
10. Wikipedia PubMed Central. http://en.wikipedia.org/wiki/PubMed_Central. Accessed 13 Jul 2013
11. Davis PM (2013) Public accessibility of biomedical articles from PubMed Central reduces journal readership: retrospective cohort analysis. FASEB J 27(7):2536–2541. doi:10.1096/fj.13-229922
12. Grefsheim SF, Rankin JA (2007) Information needs and information seeking in a biomedical research setting: a study of scientists and science administrators. J Med Libr Assoc 95(4):426–434. doi:10.3163/1536-5050.95.4.426
13. Hemminger BM, Lu D, Vaughan KTL et al (2007) Information seeking behavior of academic scientists. J Am Soc Inform Sci Tech 58(14):2205–2225
14. Kim JJ, Rebholz-Schuhmann D (2008) Categorization of services for seeking information in biomedical literature: a typology for improvement of practice. Brief Bioinform 9(6):452–465. doi:10.1093/bib/bbn032
15. PubMed Tutorial, Automatic Term Mapping. US. National Library of Medicine, National Institutes of Health. http://www.nlm.nih.gov/bsd/disted/pubmedtutorial/020_040.html
16. Embase: biomedical database. Elsevier. http://www.elsevier.com/online-tools/embase
17. Roche A-M Embase: answers to your biomedical questions. http://www.slideshare.net/rocheam/embase-introduction. Accessed 16 Jul 2013
18. Lu Z (2011) PubMed and beyond: a survey of web tools for searching biomedical literature. Database 2011:baq036. doi:10.1093/database/baq036
19. Falagas ME, Giannopoulou KP, Issaris EA et al (2007) World databases of summaries of articles in the biomedical fields. Arch Intern Med 167(11):1204–1206. doi:10.1001/archinte.167.11.1204
20. Hoskins IC, Norris WE, Taylor R (2008) Databases of biomedical literature: getting the whole picture. Arch Intern Med 168(1):113. doi:10.1001/archinternmed.2007.26, author reply 113-114
21. Bakkalbasi N, Bauer K, Glover J et al (2006) Three options for citation tracking: Google Scholar, Scopus and Web of Science. Biomed Digit Libr 3:7. doi:10.1186/1742-5581-3-7
22. Falagas ME, Pitsouni EI, Malietzis GA et al (2008) Comparison of PubMed, Scopus, Web of Science, and Google Scholar: strengths and weaknesses. FASEB J 22(2):338–342. doi:10.1096/fj.07-9492LSF
23. Bar-Ilan J (2008) Which h-index?: A comparison of WoS, Scopus and Google Scholar. Scientometrics 74(2):257–271
24. Web of science. Thomson Reuters. http://thomsonreuters.com/web-of-science/
25. Scopus: document search. Elsevier. http://www.scopus.com/home.url
26. The Thomson Reuters journal selection process. Thomson Reuters. http://wokinfo.com/essays/journal-selection-process/
27. Tuomilehto J, Lindstrom J, Eriksson JG et al (2001) Prevention of type 2 diabetes mellitus by changes in lifestyle among subjects with impaired glucose tolerance. N Engl J Med 344(18):1343–1350. doi:10.1056/NEJM200105033441801
28. CINAHL Plus with full text. EBSCO. http://www.ebscohost.com/academic/cinahl-plus-with-full-text
29. SpringerLink. Springer. http://link.springer.com/
30. ScienceDirect.com | Search through over 11 million science, health, medical journal full text articles and books. Elsevier. http://www.sciencedirect.com/
31. ScienceDirect platform brochure. Elsevier. http://www.info.sciverse.com/documents/files/content/pdf/SDPlatformBrochure_06.pdf
32. Journals. Wiley Online Library. http://olabout.wiley.com/WileyCDA/Section/id-406089.html
33. Lipman D (2012) The PubReader view: a new way to read articles in PMC. NLM Tech Bull 389:e7
34. Lu Z, Wilbur WJ, McEntyre JR et al (2009) Finding query suggestions for PubMed. AMIA Annu Symp Proc 2009:396–400
35. Neveol A, Dogan RI, Lu Z (2010) Author keywords in biomedical journal articles. AMIA Annu Symp Proc 2010:537–541
36. Islamaj Dogan R, Lu Z (2010) Click-words: learning to predict document keywords from a user perspective. Bioinformatics 26(21):2767–2775. doi:10.1093/bioinformatics/btq459
37. Lu Z, Kim W, Wilbur WJ (2008) Evaluating relevance ranking strategies for MEDLINE retrieval. AMIA Annu Symp Proc 439
38. Lu Z, Kim W, Wilbur WJ (2009) Evaluating relevance ranking strategies for MEDLINE

retrieval. J Am Med Inform Assoc 16(1):32–36. doi:10.1197/jamia.M2935

39. Errami M, Wren JD, Hicks JM et al (2007) eTBLAST: a web server to identify expert reviewers, appropriate journals and similar publications. Nucleic Acids Res 35(Web Server issue):W12–W15. doi:10.1093/nar/gkm221
40. Fontaine JF, Barbosa-Silva A, Schaefer M et al (2009) MedlineRanker: flexible ranking of biomedical literature. Nucleic Acids Res 37(Web Server issue):W141–W146. doi:10.1093/nar/gkp353
41. Ortuno FM, Rojas I, Andrade-Navarro MA et al (2013) Using cited references to improve the retrieval of related biomedical documents. BMC Bioinformatics 14:113. doi:10.1186/1471-2105-14-113
42. Tbahriti I, Chichester C, Lisacek F et al (2006) Using argumentation to retrieve articles with similar citations: an inquiry into improving related articles search in the MEDLINE digital library. Int J Med Inform 75(6):488–495. doi:10.1016/j.ijmedinf.2005.06.007
43. Poulter GL, Rubin DL, Altman RB et al (2008) MScanner: a classifier for retrieving Medline citations. BMC Bioinformatics 9:108. doi:10.1186/1471-2105-9-108
44. Soldatos TG, O'Donoghue SI, Satagopam VP et al (2012) Caipirini: using gene sets to rank literature. BioData Min 5(1):1. doi:10.1186/1756-0381-5-1
45. Kim JJ, Pezik P, Rebholz-Schuhmann D (2008) MedEvi: retrieving textual evidence of relations between biomedical concepts from Medline. Bioinformatics 24(11):1410–1412. doi:10.1093/bioinformatics/btn117
46. Nobata C, Cotter P, Okazaki N et al. (2008) Kleio: a knowledge-enriched information retrieval system for biology. Paper presented at the 31st annual international ACM SIGIR conference on research and development in information retrieval
47. Torvik VI, Smalheiser NR (2009) Author name disambiguation in MEDLINE. ACM Trans Knowl Discov Data 3(3)
48. Ohta T, Miyao Y, Ninomiya T et al (2006) An intelligent search engine and GUI-based efficient MEDLINE search tool based on deep syntactic parsing. Paper presented at the COLING/ACL Interactive presentation sessions, Sydney, Australia
49. Douglas SM, Montelione GT, Gerstein M (2005) PubNet: a flexible system for visualizing literature derived networks. Genome Biol 6(9):R80. doi:10.1186/gb-2005-6-9-r80
50. Rebholz-Schuhmann D, Kirsch H, Arregui M et al (2007) EBIMed: text crunching to gather facts for proteins from Medline. Bioinformatics 23(2):e237–e244. doi:10.1093/bioinformatics/btl302
51. Giglia E (2011) Quertle and KNALIJ: searching PubMed has never been so easy and effective. Eur J Phys Rehabil Med 47(4):687–690
52. Coppernoll-Blach P (2011) Quertle: the conceptual relationships alternative search engine for PubMed. J Med Libr Assoc 99(2):U159–U176. doi:10.3163/1536-5050.99.2.017
53. Wei CH, Kao HY, Lu Z (2013) PubTator: a web-based text mining tool for assisting biocuration. Nucleic Acids Res 41(Web Server issue):W518–W522. doi:10.1093/nar/gkt441
54. Wei CH, Kao HY, Lu Z (2012) PubTator: A PubMed-like interactive curation system for document triage and literature curation. Paper presented at the BioCreative Workshop 2012, Washington DC
55. Arighi CN, Carterette B, Cohen KB et al (2013) An overview of the BioCreative 2012 Workshop Track III: interactive text mining task. Database 2013:bas056. doi:10.1093/database/bas056
56. Arighi CN, Roberts PM, Agarwal S et al (2011) BioCreative III interactive task: an overview. BMC Bioinformatics 12(Suppl 8):S4. doi:10.1186/1471-2105-12-S8-S4
57. Lu Z, Hirschman L (2012) Biocuration workflows and text mining: overview of the BioCreative 2012 Workshop Track II. Database 2012:43. doi:10.1093/database/bas043
58. Neveol A, Wilbur WJ, Lu Z (2012) Improving links between literature and biological data with text mining: a case study with GEO, PDB and MEDLINE. Database 2012:bas026. doi:10.1093/database/bas026
59. Lu Z, Kao HY, Wei CH et al (2011) The gene normalization task in BioCreative III. BMC Bioinformatics 12(Suppl 8):S2. doi:10.1186/1471-2105-12-S8-S2
60. Van Landeghem S, Bjorne J, Wei CH et al (2013) Large-scale event extraction from literature with multi-level gene normalization. PloS One 8(4):e55814. doi:10.1371/journal.pone.0055814
61. Wei CH, Kao HY, Lu Z (2012) SR4GN: a species recognition software tool for gene normalization. PloS One 7(6):e38460. doi:10.1371/journal.pone.0038460
62. Wei CH, Harris BR, Kao HY et al (2013) tmVar: a text mining approach for extracting sequence variants in biomedical literature. Bioinformatics 29(11):1433–1439. doi:10.1093/bioinformatics/btt156
63. Leaman R, Dogan RI, Lu Z (2013) DNorm: disease name normalization with pairwise learning to rank. Bioinformatics 29:2909–2917

64. Leaman R, Khare R, Lu Z (2013) NCBI at 2013 ShARe/CLEF eHealth shared task: disorder normalization in clinical notes with DNorm. Conference and Labs of the Evaluation Forum 2013 Working Notes
65. Ding J, Hughes LM, Berleant D et al (2006) PubMed assistant: a biologist-friendly interface for enhanced PubMed search. Bioinformatics 22(3):378–380. doi:10.1093/bioinformatics/bti821
66. Schardt C, Adams MB, Owens T et al (2007) Utilization of the PICO framework to improve searching PubMed for clinical questions. BMC Med Inform Decis Mak 7:16. doi:10.1186/1472-6947-7-16
67. Richardson WS, Wilson MC, Nishikawa J et al (1995) The well-built clinical question: a key to evidence-based decisions. ACP J Club 123(3):A12–A13
68. Armstrong EC (1999) The well-built clinical question: the key to finding the best evidence efficiently. WMJ 98(2):25–28
69. Plikus MV, Zhang Z, Chuong CM (2006) PubFocus: semantic MEDLINE/PubMed citations analytics through integration of controlled biomedical dictionaries and ranking algorithm. BMC Bioinformatics 7:424. doi:10.1186/1471-2105-7-424
70. Bernstam EV, Herskovic JR, Aphinyanaphongs Y et al (2006) Using citation data to improve retrieval from MEDLINE. J Am Med Inform Assoc 13(1):96–105. doi:10.1197/jamia.M1909
71. Tanaka LY, Herskovic JR, Iyengar MS et al (2009) Sequential result refinement for searching the biomedical literature. J Biomed Inform 42(4):678–684. doi:10.1016/j.jbi.2009.02.009
72. Lin J (2008) PageRank without hyperlinks: reranking with PubMed related article networks for biomedical text retrieval. BMC Bioinformatics 9:270. doi:10.1186/1471-2105-9-270
73. Yeganova L, Comeau DC, Kim W et al (2009) How to interpret PubMed queries and Why it matters. J Am Soc Inf Sci Technol 60(2):264–274. doi:10.1002/Asi.20979
74. Yu H, Kim T, Oh J et al (2010) Enabling multi-level relevance feedback on PubMed by integrating rank learning into DBMS. BMC Bioinformatics 11(Suppl 2):S6. doi:10.1186/1471-2105-11-S2-S6
75. States DJ, Ade AS, Wright ZC et al (2009) MiSearch adaptive PubMed search tool. Bioinformatics 25(7):974–976. doi:10.1093/bioinformatics/btn033
76. Smalheiser NR, Zhou W, Torvik VI (2008) Anne O'Tate: A tool to support user-driven summarization, drill-down and browsing of PubMed search results. J Biomed Discov Collab 3:2. doi:10.1186/1747-5333-3-2
77. Yamamoto Y, Takagi T (2007) Biomedical knowledge navigation by literature clustering. J Biomed Inform 40(2):114–130. doi:10.1016/j.jbi.2006.07.004
78. Ashburner M, Ball CA, Blake JA et al (2000) Gene ontology: tool for the unification of biology. The gene ontology consortium. Nat Genet 25(1):25–29. doi:10.1038/75556
79. Doms A, Schroeder M (2005) GoPubMed: exploring PubMed with the gene ontology. Nucleic Acids Res 33(Web Server issue):W783–W786. doi:10.1093/nar/gki470
80. Perez-Iratxeta C, Bork P, Andrade MA (2001) XplorMed: a tool for exploring MEDLINE abstracts. Trends Biochem Sci 26(9):573–575
81. Perez-Iratxeta C, Perez AJ, Bork P et al (2003) Update on XplorMed: a web server for exploring scientific literature. Nucleic Acids Res 31(13):3866–3868
82. Lee EK, Lee HR, Quarshie A (2011) SEACOIN: an investigative tool for biomedical informatics researchers. AMIA Annu Symp Proc 2011:750–759
83. Mu X, Ryu H, Lu K (2011) Supporting effective health and biomedical information retrieval and navigation: a novel facet view interface evaluation. J Biomed Inform 44(4):576–586. doi:10.1016/j.jbi.2011.01.008
84. Liu F, Yu C, Meng W (2004) Personalized web search for improving retrieval effectiveness. IEEE Trans Knowl Data Eng 16(1):28–40

Chapter 3

Mapping of Biomedical Text to Concepts of Lexicons, Terminologies, and Ontologies

Michael Bada

Abstract

Concept mapping is a fundamental task in biomedical text mining in which textual mentions of concepts of interest are annotated with specific entries of lexicons, terminologies, ontologies, or databases representing these concepts. Though there has been a significant amount of research, there are still a limited number of practical, publicly available tools for concept mapping of biomedical text specified by the user as an independent task. In this chapter, several tools that can automatically map biomedical text to concepts from a wide range of terminological resources are presented, followed by those that can map to more restricted sets of these resources. This presentation is intended to serve as a guide to researchers without a background in biomedical concept mapping of text for the selection of an appropriate tool based on usability, scalability, configurability, balance between precision and recall, and the desired set of terminological resources with which to annotate the text. Only with effective automatic concept-mapping tools will systems be able to scalably analyze the biomedical literature and other large sets of documents as a fundamental part of more complex text-mining tasks such as information extraction and hypothesis evaluation and generation.

Key words Concept mapping, Concept recognition, Concept normalization, Annotation, Terminologies, Vocabularies, Ontologies

1 Introduction

One of the most fundamental tasks in biomedical text mining is the identification of mentions of specified entities of interest, such as genes/gene products, cells, chemicals, and diseases/pathologies. This task has historically been referred to in the computational linguistics community as *named-entity recognition*, where named entities are those with rigid designators such as people, organizations, and locations but also less intuitive ones such as times and quantities [1]. Partly because of the ambiguity involved in specifying what are and what are not named entities [2]—particularly in the biomedical domain—this task is now sometimes referred to more simply as *entity recognition*.

Vinod D. Kumar and Hannah Jane Tipney (eds.), *Biomedical Literature Mining*, Methods in Molecular Biology, vol. 1159, DOI 10.1007/978-1-4939-0709-0_3, © Springer Science+Business Media New York 2014

Most generally, entity recognition refers to the task of identifying textual mentions with (a typically small number of) predefined categories of interest; thus, each mention of a chemical would be marked up with a generic chemical tag without specifying the identity of the mentioned chemical, only indicating that it is a mention of a chemical. The more specific task of marking up such identified textual mentions with uniquely identifying information corresponding to entries of lexicons, terminologies, ontologies, or databases has been referred to as *term mapping* [3] or *entity normalization* [4]. The focus of this chapter is the annotation of text with elements of biomedical lexical, terminological, and ontological resources, so recognition of entities in text will always refer to this more specific sense here. Furthermore, since abstract concepts not as intuitively thought of as entities (e.g., qualities, functions, phenotypes) or explicitly not categorized as entities (e.g., processes and events, which are not typically considered to be named entities) are increasingly represented in biomedical terminologies and ontologies and important to identify in text, the more expansive *concept* is preferred over entity. *Concept mapping* will refer to the task of annotating textual mentions of any type of concept of interest with entries of lexicons, terminologies, ontologies, or databases representing those concepts, and systems that recognize textual mentions of biomedical concepts but do not link them to such specific entries are outside the scope of this chapter and will not be discussed.

Some of the obstacles to reliable concept mapping in the biomedical domain are also seen in general language and in other subdomains, namely, aspects of natural language such as synonymy (different text strings having the same meaning), polysemy (multiple meanings for a given text string), ambiguity (vagueness in the meaning of a text string), lexical variation, and abbreviation of full-length names of concepts. Additionally, in the biomedical domain, the number of named concepts in lexicons, terminologies, and ontologies is already extremely large and is rapidly growing [5, 6]. Though there are some standards, nomenclature can change quickly; furthermore, some authors instead use their own preferred terminology [7]. Biomedical terminology also tends to be longer and more complex than that in general language, and the exact lexical entries of lexicons, terminologies, and ontologies often do not commonly occur in the literature [8, 9]. Furthermore, until recently, there has been a dearth of gold-standard corpora of annotated documents [10–12] on which text-mining systems could be trained and tested for the task of biomedical concept mapping.

2 Lexicons, Terminologies, and Ontologies

A *lexicon* is essentially a collection of the words (and possibly phrases and expressions) of a language or a sublanguage, along with descriptions of how they may be used and/or how they are related to each other. A *lexical entry* for a given word, phrase, or expression may include information concerning its phonetics, morphology, syntax, frequency, semantics, or other linguistic behaviors. Depending on the types of lexical information contained within, a given lexicon may be useful for a wide range of natural language processing (NLP) tasks, including part-of-speech tagging, word-sense disambiguation, and phrasal parsing; a lexicon may be particularly useful for concept mapping, as the amount of lexical information that is the focus of lexicons may not be as thorough in corresponding terminologies and ontologies. Any typical dictionary is a lexicon, but these are created for human use and tend to be inadequate for computational work [13, 14]. The most widely used computational lexicon of English is WordNet [15], which focuses on semantic relationships between words (e.g., for a given noun, its hypernyms (semantically broader words) and hyponyms (semantically narrower words)). Prominent biomedical lexicons include the SPECIALIST Lexicon of the Unified Medical Language System (UMLS), which includes syntactic, morphological, and orthographic information for both general and biomedical terms and was developed particularly for use by its suite of NLP tools [16], and the BOOTStrep BioLexicon, in which data from large biomedical databases have been semiautomatically integrated and supplemented with morphological, syntactic, and semantic information [17].

A *terminology* (or *vocabulary*) is a collection of terms and phrases for concepts within a given domain of interest; it is often preceded by the word *controlled*, especially for a terminology that has been designed with one or more particular use cases in mind and whose entries are to serve as allowable values of fields [18–20]. Addressing the synonymy, polysemy, and complexity of natural language, terminologies are often designed to be used among distributed resources to unambiguously refer to shared content; for example, the National Cancer Institute (NCI) Thesaurus is based on a comprehensive terminology of cancer-related concepts, including genes and gene products, anatomical parts, organisms, biological processes and pathologies, clinical findings, drugs, and therapies and techniques, and was designed as a shared resource for coding, processing, and exchanging information in cancer research and care [21]. Terminologies often serve as foundations upon which knowledge of the denoted content of the terms is organized; for example, the HUGO Gene Nomenclature Committee (HGNC) oversees the terminology of official unique symbols and names of known human genes, which is used to structure the genomic, phenotypic, and proteomic information for these genes in a centralized database [22]. We broadly include uniquely represented entries of biomedical databases (e.g., the unique identifiers for entries representing species-specific proteins of the UniProt database [23]) in this category, as these essentially constitute controlled vocabularies. While inclusion of synonyms and definitions for terms is common, terminologies typically do not contain other types of lexical information that may be present in a linguistically oriented lexicon.

Ontology was originally defined in the early seventeenth century as a branch of philosophy focused on the nature of being, existence, or reality insofar as which entities can exist (and sometimes which cannot exist) and how they can be categorized and arranged within hierarchies [24]. In the last several decades, information and computer scientists have reinterpreted ontology as a philosophical area of study to *an ontology* as a (typically computational) specification that models a view of a domain of interest, i.e., a "specification of a conceptualization" [25]. Ontologies can be represented in a variety of ways, but at minimum they include classes/concepts/types that are linked via relations/properties; these may be augmented with instances/individuals of these classes, various kinds of restrictions on these relations (e.g., cardinality) at either global or specific class levels, as well as various types of axioms (e.g., rules, assertions of class disjointness). While lexicons can be hierarchically structured and terminologies are commonly developed as such, the classes of ontologies are always arranged into hierarchies. Furthermore, the hierarchical structure of a properly formed ontology should represent strict subclass–superclass relationships (i.e., every instance of a given subclass is also an instance of each of its superclasses); this contrasts to the hierarchies of some lexicons and terminologies, parts of which might be arranged by partonomy (i.e., every instance of a given subclass is a part of each of its superclasses) or by a less strict notion of semantic broadness/narrowness. Ontologies have become central resources for biomedical informatics and have been used to facilitate a wide variety of tasks, including semantic integration, knowledge inference, NLP, and data annotation, querying, and exchange [26, 27]. One of the most prominent ontological projects in the biomedical realm is the Open Biomedical Ontologies library, a collection of ontologies largely constructed in a community-driven approach, whose developers commit to a common set of desiderata including openness, shared syntax, clear versioning, demarcated content, and clear definition [28]. The most well-known among these is the Gene Ontology, which represents gene/gene product activity in terms of biological processes in which they can participate, specific molecular functionalities they can possess, and cellular locations in which they can be active [29].

We have differentiated among lexicons, terminologies, and ontologies, but in actuality the distinctions among these are much fuzzier. Though lexicons and terminologies nominally focus on words and phrases, they commonly are semantically linked to other entries, and though ontologies focus on the representation of some portion of reality (or at least a conceptualization of some portion of reality), they commonly contain linguistic information such as synonyms, alternate spellings, and abbreviations. In practice, these resources are often interchangeably referred to as lexicons, terminologies, vocabularies, or ontologies, even by developers in referring to their own resources. In the NLP task that is the subject of this chapter, these resources serve essentially the same purpose of providing a vocabulary representing biomedical types. Lexical entries of lexicons, terms of terminologies, uniquely identified database entries, and classes of ontologies will be cumulatively referred to as *concepts*, and this discussion of concept mapping will focus on identifying mentions of these formally represented concepts in natural-language documents, often as a key part of more sophisticated text-mining strategies.

3 Tools for Concept Mapping of Biomedical Text

As to the set of tools we have selected for inclusion in this discussion, we have abided by several criteria. First, they must be generically capable of analyzing English biomedical text, i.e., either an arbitrary piece of text or a very broad range of biomedical text such as an arbitrary set of PubMed abstracts. They also must be capable of not only concept recognition (i.e., marking text spans as referring to categories of interest without specifying their identity) but also mapping of text spans to specific entries of one or more lexicons, terminologies (including uniquely represented database entries), or ontologies. Additionally, this must be possible with text specified by users, not previously performed on a defined set of text with the results integrated into a wider system. Furthermore, they must be capable of concept mapping as an independent task that is not necessarily part of a more complex text-mining task (e.g., information extraction), as these more complex tasks will be discussed in subsequent chapters of this book. Finally, we only include tools that are publicly and freely available (via the Internet) and readily usable at the time of this writing; thus, published concept mapping research that is not available as a public tool but could conceivably be implemented by the user is not discussed here. We first discuss tools that map spans of text to a broad range of biomedical vocabularies and then present those that have been developed for more restricted sets of vocabularies.

3.1 Tools That Map Biomedical Text to Concepts of a Broad Range of Vocabularies

MetaMap was one of the earliest biomedical concept-mapping tools; it has been widely used and is widely regarded as a gold standard for this task [30]. It is specifically designed to map textual mentions of concepts to entries of the UMLS Metathesaurus, a large, multipurpose, multilingual thesaurus of millions of biomedical and health-related concepts compiled from more than 100 source vocabularies [31]. It is highly configurable in terms of data (e.g., choice of UMLS version), data model (e.g., types of filtering of candidate mappings), output (e.g., hiding or displaying of semantic types and/or concept unique identifiers of mappings), and processing (e.g., types of lexical variants generated to be used in mappings). MetaMap is by design tightly tied to the UMLS, which results in straightforward mapping of text to UMLS concepts; however, it relies on a dictionary in a specific format and on specific database tables, likely requiring major effort to map to other lexicons, terminologies, or ontologies. MetaMap is available as a Web interface, a Prolog program, a Java API, an Unstructured Information Management Architecture (UIMA) Annotator, or an SKR Web API.

mgrep is a command-line tool that searches text files for lines matching specified regular expressions [32]. It is similar to the popular Unix command grep and extends the latter's single-line regular expressions to multiline patterns. Because it is designed to be fast, scalable, and highly customizable, it was chosen as the concept-mapping engine behind National Center for Biomedical Ontology (NCBO) Annotator (previously named Open Biomedical Annotator (OBA)) [33]. This tool is implemented as a Web service that annotates spans of user-submitted text with concepts of user-selected terminologies and ontologies of the UMLS and/or those in the NCBO BioPortal, a Web portal providing access to biomedical ontologies along with applications relying on them [6]. In addition to direct concept mapping, the user can direct the system to also generate semantically expanded annotations, including annotations for all of a mapped concept's ancestors, annotations based on semantic distance from mapped concepts, and annotations based on existing mappings of concepts from different terminologies or ontologies. Annotations can be outputted as text, in a tab-delimited format, XML, or OWL. A caveat to consider is the fact that although the user may select any combination of UMLS and/or BioPortal vocabularies for annotation, there is no way to specify the versions of these vocabularies to use; thus, if a selected vocabulary changes between two given annotation runs, the performance and results of the annotation may change as well.

ConceptMapper [34] is designed to map spans of text to concepts of terminologies and ontologies that the user must translate into dictionaries in a specific (and fairly basic) XML format. There are many configurable parameters, including those for processing the dictionary (e.g., behavior of attributes attached to dictionary entries), processing the text (e.g., mapping case sensitively or insensitively), dictionary lookup strategies (e.g., token order-independent lookup), and output. ConceptMapper is implemented as a component of the UIMA system [35], an open-source middleware layer for text processing, and its output is in the form of UIMA annotations. Dictionary lookups are token-based and are applied within a specified context, which is typically a sentence but is configurable (e.g., to a noun phrase or a paragraph).

Peregrine is a Java-based application that annotates text with a user-supplied ontology in the form of either a text file of a specific format or a relational database conforming to a specific schema; as with ConceptMapper, the user must translate the ontology/terminology to be used into one of these specifically formatted forms if it is not already in such a form [36]. The concepts of this ontology/terminology are indexed by the system, which then splits the text into tokens, removes stop words, and attempts to match the longest possible text phrase to a concept. After the indexing engine returns results, several disambiguation procedures are attempted to determine if the results are correct in their respective contexts. Peregrine can be downloaded as .jar files or as Maven modules; its source code may also be checked out from its Subversion repository. Since this system depends on LVG Norm, an application of the Lexical Tools package of the National Library of Medicine that can standardize variations of case and plurality/singularity in English words [37], it must be downloaded and made accessible to Peregrine as well.

There are only a small number of formal evaluations of these concept-mapping tools. In an evaluation of MetaMap and mgrep, mgrep was seen to attain higher concept-mapping precision (likely at the expense of recall) and was judged to be more scalable and significantly faster; however, it generally found fewer unique concepts than MetaMap and identified many redundant concepts (i.e., concepts annotating the same text span) [38]. In another evaluation of MetaMap and mgrep, mgrep again performed with higher precision and was significantly faster but found fewer unique concepts; this study also concluded that the mapping scores of MetaMap were more useful and that leveraging these scores could result in increased performance and usability [39]. In a comprehensive evaluation involving over 1,000 parameter combinations for MetaMap, NCBO Annotator, and Concept Mapper, it was found that MetaMap attained the highest recall for five of the eight ontologies examined but performed at lower precision due to false positives generated; that NCBO Annotator attained the highest precision for four of the eight ontologies but performed at lower recall due to the fact that it cannot recognize plurals or other lexical variants of terms; and that ConceptMapper generally balanced precision and recall best of the three and attained the highest F-measure, i.e., the harmonic mean of precision and recall [40], for seven of the eight ontologies [41]. Finally, in an evaluation of mapping of biomedical text to UMLS disease concepts, the performance statistics of Peregrine and MetaMap were quite close, the former better than the latter generally by small amounts as calculated for their annotations whose text spans exactly match those of the gold-standard corpus used in the evaluation [42].

3.2 Tools that Map Biomedical Text to Concepts of More Restricted Sets of Vocabularies

In addition to tools such as those previously discussed that map text to a wide range of terminological resources, systems that can map inputted text to more restricted sets of terminological resources and export these annotations have been implemented. In particular, the centrality of macromolecular sequences in biomedical research has led to the development of several tools that map text to formally represented genes/gene products. GENO combines publicly available background knowledge and machine learning to map text mentions to the species-specific entries of the Entrez Gene database [43]; it is available as a remotely employable UIMA Analysis Engine [44]. GNAT uses background knowledge (e.g., mentions of species or cell lines in closely occurring text) to similarly map text mentions to Entrez Gene identifiers; it is available as an open-source Java library and as a remote Web service [45].

Several other publicly available systems outside the domain of genes/gene products also have been developed. LINNAEUS identifies mentions of organismal species in text and links them to NCBI Taxonomy entries [46]; it is available as a stand-alone software system or as a server, and the annotations can be outputted as XML, HTML, and tab-delimited files or to a database [47]. The Open-Source Chemistry Analysis Routines (OSCAR) toolkit attempts to map all identified mentions of chemicals to concepts of the Chemical Entities of Biological Interest (ChEBI) ontology [48]; the most recent version, OSCAR4, is available as a Java library [49]. ChemSpot, also available as a Java library, also identifies mentions of chemicals and specifies those that were extracted by its dictionary component with identifiers of the Chemical Abstracts Service (CAS) Registry [50]; additionally, methods returning identifiers from other terminological resources, including ChEBI, ChemIDplus [51], and PubChem [52], can be invoked for those annotations linked to CAS Registry identifiers [53]. Whatizit is a suite of Web service modules that can be invoked to annotate text with entries of ChEBI (via OSCAR3), UMLS diseases (via MetaMap), DrugBank [54], the Gene Ontology, the NCBI Taxonomy, or UniProt/Swiss-Prot; input may be in the form of either Unicode-encoded text or a list of identifiers of PubMed articles, whose abstracts are analyzed [55].

Additionally, there exist several publicly available tools with which text of Web pages can be mapped to biomedical concepts and the annotations displayed to the user, principally for browsing purposes. The aforementioned Whatizit also offers a Web interface in which the user can specify text or PubMed article abstracts to be analyzed and have concept annotations subsequently rendered. GoPubMed is a Web server that is designed to allow the user to navigate PubMed abstracts by category; it creates and displays

Gene Ontology (GO) annotations of the text and presents the ontology, with which the user may explore abstracts with semantically related GO annotations [56]. Reflect can identify genes/gene products and small molecules in the text of an inputted URL; clicking on an annotated entity opens a pop-up window containing information gathered about the entity, including links to its representation in prominent databases [57].

3.3 Getting Started with NCBO Annotator

As an introduction for the computational biologist without a background in concept mapping of biomedical text, we present a brief guide to making use of NCBO Annotator. We briefly focus on this tool here because (a) it allows the user to submit arbitrary pieces of text for automatic annotation; (b) it allows the users to annotate their submitted documents with any selected combination of a large number of ontologies and terminologies (any OBO or UMLS ontology or terminology, amounting to 320 at the time of this writing); and (c) it requires the least overhead of any of the systems that can automatically annotate text with a wide range of ontologies and terminologies (as presented in Subheading 3).

The most straightforward way to make use of NCBO Annotator is through its Web interface at the NCBO BioPortal (http://bioportal. bioontology.org/annotator). The user can paste up to 500 words directly into a text box and select any combination of OBO and/or UMLS ontologies and terminologies and the level of ancestor concept annotations to include, if any. For each annotation created, the tool indicates the concept used and its source ontology/terminology (along with hyperlinks to the BioPortal ontology browser), type of annotation (i.e., direct or ancestor), and textual context. These results can be outputted as a text document, a file of comma-separated values, or an XML document.

Alternately, users can programmatically access NCBO Annotator through its API, with extensive documentation at http://data.bioontology.org/documentation#nav_home; included here are definitions of and possible values for all of the configurable annotation parameters as well as a description of the content of the Web service response. Additionally, the developers have provided examples for HTML, Java, Perl, Python, R, and Ruby clients; these are accessible from http://www.bioontology.org/wiki/index.php/Annotator_Client_Examples. These are simple, working examples in which the user needs to only replace values for the text to be annotated, the set of ontologies/terminologies with which to annotate the text, and, if desired, values for the annotation parameters if different from the default settings. Note that to use this Web service, the user must have a BioPortal account; users may sign up (for free) at http://bioportal.bioontology.org/accounts/new.

4 Conclusions

Though there has been a significant amount of research, there are still a limited number of practical, publicly available tools for the mapping of biomedical text to specific concepts in lexicons, terminologies, and ontologies.

For the annotation of text to a wide range of biomedical terminological resources, the user can choose from MetaMap (available as a Web interface, a Prolog program, a Java API, a UIMA Annotator, or an SKR Web API), mgrep (via NCBO Annotator, available as a Web interface or through its API), ConceptMapper (available as a UIMA component), and Peregrine (available as a Java-based application). ConceptMapper and Peregrine can map text to any terminological resource provided the user converts it into a dictionary of a specified format (or, for Peregrine, a database of a specified schema), while NCBO Annotator can annotate with any vocabulary in the UMLS or the NCBO BioPortal and MetaMap only to those vocabularies within the UMLS (without significant additional effort). Among MetaMap and mgrep, mgrep has been shown to be more precise at the expense of recall, while MetaMap with its higher recall finds more correct unique concepts. ConceptMapper has been found to generally balance precision and recall better and therefore attain higher F-measure scores than MetaMap and mgrep. For projects for which speed and/or scalability are concerns, mgrep is significantly faster than MetaMap.

A small number of publicly available tools that map text to more restricted sets of lexicons, terminologies, and ontologies have also been developed. Both GNAT, available as either a Java library or a Web service, and GENO, available as a UIMA Analysis Engine, map mentions of genes/gene products to entries of the Entrez Gene database. LINNAEUS, available as a stand-alone software system or as a server, maps species mentions to NCBI Taxonomy entries. OSCAR4 attempts to map chemical mentions to concepts of the ChEBI ontology, while ChemSpot annotates chemical mentions with CAS Registry identifiers (and indirectly to other resources as well, including ChEBI, ChemIDplus, and PubChem); both are available as Java libraries. Whatizit is a suite of Web service modules that can be invoked to annotate text with entries of ChEBI (via OSCAR3), UMLS diseases (via MetaMap), DrugBank, the Gene Ontology, the NCBI Taxonomy, and UniProt/Swiss-Prot. Finally, several tools can render concept mappings of text primarily for Web browsing; in addition to Whatizit, these include GoPubMed, which creates and displays Gene Ontology annotations of PubMed abstracts, and Reflect, which links identified genes/gene products and small molecules to prominent databases in Web text.

In addition to the limited number of publicly available ready-to-use tools, there is ongoing research in the annotation of biomedical text with concepts from lexicons, terminologies, and ontologies. Researchers can take advantage of recent gold-standard corpora annotated with formally represented biomedical concepts such as the CRAFT Corpus to guide the development of such systems, though it is likely that different strategies will be needed to maximize performance of mapping to vocabularies of different domains. Only with effective automatic concept-mapping tools will systems be able to scalably analyze the biomedical literature and other large sets of documents as a fundamental part of more complex text-mining tasks such as information extraction and hypothesis evaluation and generation.

References

1. Nadeau K, Sekine S (2007) A survey of named entity recognition and classification. Lingvisticae Investigationes 30(1):3–26
2. Tanabe L, Xie N, Thom LH, Matten W, Wilbur WJ (2005) GENETAG: a tagged corpus for gene/protein named entity recognition. BMC Bioinform 6(Suppl I):S3
3. Krauthammer M, Nenadic G (2004) Term identification in the biomedical literature. J Biomed Inform 37:512–526
4. Morgan AA, Lu Z, Wang X, Cohen AM, Fluck J, Ruch P, Divoli A, Fundel K, Leaman R, Hakenburg J, Sun C, Liu H-H, Torres R, Krauthammer M, Lau WM, Liu H, Hsu C-N, Schuemie M, Cohen KB, Hirschman L (2008) Overview of BioCreative II gene normalization. Gen Biol 9(Suppl 2):S3
5. Bales ME, Lussier YA, Johnson SB (2007) Topological analysis of large-scale biomedical terminology structures. J Am Med Inform Assoc 14:788–797
6. Whetzel PL, Noy NF, Shah NH, Alexander RR, Nyulas C, Tudorache T, Musen MA (2011) BioPortal: enhanced functionality via new Web services from the National Center for Biomedical Ontology to access and use ontologies in software applications. Nucleic Acids Res 39(Web Server issue):W541–W545
7. Chen L, Liu H, Friedman C (2005) Gene name ambiguity of eukaryotic nomenclatures. Bioinformatics 21:248–255
8. Hirschman L, Morgan AA, Yeh AS (2002) Rutabaga by any other name: extracting biological names. J Biomed Inform 35(4): 247–259
9. McCray AT, Browne AC, Bodenreider O (2002) The lexical properties of the gene ontology. Proc AMIA Annual Symp, 504–508
10. Kim JD, Ohta T, Tateisi Y, Tsujii J (2003) GENIA corpus: a semantically annotated corpus for bio-text mining. Bioinformatics 19(Suppl 1):i180–i182
11. Pyysalo S, Ginter F, Heimonen J, Björne J, Boberg J, Järvinen J, Salakoski T (2007) BioInfer: a corpus for information extraction in the biomedical domain. BMC Bioinform 8:50
12. Bada M, Eckert M, Evans D, Garcia K, Shipley K, Sitnikov D, Baumgartner Jr. WA, Cohen KB, Verspoor V, Blake JA, Hunter LE (2012) Concept annotation in the CRAFT corpus. BMC Bioinform 13:161
13. Briscoe T (1991) Lexical issues in natural language processing. In: Klein E, Veltman F (eds) Natural language and speech. Springer, Berlin
14. Hirst G (2009) Ontology and the Lexicon. In: Staab S, Studer S (eds) Handbook on ontologies. Springer, Berlin, pp 269–292
15. Fellbaum C (1998) WordNet: an electronic lexical database. MIT Press, Cambridge, MA
16. McCray AT, Srinavasan S, Browne AC (1994) Lexical methods for managing variation in biomedical terminologies. Proc Annu Symp Comput Appl Med Care, 235–239
17. Quochi V, Monachini M, Del Gratta R, Calzolari N (2008) A lexicon for biology and bioinformatics: the BOOTStrep experience. Proceedings international conf on language resources and evaluation (LREC) 2008, Marrakech, Morocco
18. Chute C (2000) Clinical classification and terminology: some history and current observations. J Am Med Informatics Assoc 7(3): 298–303
19. Svenonius E (2003) Design of controlled vocabularies. In: Drake M (ed) Encyclopedia of library and information science. Marcel Dekker, New York, NY, pp 822–838

20. Ingenerf J, Pöppl S (2007) Biomedical vocabularies: the demand for differentiation. Proc Internat Conf Med Informatics (MEDINFO) 2007, Brisbane
21. Sioutos N, de Coronado S, Haber MW, Hartel FW, Shaiu W-L, Wright LW (2007) NCI thesaurus: a semantic model integrating cancer-related clinical and molecular information. J Biomed Inform 40:30–43
22. Gray KA, Daugherty LC, Gordon SM, Seal RL, Wright MW, Bruford EA (2013) Genenames.org: the HGNC resources in 2013. Nucl Acids Res 41(Database issue):D545–D552
23. The UniProt Consortium (2012) Reorganizing the protein space at the Universal Protein Resource (UniProt). Nucleic Acids Res 40(D1): D71–D75
24. Smith B (2003) Ontology. In: Floridi L (ed) Blackwell guide to the philosophy of computing and information. Blackwell, Oxford, pp 155–166
25. Gruber TR (1995) Toward principles for the design of ontologies used for knowledge sharing. Int J Hum Comp Stud 43(5/6):907–928
26. Bodenreider O, Stevens R (2006) Bio-ontologies: current trends and future directions. Brief Bioinform 7(3):256–274
27. Rubin DL, Shah NH, Noy NF (2007) Biomedical ontologies: a functional perspective. Brief Bioinform 9(1):75–90
28. Smith B, Ashburner M, Rosse C, Bard C, Bug W, Ceusters W, Goldberg LJ, Eilbeck K, Ireland A, Mungall CJ, The OBI Consortium, Leontis N, Rocca-Serra P, Ruttenberg A, Sansone SA, Scheuermann RH, Shah N, Whetzel PL, Lewis S (2007) The OBO Foundry: coordinated evolution of ontologies to support biomedical data integration. Nat Biotechnol 25:1251–1255
29. The Gene Ontology Consortium (2000) Gene Ontology: tool for the unification of biology. Nat Genet 25:25–29
30. Aronson AR, Lang F-M (2010) An overview of MetaMap: historical perspective and recent advances. J Am Med Inform Assoc 17:229–236
31. Schuyler PL, Hole WT, Tuttle MS, Sherertz DD (1993) The UMLS Metathesaurus: representing different views of biomedical concepts. Bull Med Libr Assoc 81(2):217–222
32. Dai M, Shah NH, Xuan W, Musen MA, Watson SJ, Athey BD, Meng F (2008) An efficient solution for mapping free text to ontology terms. Proc AMIA Summit Translat Bioinform
33. Jonquet C, Shah NH, Musen MA (2009) The open biomedical annotator. Proc AMIA Summit Translat Bioinform
34. Tanenblatt M, Coden A, Saminsky I (2010) The ConceptMapper approach to named entity recognition. Proc 7th Internat Conf Lang Resources and Eval (LREC)
35. Ferrucci D, Lally A (2004) UIMA: An architectural approach to unstructured information processing in the corporate research environment. Nat Lang Eng 10(3–4):327–348
36. Schuemie MJ, Jelier R, Kors JA (2007) Peregrine: lightweight gene name normalization by dictionary lookup. Proc 2nd BioCreative Challenge Evaluation Workshop, 131–133
37. Browne AC, Divita G, Lu C, McCreedy L, Nace D (2003) Lexical systems; a report to the board of scientific counselors. Lister Hill National Center for Biomedical Communications Technical Report LHNCBC-TR-2003-003
38. Shah NH, Bhatia N, Jonquet C, Rubin D, Chiang AP, Musen MA (2009) Comparison of concept recognizers for building the open biomedical annotator. BMC Bioinform 10 (Suppl 9):S14
39. Stewart SA, von Maltzahn ME, Abidi SSR (2012) Comparing MetaMa to MGrep as a tool for mapping free text to formal medical lexicons. Proc 1st international workshop on knowledge extraction and consolidation from social media (KECSM)
40. Hripcsak G, Rothschild AS (2005) Agreement, the F-measure, and reliability in information retrieval. J Am Med Inform Assoc 12:296–298
41. Funk C, Baumgartner Jr. W, Garcia B, Roeder C, Bada M, Cohen KB, Hunter LE, Verspoor K (2013) Large-scale biomedical concept recognition: an evaluation of current automatic annotators and their parameters. BMC Bioinform
42. Kang N, Singh B, Afzal Z, van Mulligen EM, Kors JA (2013) Using rule-based natural language processing to improve disease normalization in biomedical text. J Am Med Inform Assoc 0:1–6
43. Maglott D, Ostell J, Pruitt KD, Tatusova T (2011) Entrez Gene: gene-centered information at NCBI. Nucl Acids Res 39(Database Issue):D52–D57
44. Wermter J, Tomanek K, Hahn U (2009) High-performance gene name normalization with GENO. Bioinformatics 25(6):815–821
45. Hakenberg J, Gerner M, Haeussler M, Solt I, Plake C, Schroeder M, Gonzalez G, Nenadic G, Bergman CM (2011) The GNAT library for local and remote gene mention normalization. Bioinformatics 27(19):2769–2771
46. Sayers EW, Barrett T, Benson DA, Bryant SH, Canese K, Chetvernin V, Church DM, DiCuccio M, Edgar R, Federhen S, Feolo M, Geer LY, Helmberg W, Kapustin Y, Landsman D, Lipman DJ, Madden TL, Maglott DR, Miller V, Mizrachi I, Ostell J, Pruitt KD,

Schuler GD, Sequeria E, Sherry ST, Shumway M, Sirotkin K, Souvarov A, Starchenko G, Tatusova TA, Wagner L, Yaschenko E, Ye J (2009) Database resources of the National Center for Biotechnology Information. Nucl Acids Res 37(Database Issue):D5–D15

47. Gerner M, Nenadic G, Bergman CM (2010) LINNAEUS: a species name identification system for biomedical literature. BMC Bioinform 11:85
48. Degtyarenko K, de Matos P, Ennis M, Hastings J, Zbinden M, McNaught A, Alcantara R, Darsow M, Guedj M, Ashburner M (2008) ChEBI: a database and ontology for chemical entities of biological interest. Nucl Acids Res 36(Database Issue):D344–D350
49. Jessop DM, Adams SE, Willighagen EL, Hawizy L, Murray-Rust P (2011) OSCAR4: a flexible architecture for chemical text-mining. J Cheminform 3:41
50. Weisgerber DW (1997) Chemical abstracts service chemical registry system: history, scope, and impacts. J Am Soc Inform Sci 48(4): 349–360
51. Tomasulo P (2002) ChemIDplus: super source for chemical and drug information. Med Ref Serv Q 21(1):53–59
52. Li Q, Cheng T, Wang Y, Bryant SH (2010) PubChem as a public resource for drug discovery. Drug Discov Today 15(23–24): 1052–1057
53. Rocktäschel T, Weidlich M, Leser U (2012) ChemSpot: a hybrid system for chemical named entity recognition. Bioinformatics 28(12): 1633–1640
54. Knox C, Law V, Jewison T, Liu P, Ly S, Frolkis A, Pon A, Banco K, Mak C, Neveu V, Djombou Y, Eisner R, Guo AC, Wishart DS (2011) DrugBank 3.0: a comprehensive resource for "Omics" research on drugs. Nucl Acids Res 39(Database Issue): D1035–D1041
55. Rebholz-Schuhmann D, Arregui M, Gaudan S, Kirsch H, Jimeno A (2008) Text processing through Web services: calling Whatizit. Bioinformatics 24(2):296–298
56. Doms A, Schroeder M (2005) GoPubMed: exploring PubMed with the gene ontology. Nucl Acids Res 33(Web Server Issue): W783–W786
57. Pafilis E, Donoghue SI, Jensen LJ, Horn H, Kuhn M, Brown NP, Schneider R (2009) Reflect: augmented browsing for the life scientist. Nat Biotechnol 27:508–510

Chapter 4

Text Mining for Drug–Drug Interaction

Heng-Yi Wu, Chien-Wei Chiang, and Lang Li

Abstract

In order to understand the mechanisms of drug–drug interaction (DDI), the study of pharmacokinetics (PK), pharmacodynamics (PD), and pharmacogenetics (PG) data are significant. In recent years, drug PK parameters, drug interaction parameters, and PG data have been unevenly collected in different databases and published extensively in literature. Also the lack of an appropriate PK ontology and a well-annotated PK corpus, which provide the background knowledge and the criteria of determining DDI, respectively, lead to the difficulty of developing DDI text mining tools for PK data collection from the literature and data integration from multiple databases.

To conquer the issues, we constructed a comprehensive pharmacokinetics ontology. It includes all aspects of in vitro pharmacokinetics experiments, in vivo pharmacokinetics studies, as well as drug metabolism and transportation enzymes. Using our pharmacokinetics ontology, a PK corpus was constructed to present four classes of pharmacokinetics abstracts: in vivo pharmacokinetics studies, in vivo pharmacogenetic studies, in vivo drug interaction studies, and in vitro drug interaction studies. A novel hierarchical three-level annotation scheme was proposed and implemented to tag key terms, drug interaction sentences, and drug interaction pairs. The utility of the pharmacokinetics ontology was demonstrated by annotating three pharmacokinetics studies; and the utility of the PK corpus was demonstrated by a drug interaction extraction text mining analysis.

The pharmacokinetics ontology annotates both in vitro pharmacokinetics experiments and in vivo pharmacokinetics studies. The PK corpus is a highly valuable resource for the text mining of pharmacokinetics parameters and drug interactions.

Key words Pharmacokinetics, Pharmacodynamics, Drug–drug interaction, Text mining, Corpus, Ontology, Relation extraction, Enzyme, Transporter

1 Introduction

Adverse drug reaction (ADR) is one of the major causes of morbidity and mortality occurring in clinical care every year. To investigate the crucial problem, the US Food and Drug Administration (FDA) found that more than 40 % of the US population is prescribed more than four medications at a single time, which makes them more susceptible to ADR [1]. A literature search in Medline and Embase database from 1990 to 2006

Vinod D. Kumar and Hannah Jane Tipney (eds.), *Biomedical Literature Mining*, Methods in Molecular Biology, vol. 1159, DOI 10.1007/978-1-4939-0709-0_4,

showed that drug–drug interactions (DDIs) were held responsible for 0.054 % of the emergency department (ED) visits, 0.57 % of the hospital admissions, and 0.12 % of the re-hospitalizations [2]. It is possible that drug interaction can be beneficial or detrimental. The use of multiple drugs might provide synergism such as increasing the efficacy of therapeutic effect, decreasing dosage but holding the same efficacy to avoid toxicity, or minimizing the drug resistance [3]. However, we have more interests in the investigation of negative interaction because pathological significance is often unexpected and hard to be diagnosed. To predispose DDI, the importance of high-risk factors like age, polypharmacy, and genetic polymorphisms should be carefully evaluated [4]. In the elder population, DDIs account for 4.8 % of the hospital admissions, which is much higher than the proportion of DDI victims within the total population. The reason is directed to the abatement of liver metabolism or kidney function [5, 6]. Genetic polymorphism has profound influence on enzyme function, which might result in increased drug metabolism and absence of drug response. Evidences [7] suggested that patients affected by genetic polymorphisms will experience severe toxicities upon drug intake.

For economic aspect, the problem of DDI effect or co-medication effect has scaled such heights that it has even led to withdrawing of drugs from the market after approval. The 1990s saw the withdrawal of more than 11 drugs as shown in ref. 8. In 2007, the biopharmaceutical industry invested roundabout $58.8 billion for the research and development as the withdrawing of drugs [9] is a major setback to the industry as the deployment of a single drug compound is estimated at $200 million.

1.1 Drug–Drug Interaction Mechanisms and In Vitro and In Vivo Drug Interaction Studies

DDI can result when a substance affects the activity of a drug or its metabolites when these two drugs are administrated at the same time. The simultaneous administration of two drugs, which causes synergistic or antagonistic effect, might lead to the alternation of medication effectiveness or some harming effects on patient body. Those potential influences on human body should be noticed to prevent from a high risk of multiple interactions because the number of approved drugs increases. To preclude the possibility of hazardous interaction, understanding the significant scientific principles or mechanisms of DDI is important.

Due to the continued growth in drug development and the insight into molecular biology, we come to realize that transporter and enzyme played an important role in drug elimination, which inspired a clue to dig the mechanisms surrounding DDI. In brief, there are two major molecular mechanisms of DDI, enzyme-based drug metabolism and transporter-based drug transportation [10].

If an enzyme that is responsible for the metabolism of one drug is induced or inhibited by another drug, then the clearance of original drug will be changed, which might result in being toxic or less effective. For transporter-based drug transportation, transporter is important to drug deposition. Drugs can be metabolized only after they are transported into liver cells. To understand how a transporter-mediated DDI happens, the knowledge of the transporter substrates and inhibitors can suggest potential DDIs [11].

There are two basic types of drug interaction, pharmacokinetics (PK) and pharmacodynamics (PD). In short, PK investigates the activity of drug combinations with drug absorption, disposition, metabolism, excretion, and transportation (ADMET), which describes how these five criteria influence drug level (concentration). Pharmacokinetically speaking, potentiative or reductive combinations are, respectively, correlated to positive or negative modulation of drug transport, permeation, distribution, localization, or metabolism. Potentiative modulation of drug transport will enhance drug absorption via the disruption of transport carrier, increase drug concentration in plasma by inhibiting metabolic process, and stimulate or inhibit the metabolism of drugs into active or inactive form. On the other hand, reductive modulation provides contrasting perspectives to potentiative modulation. The reductive modulation of drug transport typically blocks drug absorption, decreases drug concentration in plasma, and reduces drug metabolism activity [12]. Those information brings to systematically investigate the physiological and biochemical mechanisms of drug exposure in multiple tissue types, cells, animals, and human subjects [13], which links preclinical and clinical phase of drug development. If the PK can be interpreted as the dose–concentration relationship, pharmacodynamics (PD) can be defined as the mechanism of drug action and relationship between drug concentration and effect. A drug's pharmacodynamics effect ranges widely from the molecular signals (such as its targets or downstream biomarkers) to clinical symptoms (such as the efficacy or side effect endpoints). Classification of its therapeutic effects: It can be synergistic, additive, or antagonistic if the effect is greater than, equal to, or less than the summed effects of drug combinations [12].

As stated in the previous section, the complicated transporter–enzyme interplay in the deposition of drug leads to the difficulty for the identification of DDIs in drug administration and drug development. Thus, understanding the molecular mechanism underlying different types of drug interaction could facilitate the discovery of novel DDI. Recently, in vitro technologies can qualitatively provide an insight into the potential DDI based on the observation of enzyme kinetic parameters. Via ADME screening efforts as well as the assessment of CYP inhibition, the choice of test compound inhibiting the metabolism of one probe substrate for an enzyme in the in vitro experiment can be fulfilled to carry out the prediction of in vivo DDI. Wienkers and Heath [14] addressed the basic principles of in vitro inhibition prediction underlying the generation of in vitro drug metabolism data and suggested several factors that introduced error or uncertainty into a quantitative prediction of in vivo DDI based on in vitro-derived PK parameters. In ref. 15, three factors authors recommended for the ideal model to predict metabolic drug–drug interaction (M-DDI) should be an accurate measurement of the average increase in the area under the plasma concentration–time curve (AUC) of a victim drug following administration of a perpetrator drug, the plasma binding displacement interaction, and the impact of the concentration–time profile of the inhibitor. To evaluate the potential for M-DDI [15] developed an in silico software SIMCYP, which incorporates extensive data on demographics; disease states; anatomical, physiological, genetic, and biochemical variables; and input of information on in vitro drug metabolism and transport.

1.2 Computational Drug Interaction Prediction and Drug Interaction Text Mining

1.2.1 Overview of Computational Drug Interaction Prediction

The evaluation of the potential risk of DDI is of importance in patient safety since DDIs can raise the danger of patients and the cost of healthcare system. According to the guidance for industry from the Food and Drug Administration [16], study design, data analysis, and implication for dosing and labeling are suggested to deal with drug interaction studies. When studying DDI for a new drug, it usually begins with in vitro study to determine whether a drug is a substrate, inhibitor, or inducer of metabolizing enzymes. The consequence of in vitro investigations can serve as an evidence to screen out the candidate potential drug pairs for additional in vivo study. To conduct an in vivo DDI study for an investigating drug, a quantitative analysis to mathematically describe the kinetics of drug metabolism involved in ADME process is needed. The basic model for the initial assessment of DDI based on in vitro and in vivo studies can be achieved by physiologically based pharmacokinetics (PBPK) modeling. From published in vitro experiments and in vivo studies [17–24] had developed Bayesian models and computational algorithms to construct PBPK models for DDI prediction.

Another common way to explore novel DDI is literature-based discovery. The hidden knowledge among information embedded in publications can be dug out through finding connections between articles. To this end, many researchers took advantage of some commercial or public databases as resource, such as Metabolism and Transport Drug Interaction Database (DIDB) [25], PharGKB [26], and DrugBank [27] which provided extensive lists of DDI information published in articles, clinical files, or biomedical research reports. Gottlieb et al. [28] proposed a computational framework INDI to infer and explore DDI by calculating similarity measurement between drug pair via diverse feature measurements, i.e., chemical based, ligand based, side effect based, annotation based, and sequence based. However, the problem of data inconsistency arose when using different databases. Some significant scientific evidences associated with DDI are limited or lacking in some existing databases. This deficiency is hard to prevent because the tasks of data collections are manually accomplished by different research groups or professional experts. To conquer this problem, employing the technologies from information retrieval (IR) or natural language processing (NLP) can be a solution to help extract data more efficiently and consistently.

1.2.2 Biomedical Text Mining

Text mining refers to the process of deriving high-quality information from text, which relies on NLP. To translate the text into computer-readable language, there are some basic steps of NLP [29], including sentence splitting, tokenization, part of speech, named entity recognition (NER), shallow parsing, and syntactic parsing. In this section, we do not go into the details of techniques for NLP tools. The attentions will be paid more on the tasks of corpus construction, IR, or information extraction (IE), which employs highly scalable statistics-based techniques to index and search large volume of text efficiently.

Extracting facts from texts is the goal of text-mining systems. The range of extraction tasks can be narrow from retrieving potentially relevant articles by sophisticated keyword search or classifying papers into different ontological types (IR), recognizing biological entities or concepts in text, and detecting relations between biological entities (IE) and broader to document summarization or question answering (beyond IE) [30]. To fulfill those tasks in biomedical domain, NER is an initial processing step because the significant knowledge is usually centered on the mechanism of biological activities which are described by nominalized verbs and nouns within sentences. Therefore, identifying text that satisfies various types of information needs is an important first step toward accurate text mining. But how to utilize the identified entities for improving text mining is challenging. One solution to this problem is an annotated corpus. The corpus annotated with such information allows real usage within text to be taken into account. The annotated sentence then can be represented in syntactic and semantic format, which shows the different levels of scientific characteristics. However, the strategy of constructing corpus is diverse. It differs with the purpose of text mining task and the methodology we used in extracting information. Kim et al. [31] introduces GENIA corpus with linguistically rich annotations for biomedical articles. The value of GENIA corpus comes from its annotations. All biologically meaningful terms are semantically annotated with descriptors from GENIA ontology. Wilbur et al. [32] suggest the basic guideline and criteria of corpus construction and annotation task for facilitating the training components of IE system by using machine learning method. Another value of annotated corpus is being a gold standard that facilitates the evaluation of approach. The success of practical applications crucially depends on the quality of extraction results, which is against the access of gold standard reference.

1.2.3 Relationship Extraction

Within IE methods, we are more interested in relationship extraction. The goal of relationship extraction is to detect the prespecified type of relation between a pair of entities of given types. A relation is typically represented as a pair of entities, linked by an arc that is either directed or undirected. The arc is given a label usually corresponding to a semantic type. In biology, the type of entities can be very specific such as gene, protein, or drug, while the type of relationship can be referred from some particular verbs, including transcribe, repress, or inhibit.

To effectively extract relationship, analysis of sentence structure is necessary. The use of semantic processing or deep parsing techniques that analyze both the syntactic and semantic structure of texts can benefit relation extraction. Several approaches had been reported in literature to extract the relation of interest. Generally, there are three main approaches for relationship extraction: co-occurrence-based, rule-based, and machine learning based approaches. Muller et al. [33] employ co-occurrence-based method, which is the simplest way to capture relationships relying on co-occurrence of two entities to derive a relation. Rule-based approaches [34, 35] are to take advantage of linguistic technology to grasp syntactic structure or semantic meaning for understanding the relationship from the unstructured text. Feldman et al. [34] employed an NP1–verb–NP2 template to identify the relation between two domain-specific entities. Fundel et al. [35] constructed a set of domain-specific rules and apply them to dependency parse tree to capture different forms of expressing a given relationship. Finally, classifiers using machine learning approaches such as support vector machines (SVM) [36] are often used for relation extraction. This method needs laborious efforts to define grammars or rules, and text in training dataset is manually tagged by a human expert. This text mining method uses the training data to automatically learn the "rules" so it can mine wanted information or identify the necessary knowledge [37–40].

The comparison among different methods is not easy because each method obtains its inherited pros and cons. Co-occurrence method provides the highest recall but poor precision among three. A large amount of false-positive relations are returned whenever the sentence is sophisticated with more than two entities or two key entities co-occurred in each single sentence but it does not state their relationship. Thus, co-occurrence method is more suitable to use as a simple baseline method for performance comparison.

Rule-based method achieves better precision in extracting binary relationships due to the more precise rule conditions for defining relationship. But when it meets the complex sentence with various coordinates and relational clauses, the performance turns down obviously [41]. In general, machine learning-based method performs the best among methods. As an evidence in BioCreative challenge [42], the frameworks using supervised machine learning algorithm outperformed the existing methods in detecting protein–protein interaction (PPI). One important advantage is that system can predict categories for unseen samples. However, this advantage is heavily relying on annotated corpus [43]. Therefore, it can also be a big disadvantage because of the need for huge learning set.

1.2.4 Literature Review for Extracting Drug–Drug Interaction

Different approaches had been developed for extracting biomedical relationships such as PPI. From the experience of previous researches centered on PPI [36–40], few approaches have been proposed to the problem of detecting DDI. To promote the development of DDI extraction tools, DDIExtraction 2011, the first challenge task on DDI extraction, was held in 2011 at Spain. In this workshop, they provided evidence for the most effective methods available to solve specific problems and reveal the performance on these problems. In competition, most participants proposed systems using classifiers SVM or RLS. Their choices verified that machine learning can outperform other methods in relation extraction. Observed from results, approaches based on kernel methods achieved better performance than the classical feature-based methods [44]. Thus, the advantages of kernel-based method using machine learning classifier are spotlighted in this workshop.

In literature, some articles are outstanding in DDI extraction. The co-chairs of DDIExtraction 2011 [43] proposed a hybrid approach, which combines shallow parsing and pattern matching to extract relation between drugs based on annotated corpus. It utilizes the proposed syntactic patterns to split the sentence into clauses from which relations are extracted by matching patterns. The ability of dealing with complicated sentence is the advantage of this method. Complexity can be diminished by separating a long sentence into simplified clauses and by the detection of the apposition and coordinate structure. But there is one gap in the extraction of DDI information if used in pharmacokinetics or pharmacogenetics articles. Only exploring DDI based on literal denotation will lead to the missing detection of actual DDI information due to the lack of scientific knowledge. In ref. 45, DDIs are identified by aggregating gene–drug interactions which are extracted via rule-based method. The extracted interactions are then normalized and mapped into their standardized ontology to form the semantics network. The network could be useful to find potential DDIs. Differed from Percha et al. [45] who extracted DDI via the perspective of pharmacogenetics, Teri et al. [46] developed a method that combined text mining and automated reasoning to predict enzyme specific DDIs. In most situations, the extracted relations from the results of conventional relation extraction are not sufficient to derive DDI. By representing the general knowledge related to metabolism and interaction with the form of logic rules, DDI can be acquired in the reasoning phase.

2 Materials

For PK DDI text mining, the materials for the construction of PK ontology are prepared. A descriptor from specialized ontology can be used to describe the environment of PK experiments (in vivo and in vitro) and the nature of drug mechanisms (all drug metabolism and transportation enzymes.

For drug name, the dictionary is created using drug names from DrugBank 3.0 [27]. DrugBank consists of 6,829 drugs which can be grouped into different categories of FDA-approved, FDA-approved biotech, nutraceuticals, and experimental drugs. The drug names are mapped to generic names, brand names, and synonyms. The environment condition-specific in vitro PK experiment and their associated PK parameters are referred from [47–50]. The materials for in vivo study are summarized from two textbooks [13, 51]. The information of tissue-specific transporters and enzymes with all their probe inhibitors, inducers, and substrates were collected from industry standard (http://www.fda.gov/Drugs/GuidanceComplianceRegulatoryInformation/Guidances/ucm064982.htm), reviewed in the top pharmacology journal [16].

3 Methods

To extract PK DDI by text-mining system, there are three noteworthy issues we should carefully deal with. (1) Recognition of drug name is one of the most salient issues in DDI text mining. Without satisfied performance in tagging drug name, false-positive or missing detection eliminates the accuracy of DDI results. Unlike gene's or protein's name, the representations of drug name are more sophisticated. The same drug may show in different documents with a number of ways, especially for metabolites of a compound [52]. The diversity of naming conventions perplexes the identification of drug names in pharmacokinetics articles. (2) Ontology is the main repository of formally represented knowledge for DDI text-mining system. The hierarchical repository provides a framework for knowledge integration and sharing, which give machine-readable descriptors of biomedical concepts and their relations. The challenge for ontology construction is to develop appropriate ontology resources and link them to adequate terminological lexicons [53]. (3) Corpus construction is essential to make text mining successful. It is not possible for a machine to capture useful information from text data written in natural language directly. To bridge the gap between text data and machine, corpus creates the accessibility for computer to read text data precisely [31, 54]. Another important issue within corpus is the scheme of biological annotation. The task of annotation can be regarded as identifying and classifying entities or sentences according to predefined categories. A well-defined scheme for annotation task is indispensable to corpus construction.

An ideal system for PK DDI extraction should provide not only a comprehensive list of DDIs in a cost-efficient manner but also the mechanism behind interactions. In current DDI extraction methods, most researches extract DDIs centering on exploring the semantics of sentence. Given a sentence with at least two drugs, they analyze sentence structure and identify drug entities and trigger words (e.g., verbs like inhibit or induce) to accomplish this task. However, in most situations, complete DDI information is presented in complicated ways with more than a single sentence. More concrete DDI conditions such as experimental measurements might be mentioned in those sentences which only have a single drug. For instance, the way to express DDI information in pharmacokinetics articles is quite different from that in pharmacodynamics articles. The sentence only with a single drug frequently mentions its corresponding PK parameters or other measurements, which show the practical conditions for drug metabolism. The merit of those parameters gives the clue to determine inhibition or induction of DDI as well as provides a criterion to exam the reliability of found DDIs.

To meet the abovementioned issues, this chapter tries to propose a system to detect DDI information not only from narrative sentences but also in those sentences with a single drug, which contains possible DDI candidate. Besides detecting DDI pairs from sentence structure, considering PK parameter as an evidence to determine DDI is important in our strategy. In the following sections, we carefully discuss how the task of drug name mapping works, the construction of an integrated pharmacokinetics ontology and corpus for text-mining system, and finally how to apply them in the text-mining system.

3.1 Drug Name Mapping

To detect the name by using NER, the performance of DDI extraction matters if the accuracy of drug name identification is not satisfied [52].

Drug names were created using the drug names from DrugBank 3.0 [27]. DrugBank consists of 6,829 chemicals with unique DrugBank ID which can be grouped into different categories of FDA-approved, FDA-approved biotech, nutraceuticals, and experimental drugs. The chemicals are mapped to generic names, brand names, and synonyms which results in 36,433 unique DrugBank ID–name pairs. 315 names in DrugBank have less than 4 letters such as chloramphenicol, DB0046 has a synonym CAP, and cholecalciferol, DB00169 has a synonym CC. The words with less than four letters may cause bad NER; therefore, they were removed.

In addition, drug metabolites were also tagged, because they are important in in vitro studies. The metabolites were judged by either prefix or suffix: oxi, hydroxyl, methyl, acetyl, N-dealkyl, N-demethyl, nor, dihydroxy, O-dealkyl, and sulfo. These prefixes and suffixes are due to the reactions due to phase I metabolism (oxidation, reduction, hydrolysis) and phase II metabolism (methylation, sulfation, acetylation, glucuronidation) [55].

3.2 PK Ontology Construction

The motivation for ontology in biomedical text mining is to make sense of raw text. According to the defined concepts, properties, relationships, instances, and axioms for a given domain, raw text can be interpreted by the descriptors of ontology with a standardized format and organized into hierarchical structure. Such advantages allow complex text to be represented with semantic and consistent manner [56].

The process of building ontology is a complex and tedious process. Various domain-specific resources and lexicons are required to satisfy the needs of a text-mining system using in a specific scope. According to the introduction of DDI mechanism we mentioned in Subheading 1.1, the domain of PK DDI is concerned with the process of drug disposition within the organism, the response of drug level, and the kinetics of drug exposure to different tissue types. Even in different experimental studies, DDI is defined with distinct measurements. However, no single system is currently capable of covering a complete domain for all aspects. For this reason, we introduce an integrated PK ontology which is composed of several components: experiment, metabolism enzyme, transporter, drug, and subject. In this work, the primary contribution is the ontology development for the PK experiment and integration of the PK experiment ontology with other PK-related ontologies.

Experiment specifies in vitro and in vivo PK studies and their associated PK parameters. The definitions and units for both in vitro or in vivo PK parameters and their corresponding experiment conditions should be included.

Within different types of in vitro PK experiments, different in vitro *PK parameters* are employed.

- *Single-drug metabolism experiment* includes Michaelis–Menten constant (K_m), maximum velocity of the enzyme activity (V_{max}), intrinsic clearance (CL_{int}), metabolic ratio, and fraction of metabolism by an enzyme (fm_{enzyme}) [47].

- *Single-drug transporter experiment*: PK parameters include apparent permeability (Papp), ratio of the basolateral to apical permeability and apical to basolateral permeability (Re), radioactivity, and uptake volume [57].
- *Drug interaction experiment*: IC_{50} is the inhibition concentration that inhibits to 50 % enzyme activity; it is substrate dependent; and it does not imply the inhibition mechanism. K_i is the inhibition rate constant for competitive inhibition, noncompetitive inhibition, and uncompetitive inhibition. It represents the inhibition concentration that inhibits to 50 % enzyme activity, and it is substrate concentration independent. K_{deg} is the degradation rate constant for the enzyme. K_I is the concentration of inhibitor associated with half maximal inactivation in the mechanism-based inhibition; and K_{inact} is the maximum degradation rate constant in the presence of a high concentration of inhibitor in the mechanism-based inhibition. E_{max} is the maximum induction rate, and EC_{50} is the concentration of inducer that is associated with the half maximal induction [15].
- *Type of drug interaction*: There are multiple drug interaction mechanisms, including competitive inhibition, noncompetitive inhibition, uncompetitive inhibition, mechanism-based inhibition, and induction [15].

For *in vitro experiment conditions*, metabolism enzyme, transporter, and some other factors should be considered.

- *Metabolism enzyme* experiment conditions include buffer, NADPH sources, and protein sources. In particular, protein sources include recombinant enzymes, microsomes, and hepatocytes. Sometimes, genotype information is available for the microsome or the hepatocyte samples.
- *Transporter* experiment conditions include bidirectional transporter, uptake/efflux, and ATPase.
- *Other factors* of in vitro experiments include preincubation time, incubation time, quantification methods, sample size, and data analysis methods.

All these information can be found in the FDA website (http://www.abclabs.com/Portals/0/FDAGuidance_DraftDrugInteractionStudies2006.pdf).

Differed from in vitro study, in vivo refers to experimentation using a whole, living organism such that its experiment condition and parameters are quite different. Within in vivo study, in vivo PK parameters, pharmacokinetics models, study designs, and quantification methods are the key components to investigate an in vivo experiment.

- All of the information for in vivo *PK parameters* is summarized from two text books [13, 51]. There are several main classes of PK parameters. Area under the concentration curve parameters are AUC_{inf}, AUC_{SS}, AUC_t, and AUMC; drug clearance parameters are CL, CL_b, CL_u, CL_H, CL_R, CL_{po}, CL_{IV}, CL_{int}, and CL_{12}; drug concentration parameters are C_{max} and C_{SS}; extraction ratio and bioavailability parameters are E, E_H, F, F_G, F_H, F_R, f_e, and f_m; rate constants include elimination rate constant k, absorption rate constant k_a, urinary excretion rate constant k_e, Michaelis–Menten constant K_m, distribution rate constants (k_{12}, k_{21}), and two rate constants in the two-compartment model (λ_1, λ_2); blood flow rate (Q, Q_H); time parameters (t_{max}, $t_{1/2}$); volume distribution parameters (V, V_b, V_1, V_2, V_{ss}); maximum rate of metabolism, V_{max}; and ratios of PK parameters that present the extent of the drug interaction (AUCR, CL ratio, C_{max} ratio, C_{ss} ratio, $t_{1/2}$ ratio).

- Two types of *pharmacokinetics models* are usually presented in the literature: non-compartment model and one- or two-compartment models.
- The *design strategies* are very diverse: single arm or multiple arms, crossover or fixed-order design, with or without randomization, with or without stratification, pre-screening or no pre-screening based on genetic information, prospective or retrospective studies, and case reports or cohort studies. The sample size includes the number of subjects and the number of plasma or urine samples per subject. The time points include sampling time points and dosing time points. The sample type includes blood, plasma, and urine. The hypotheses include the effect of bioequivalence, drug interaction, pharmacogenetics, and disease conditions on a drug's PK.
- The drug *quantification methods* include HPLC/UV, LC/MS/MS, LC/MS, and radiographic.

Metabolism enzyme: The cytochrome P450 (officially abbreviated as CYP) enzymes predominantly exist in the gut wall and liver. The CYP450 superfamily is a large and diverse group of enzymes that catalyze the oxidation of organic substances. The substrates of CYP enzymes include metabolic intermediates such as lipids and steroidal hormones as well as xenobiotic substances such as drugs and other toxic chemicals. CYPs are the major enzymes involved in drug metabolism and bioactivation, accounting for about 75 % of the total number of different metabolic reactions [58]. CYP enzyme names and genetic variants were mapped from the Human Cytochrome P450 (CYP) Allele Nomenclature Database (http://www.cypalleles.ki.se/). This site contains the CYP450 genetic mutation effect on the protein sequence and enzyme activity with associated references.

In the pharmacology research, probe drug is another important concept. An enzyme's probe substrate means that this substrate is primarily metabolized or transported by this enzyme. In order to experimentally prove whether a new drug inhibits or induces an enzyme, its probe substrate is always utilized to demonstrate this enzyme's activity before and after inhibition or induction. An enzyme's probe inhibitor or inducer means that it inhibits or induces this enzyme primarily. Similarly, an enzyme's probe inhibitor needs to be utilized if we investigate whether a drug is metabolized by this enzyme. Due to its importance, all the probe inhibitors, inducers, and substrates of CYP enzymes are also included in our PK ontology. All this information was collected from industry standard (http://www.fda.gov/Drugs/GuidanceComplianceRegulatoryInformation/Guidances/ucm064982.htm), reviewed in the top pharmacology journal [16].

Transporters are tissue specific. With different aliases, their tissue-specific transports and corresponding functions are different. *Transport proteins* are proteins which serve the function of moving other materials within an organism. Transport proteins are vital to the growth and life of all living things. Transport proteins are involved in the movement of ions, small molecules, or macromolecules, such as another protein, across a biological membrane. They are integral membrane proteins; that is, they exist within and span the membrane across which they transport substances. Their names and genetic variants were mapped from the Transporter Classification Database (http://www.tcdb.org). In addition, we also added the probe substrates and probe inhibitors and inducers to each one of the metabolism and transportation enzymes [16].

Drug names were created using the drug names from DrugBank 3.0 [27]. DrugBank consists of 6,829 drugs which can be grouped into different categories of FDA-approved, FDA-approved biotech, nutraceuticals, and experimental drugs. The drug names are mapped to generic names, brand names, and synonyms.

Subject included the existing ontologies for human disease ontology (DOID), suggested Ontology for Pharmacogenomics (SOPHARM), and mammalian phenotype (MP) from http://bioportal.bioontology.org.

The PK ontology was implemented with Protégé [59] and uploaded to the BioPortal ontology platform.

3.3 PK Corpus

Corpus is the key component to make NLP technologies successfully applied to text. To materialize text into computer-readable format, two types of annotations are needed, biological annotation and linguistic annotation [54]. Biological annotation belongs to event annotation, which identifies the location of biological information in the article. The scope of biological annotation can be narrowed down at single biological terms or broadened to include a whole sentence, which describes a biological event. Practically, event annotations are more complicated than term annotations. Term annotation only needs the terms to be annotated and hierarchically organized into categories. Unlike term annotation, an event has its own internal structure and it also involves biological entities (from term annotation) as its participants. Therefore, well-defined conditions to call biological events are required. On the other hand, linguistic annotation gives linguistic parsing such as POS or syntactic trees to know the type and role of term in natural language. The main purpose of linguistic annotation is to use it in the study of language through analysis of natural-occurring data. It involves computational methods and tools for analyzing linguistic pattern IR based on annotated corpora.

Most existing DDI extraction methods are designed to capture pairs of drugs that have the relation of interaction via semantic interpretation. There is one gap if we continue to use the same method for extracting DDI information from a pharmacokinetics perspective. The gap comes from the lack of knowledge to define a PK DDI. Pharmacokinetics parameters and knowledge from in vitro and in vivo DDI experimental designs, especially the selection of enzyme-specific probe substrates and inhibitors, should be considered. For instance, important pharmacokinetic parameters such as K_i, IC_{50}, and AUCR have not been included in the existing text mining approaches to DDI. This kind of pharmacokinetic information may be particularly relevant when seeking evidence of causal mechanisms behind DDIs and as a complement to DDI text mining of patient records.

3.3.1 Corpus Construction

A PK abstract corpus was constructed to cover four primary classes of PK studies: clinical PK studies ($n=56$); clinical pharmacogenetic studies ($n=57$); in vivo DDI studies ($n=218$); and in vitro drug interaction studies ($n=210$). The PK corpus construction is a manual process. The abstracts of clinical PK studies related to the most popular CYP3A substrate, midazolam, were investigated [60]. The clinical pharmacogenetic abstracts were selected based on the most polymorphic CYP enzyme, CYP2D6. We think that these two selection strategies represent very well all the in vivo PK and PG studies. In searching for the drug interaction studies, the abstracts were randomly selected from a PubMed query, which used probe substrates/inhibitors/inducers for metabolism enzymes.

Once the abstracts have been identified in four classes, their annotation is a manual process (Fig. 1). The annotation was firstly carried out by three master-level annotators (Shreyas Karnik, Abhinita Subhadarshini, and Xu Han) and one Ph.D. annotator (Lang Li). They have different training backgrounds: computational science, biological science, and pharmacology. Any differentially annotated terms were further checked by Sara K. Quinney and David A. Flockhart, one Pharm D. scientist and one M.D. scientist with extensive pharmacology training background. Among the disagreed annotations between these two annotators, a group review was conducted (Drs Quinney, Flockhart, and Li) to reach the final agreed annotations. In addition a random subset of 20 % of the abstracts that had consistent annotations among four annotators (three masters and one Ph.D.) were double checked by two Ph.D.-level scientists.

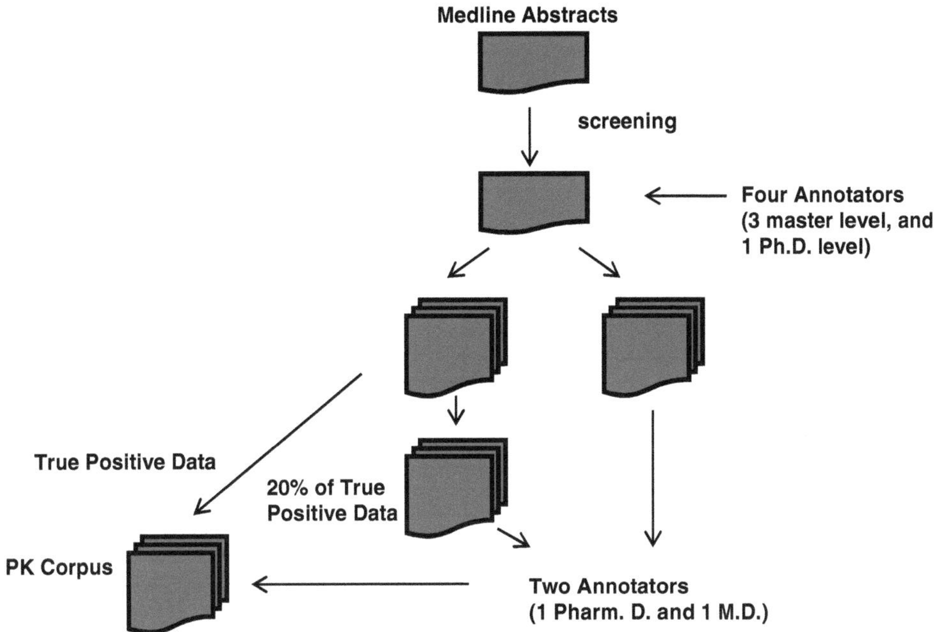

Fig. 1 PK corpus annotation flow chart

3.3.2 DDI Annotation Scheme

A structured annotation scheme was implemented to annotate three layers of pharmacokinetics information: key terms, DDI sentences, and DDI pairs (Fig. 2). DDI sentence annotation scheme depends on the key terms; and DDI annotations depend on the key terms and DDI sentences. Their annotation schemes are described as follows.

Term-level annotation: Key terms include drug names, enzyme names, PK parameters, numbers, mechanisms, and change. The boundaries of these terms among different annotators were judged by the following standard.

- *Drug names* were defined mainly on DrugBank 3.0 [27]. In addition, drug metabolites were also tagged, because they are important in in vitro studies. The metabolites were judged by either prefix or suffix: oxi, hydroxyl, methyl, acetyl, N-dealkyl, N-demethyl, nor, dihydroxy, O-dealkyl, and sulfo. These prefixes and suffixes are due to the reactions due to phase I metabolism (oxidation, reduction, hydrolysis) and phase II metabolism (methylation, sulfation, acetylation, glucuronidation) [55].
- *Enzyme names* covered all the CYP450 enzymes. Their names are defined in the Human Cytochrome P450 Allele Nomenclature Database, http://www.cypalleles.ki.se/. The variations of the enzyme or the gene names were considered.
- *PK parameters* were annotated based on the defined in vitro and in vivo PK parameter ontology. In addition, some PK parameters

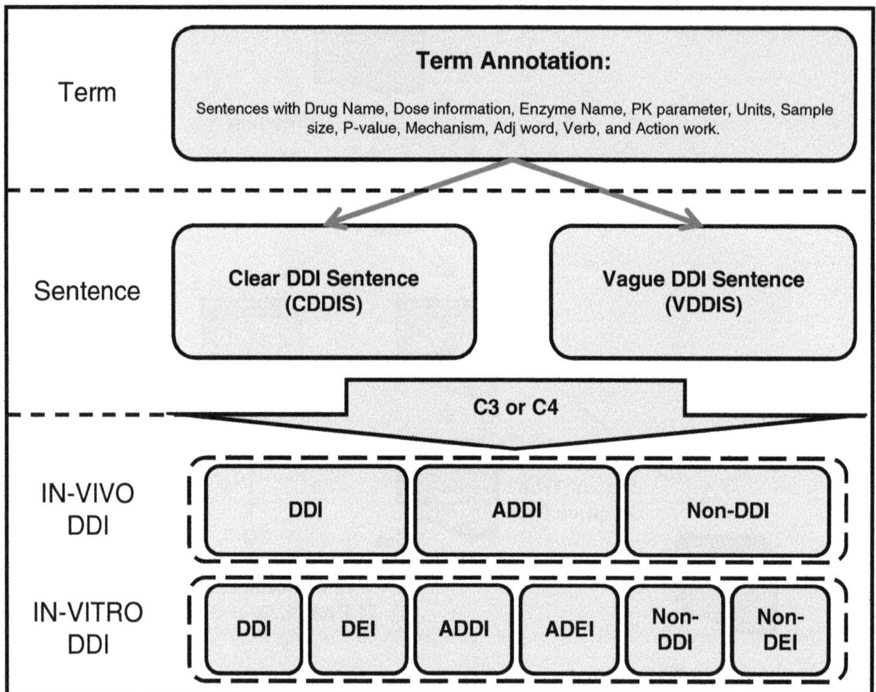

Fig. 2 A three-level hierarchical PK and DDI annotation scheme

have different names, such as CL = clearance, $t_{1/2}$ = half-life, AUC = area under the concentration curve, and AUCR = area under the concentration curve ratio. Those terms need to be handled carefully because their formats are varied.

- *Numbers* such as dose, sample size, values of PK parameters, and *p*-values were all annotated. If presented, their units were also covered in the annotations.
- *Mechanisms* denote the drug metabolism and interaction mechanisms. Linguistic realization of those terms is usually presented in various contexts. The nominalization of the following terms, inhibit, catalyze, correlate, metabolize, induce, form, stimulate, activate, and suppress, is annotated with regular expression patterns.
- *Change* describes the change of PK parameters. The following words and its nominalizations were annotated in the corpus to denote the change: strong, moderate, high, slight, significant, obvious, marked, great, pronounced, modest, probably, may, might minor, little negligible, doesn't interact, affect, reduce, and increase.

Sentence-level annotation: The middle-level annotation focused on the drug interaction sentences. Because two interaction drugs were not necessary all presented in the sentence, sentences were categorized into two classes:

- *Clear DDI sentence* (*CDDIS*): Two drug names (or drug–enzyme pair in the in vitro study) are in the sentence with a clear interaction statement, i.e., either "interaction" or "non-interaction", or ambiguous statement (i.e., such as "possible interaction" or "might interact").
- *Vague DDI sentence* (*VDDIS*): One drug or enzyme name is missed in the DDI sentence, but it can be inferred from the context. Clear interaction statement also is required.

DDI-level annotation: Once DDI sentences were labeled, the DDI pairs in the sentences were further annotated. Because of the fundamental difference between in vivo DDI studies and in vitro DDI studies, their DDI relationships were defined differently. In in vivo studies, three types of DDI relationships were defined (Table 1): DDI, ambiguous DDI (ADDI), and non-DDI (NDDI). Four conditions are specified to determine these DDI relationships. Condition 1 (C1) requires that at least one drug or enzyme name has to be contained in the sentence; condition 2 (C2) requires that the other interaction drug or enzyme name can be found from the context if it is not from the same sentence; condition 3 (C3) specifies numeric rules to define the DDI relationships based on the PK parameter changes; and condition 4 (C4) specifies the language expression patterns for DDI relationships. Using the rules summarized in Table 1, DDI, ADDI, and NDDI can be defined by $C1 \wedge C2 \wedge (C3 \vee C4)$. The priority rank of in vivo PK parameters is $AUC > CL > t_{1/2} > C_{max}$. In in vitro studies, six types of DDI relationships were defined (Table 1). DDI, ADDI, and NDDI were similar to in vivo DDIs, but three more drug–enzyme relationships were further defined: DEI, ambiguous DEI (ADEI), and non-DDI (NDEI). C1, C2, and C4 remained the same for in vitro DDIs. The main difference is in C3, in which either K_i or IC_{50} (inhibition) or EC_{50} (induction) was used to defined DDI relationship quantitatively. The priority rank of in vitro PK parameters is $K_i > IC_{50}$. Table 2 presents eight examples of how DDIs or DEIs were determined in the sentences.

Corpus evaluation: Agreement measurement is one of the important steps in corpus construction, which carries out the assessment of reference standard quality. If there is little agreements among annotators, that means that the task of annotation is not reliable and the quality of reference standard is suspected. In this work, Krippendorff's alpha [61] was calculated to evaluate the reliability of annotations from four annotators. The frequencies of key terms, DDI sentences, and DDI pairs are presented in Table 3. Their Krippendorff's alphas are 0.953, 0.921, and 0.905, respectively. Please note that the total DDI pairs refer to the total pairs of drugs within a DDI sentence from all DDI sentences.

The PK corpus was constructed by the following process. Raw abstracts were downloaded from PubMed in XML format. Then XML files were converted into GENIA corpus format following

Table 1
DDI definitions in corpus

DDI relationship	C1	C2	C3[b]	C4[b]
In vivo study				
DDI	Yes	Yes	The PK parameter with the highest priority[a] must satisfy p-value <0.05 and FC > 1.50 or FC < 0.67	Significant, obviously, markedly, greatly, pronouncedly, etc.
Ambiguous DDI (ADDI)			The PK parameter with the highest priority[a] in the conditions of p-value <0.05 but 0.67 < FC < 1.50; or FC > 1.50 or FC < 0.67, but p-value >0.05	Modestly, moderately, probably, may, might, etc.
Non-DDI (NDDI)			The PK parameter with the highest priority[a] is in the condition of p-value >0.05 and 0.67 < FC < 1.50	Minor significance, slightly, little or negligible effect, does not interact, etc.
In vitro study				
DDI DEI	Yes	Yes	($0 < K_i < 10$ or $0 < EC_{50} < 10$ μM, and p-value <0.05)	Significant, obviously, markedly, greatly, pronouncedly, etc.
ADDI Ambiguous DEI (ADEI)			($10 < K_i < 100$ or $10 < EC_{50} < 100$ μM, and p-value <0.05 or vice versa)	Modestly, moderately, probably, may, might, etc.
NDDI Non-DEI (NDEI)			($K_i > 100$ μM or $EC_{50} > 100$ μM, and p-value >0.05)	Minor significance, slightly, little or negligible effect, does not interact, etc.

Note:
C1: At least one drug or enzyme name has to be contained in the sentence
C2: Need to label the drug name if it is not from the same sentence
C3: PK parameter and value dependent
C4: Significance statement
[a]For the priority of PK parameters: AUC > CL > $t_{1/2}$ > C_{max}; the priority of in vitro PK parameters: $K_i > IC_{50}$. [b]Priority issue: When C3 and C4 occur and conflict, C3 dominates the sentence

Table 2
Examples of DDI definitions

PMID	DDI sentence	Relationship and commend
20012601	The pharmacokinetic parameters of ***verapamil*** were ***significantly*** altered by the co-administration of ***lovastatin*** compared to the control	Because of the words, "significantly," (verapamil, lovastatin) is a *DDI*
20209646	The ***clearance*** of ***mitoxantrone*** and ***etoposide*** was ***decreased*** by *64* and *60 %*, respectively, when combined with ***valspodar***	Because the fold changes were less than 0.67 (*mitoxantrone, valspodar*) and (*etoposide, valspodar*) are *DDIs*
20012601	The *(AUC (0-infinity))* of ***norverapamil*** and the terminal ***half-life*** of ***verapamil did not significantly changed*** with ***lovastatin*** co-administration	Because of the words, "not significantly changed," *(verapamil, lovastatin)* is an *NDDI*
17304149	Compared with placebo, ***itraconazole*** treatment ***significantly increases*** the peak plasma concentration (*Cmax*) of paroxetine by *1.3-fold* (6.7 ± 2.5 versus 9.0 ± 3.3 ng/mL, $p \leq 0.05$) and the area under the plasma concentration–time curve from zero to 48 h (*AUC*(0–48)) of ***paroxetine*** by *1.5-fold* (137 ± 73 versus 199 ± 91 ng × *h/mL, $p \leq 0.01$)	AUC has a higher rank than Cmax, and it had a 1.5-fold change and less than 0.05 *p*-value; thus, (*itraconazole, paroxetine*) is a *DDI*
13129991	The mean (SD) ***urinary ratio*** of ***dextromethorphan*** to its metabolite was *0.006* (0.010) at baseline and *0.014* (0.025) after ***St John's wort*** administration ($p = 0.26$)	The change in PK parameter is more than 1.5-fold but *p*-value is >0.05. Thus, (dextromethorphan, St John's wort) is an *ADDI*
19904008	The obtained results show that ***perazine*** at its therapeutic concentrations is a ***potent inhibitor*** of human *CYP1A2*	Because of the word, "potent inhibitor," (perazine, CYP1A2) is a *DEI*
19230594	After human hepatocytes were exposed to 10 μM *YM758*, microsomal activity and mRNA level for *CYP1A2* were ***not induced*** while those for *CYP3A4* were ***slightly induced***	Because of the words, "not induced" and "slightly induced," (YM758, CYP1A2) and (YM758, CYP1A2) are *NDEIs*
19960413	From these results, *DPT* was characterized to be a competitive ***inhibitor*** of *CYP2C9* and *CYP3A4*, with *K(i)* values of *3.5* and *10.8 μM* in HLM and *24.9* and *3.5*μM in baculovirus–insect cell-expressed human CYPs, respectively	Because *K* was larger than 10 μM, (DPT, CYP2C9) and (DPT, CYP3A4) are *ADEIs*

Table 3
Annotation performance evaluation

Key terms	Annotation categories	Frequencies	Krippendorff's alpha
	Drug	8,633	0.953
	CYP	3,801	
	PK parameter	1,508	
	Number	3,042	
	Mechanism	2,732	
	Change	1,828	
	Total words	97,291	
DDI sentences	CDDI sentences	1,191	0.921
	VDDI sentences	120	
	Total sentences	4,724	
DDI pairs	DDI	1,239	0.905
	ADDI	300	
	NDDI	294	
	DEI	565	
	ADEI	95	
	NDEI	181	
	Total drug pairs	12,399	

the gpml.dtd from the GENIA corpus [31]. The sentence detection in this step is accomplished by using the Perl module Lingua:: EN::Sentence, which was downloaded from the Comprehensive Perl Archive Network (CPAN, www.cpan.org). GENIA corpus files were then tagged with the prescribed three levels of PK and DDI annotations. Finally, a cascading style sheet (CSS) was implemented to differentiate colors for the entities in the corpus. This feature allows the users to visualize annotated entities. We would like to acknowledge that a DDI corpus was recently published as part of a text-mining competition DDIExtraction 2011 (http://labda.inf.uc3m.es/DDIExtraction2011/dataset.html). Their DDIs were clinical outcome oriented, not PK oriented. They were extracted from DrugBank, not from PubMed abstracts. Our PK corpus complements to their corpus very well.

3.4 DDI Text Mining

We implemented the approach described by [37] for the DDI extraction. Prior to performing DDI extraction, the testing and validation DDI abstracts in our corpus were preprocessed and converted into the unified XML format [37]. The following steps were conducted:

- Drugs were tagged in each of the sentences using dictionary based on DrugBank. This step revised our prescribed drug name annotations in the corpus. One purpose is to reduce the redundant synonymous drug names. The other purpose is only

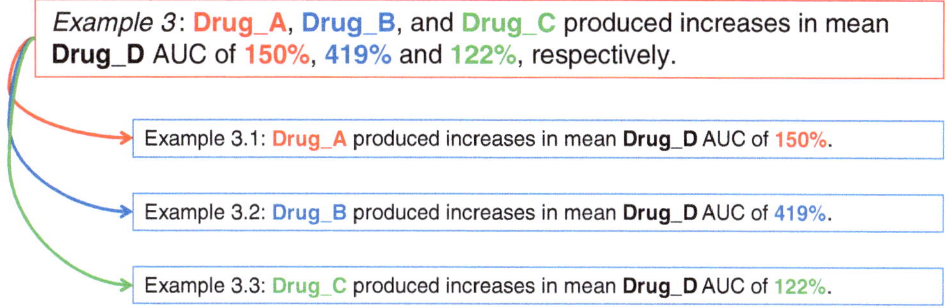

Fig. 3 Sentence separation

to keep the parent drugs and remove the drug metabolites from the tagged drug names from our initial corpus, because parent drugs and their metabolites rarely interact. In addition, enzymes (i.e., CYPs) were also tagged as drugs, since enzyme–drug interactions have been extensively studied and published. The regular expression of enzyme names in our corpus was used to remove the redundant synonymous gene names.

- Each of the sentences was subjected to tokenization, POS tags, and dependency tree generation using the Stanford parser [62].
- C_2^n drug pairs from the tagged drugs in a sentence were generated automatically, and they were assigned with default labels as no-drug interaction. Please note that if a sentence had only one drug name, this sentence did not have a DDI. This setup limited us to consider only CDDI sentence in our corpus.
- The drug interaction labels were then manually flipped based on their true drug interaction annotations from the corpus. Please note that our corpus had annotated DDIs, ADDIs, NDDIs, DEIs, ADEIs, and NDEIs. Here only DDIs and DEIs were labeled as true DDIs. The other ADDIs, NDDIs, DEIs, and ADEIs were all categorized into the no-drug interactions.

Then sentences were represented with dependency graphs using interacting components (drugs) (Fig. 3). The graph representation of the sentence was composed of two items: (1) one dependency graph structure of the sentence and (2) a sequence of POS tags (which was transformed to a linear order "graph" by connecting the tags with a constant edge weight). We used the Stanford parser [62] to generate the dependency graphs. Airola et al. proposed to combine these two graphs to one weighted, directed graph. This graph was fed into a SVM for DDI/non-DDI classification. More details about the all paths graph kernel algorithm can be found in [37].

DDI extraction was implemented in the in vitro and in vivo DDI corpus separately. Table 4 presents the training sample size and testing sample size in both corpus sets. Then Table 5 presents the DDI extraction performance. In extracting in vivo DDI pairs,

Table 4
DDI data description

Datasets	Abstracts	Sentences	DDI pairs	True DDI pairs
In vivo DDI training	174	2,112	2,024	359
In vivo DDI testing	44	545	574	45
In vitro DDI training	168	1,894	7,122	783
In vitro DDI testing	42	475	1,542	146

Table 5
DDI extraction performance

Datasets	Precision	Recall	*F*-measure
In vivo DDI training	0.67	0.78	0.72
In vivo DDI testing	0.67	0.79	0.73
In vitro DDI training	0.51	0.59	0.55
In vitro DDI testing	0.47	0.58	0.52

the precision, recall, and *F*-measure in the testing set are 0.67, 0.79, and 0.73, respectively. In the in vitro DDI extraction analysis, the precision, recall, and *F*-measure are 0.47, 0.58, and 0.52, respectively, in the in vitro testing set. In our early DDI research published in the DDIExtract 2011 Challenge [63], we used the same algorithm to extract both in vitro and in vivo DDIs at the same time, and the reported *F*-measure was 0.66. This number is in the middle of our current in vivo DDI extraction *F*-measure 0.73 and in vitro DDI extraction *F*-measure 0.52.

Error analysis was performed in testing samples. Table 6 summarizes the results. Among the known reasons for the false positives and false negatives, the most frequent one is that there are multiple drugs in the sentence or the sentence is long. The other reasons include that there is no direct DDI relationship between two drugs, but the presence of some words, such as dose and increase, may lead to a false-positive prediction; or DDI is presented in an indirect way; or some NDDIs are inferred due to some adjectives (little, minor, negligible).

4 Notes (Challenges and Possible Solutions)

As we have seen, there had been a number of approaches for DDI extraction research. Nonetheless, there are significant unsolved problems or difficulties when we apply those approaches in PK

Table 6
DDI extraction error analysis from testing DDI sets

No.	Error categories	Error type	Frequency		Examples
			In vivo	In vitro	
1	There are multiple drugs and PK parameters in the sentence, and the sentence is long	FP	6	34	PMID: 12426514: In three subjects with measurable concentrations in the single-dose study, rifampin significantly decreased the mean maximum plasma concentration (C(max)) and area under the plasma concentration–time curve from 0 to 24 h (AUC(0–24)) of praziquantel by 81 % ($p<0.05$) and 85 % ($p<0.01$), respectively, whereas rifampin significantly decreased the mean C(max) and AUC(0–24) of praziquantel by 74 % ($p<0.05$) and 80 % ($p<0.01$), respectively, in five subjects with measurable concentrations in the multiple-dose study
		FN	2	17	PMID: 10608481: Erythromycin and ketoconazole showed a clear inhibitory effect on the 3-hydroxylation of lidocaine at 5 μM of lidocaine (IC_{50} 9.9 μM and 13.9 μM, respectively) but did not show a consistent effect at 800 μM of lidocaine ($IC_{50}>250$ μM and 75.0 μM, respectively)
2	There is no direct DDI relationship between two drugs, but the presence of some words, such as dose and increase, may lead to a false-positive prediction	FP	6	14	PMID: 17192504: A significant fraction of patients to be treated with HMR1766 is expected to be maintained on warfarin
3	DDI is presented in an indirect way	FN	2	19	PMID: 11994058: In CYP2D6 poor metabolizers, systemic exposure was greater after chlorpheniramine alone than in extensive metabolizers, and administration of quinidine resulted in a slight increase in CLoral
4	Design issue: Some NDDIs are inferred due to some adjectives (little, minor, negligible)	FP	1	3	PMID: 10223772: In contrast, the effect of ranitidine or ebrotidine on CYP3A activity in vivo seems to have little clinical significance
5	Unknown	FP	5	44	PMID: 10383922: CYP1A2, CYP2A6, and CYP2E1 activities were not significantly inhibited by azelastine and the two metabolites
		FN	6	26	PMID: 10681383: However, the most unusual result was the interaction between testosterone and nifedipine

DDI text mining. According to our annotation scheme, the three-level annotation is designed to identify key terms, DDI sentences, and DDI pairs. From our DDI extraction error analysis, we found that major errors come from the challenges of annotations. Most missing detections result from the issue of drug name mapping in term annotation level. The reason to cause the errors classified into the third category is that the approach we use to extract DDI lacks the ability of co-reference resolution. Due to the omission of VDDIS in DDI sentence level, these kinds of errors happen. Finally, a major part of failure resulted largely from the long sentences with multiple drugs and PK parameters. To meet these three issues, we discuss the problem of errors and try to explore their possible solution in the following three subsections.

4.1 Issues in Drug Name Mapping

In term annotation level, most biological terms can be annotated with satisfied performance by using NER, except for drug name. The representations of drug names are diverse in pharmacology articles. The main reason to this issue comes from the naming convention of different drug companies. Each drug with the same generic name might have multiple brand names or synonym name. Due to the different backgrounds of authors, the preferences of name adoption are quite different. Among drug names, some really confuse NER tools by its confliction with other terms. For example, one of ketoconazole's synonyms (DB01026) is "2 %" and a small molecule (db03951) is denominated with "16 g." For the possible solution for this issue, we recommend to remove those terms from your dictionaries because few authors use those peculiar terms as drug names. Another issue in NER is to recognize acronym and abbreviation of drugs or other terms. There are no rules or exact patterns for the creation of acronym and abbreviation from their full form. To meet this problem, there are two possible solutions. First, parenthetical expression might be the solution to distinguish acronyms. By using Schwartz and Hearst's algorithm, it searches for parentheses in text and limits context around brackets as a mark of term, such as single or more words, e.g., nevirapine (NVP) or human liver microsomes (HLMs). Otherwise, using FDA-provided acronym and abbreviation database as another dictionary can be the second solution. This database can be downloaded at the following link: http://www.fda.gov/AboutFDA/FDAAcronymsAbbreviations/ucm070296.htm.

4.2 Vague DDI Sentence Problem

In most DDI extraction approaches, CDDIS are considered to be candidates for the analysis of DDI extraction. Nevertheless, we found that the number of VDDISs amounts to one-tenth of CDDISs' quantity in Table 3. If we omit investigating those sentences, it means that up to 10 % of true information is possibly missed. Although this problem also happens in many articles related to protein–protein or protein–gene interactions, it harms PK DDI articles more. It is because the interactions between proteins or genes are more often expressed with narrative ways while many of PK DDIs in text can be determined only with the measurements of ADME activities. Such omissions will highly increase the chance of missing detections, especially for the task of PK curation.

To retrieve VDDIS, human beings can easily recognize DDI information from VDDIS via the reference to other sentences. The process of determining the pronoun or the antecedent from its context is called co-reference resolution [64]. Some previous works [65, 66] had considered this problem on a pre-sentence basis and used it to explore neglected useful information in the same article. Grosz et al. [67] considered the feature of significant entities which are mentioned multiple times in context and its transitivity property to extract event–argument relations. But no one has yet considered using it to improve the performance of DDI extraction. Here we would like to choose one appropriate approach among published co-reference resolution method to transform the VDDIS into CDDIS.

Bridging references arise when a reference to a noun phrase that is not directly mentioned is made. For an example sentence in PMID-17518508 (example 1), it does not mention that which drug is the CYP3A4 inhibitor in the sentence, but readers can figure out that it is ketoconazole from few sentences before. Another type of VDDIS is more challenging to determine because the noun phrase and pronoun are not even mentioned in the sentence of PMID-17909805 (example 2). The pronoun or the antecedent for inhibitor drug even does not show in this sentence, and its argument is located in few sentences behind, which makes it more difficult to find its co-reference.

Example 1
Co-administration of *a potent CYP3A4 inhibitor* moderately increased cinacalcet exposure in study subject.

Example 2
The plasma clearances of docetaxel and midazolam were reduced by 1.7- and 6-fold, respectively.

Centering theory [66, 68, 69] is a method to model the relations among focus of attention, choice of referring expression, and perceived coherence of utterances within a discourse segment. This approach should conquer the problem of example 1. As for the second example, it cannot be answered by only finding the co-reference of pronoun. Finding the relation between an event and its argument across co-reference relations will really help find the argument of events. To achieve the cross-sentence event–argument relation, some previous works [67, 70] had been capable of identifying the event for intra-sentence argument. To handle the challenges in both examples, we are eager to look for a method which reaches the best performance.

4.3 Multiple Drug Pairs and PK Parameters

The purpose of DDI-level annotation is to label drug pairs and PK parameters in text and conduct the relationships for DDI pairs. From the experience of evaluating corpus and the error analysis for DDI results, there are two challenges when extracting DDI. (1) Long sentences with multiple drugs significantly complicated syntactic structure and led to the most frequent faults of first category in Table 6. In fact, such sentences often occurred in the articles related to in vivo and in vitro experiments. Authors try to compare the intensity of drug interactions among different drugs and place their PK parameters as well as dose conditions after. (2) How to take advantage of PK parameters for DDI extraction is another challenge. In the previous works, machine learning-based approach deals with the task of extracting relations by classifying pairs of drug with/without DDI categories, while rule-based or pattern-based method locates drug pairs as well as trigger words to build up a tree for determining their relationship. But, no one has yet considered using it to improve the performance of DDI extraction.

To overcome both the problem of multiple drugs and PK parameters and the challenge of utilizing PK parameters, simplifying sentence is an idea that came from Segura-Bedmar's method [43], which split the long sentences into clauses from which relations are extracted by a pattern matching algorithm. Such a simplification significantly improves the performance of dealing with long sentences. This inspires us to split a long sentence with different way. According to the characteristics of utterances in PK articles, the orders and locations of drug names and their corresponding PK parameters are parallelly located. The example in Fig. 3 shows that there are three different drugs interacting one drug followed by the corresponding fold change of AUC value. But when looking into its structure of dependent graph tree which is often used for machine learning- or rule-based pattern (Fig. 4), we found that both drugs and PK values are connected with *conj_and* edge. It is not possible to differentiate which PK value is belonging to which drug. Thus, splitting the sentence according to drugs and PK parameters before machine learning- or rule-based pattern matching is necessary. Using example 3 as an instance, we hope that the sentence can break down into three sentences (examples 3.1, 3.2, and 3.3 in Fig. 4). This separation greatly simplifies the sentences' complexity and resolves the problem of matching PK parameters.

Example 3
Drug_A, Drug_B, and Drug_C produced increases in mean Drug_D AUC of 150, 419, and 122 %, respectively.

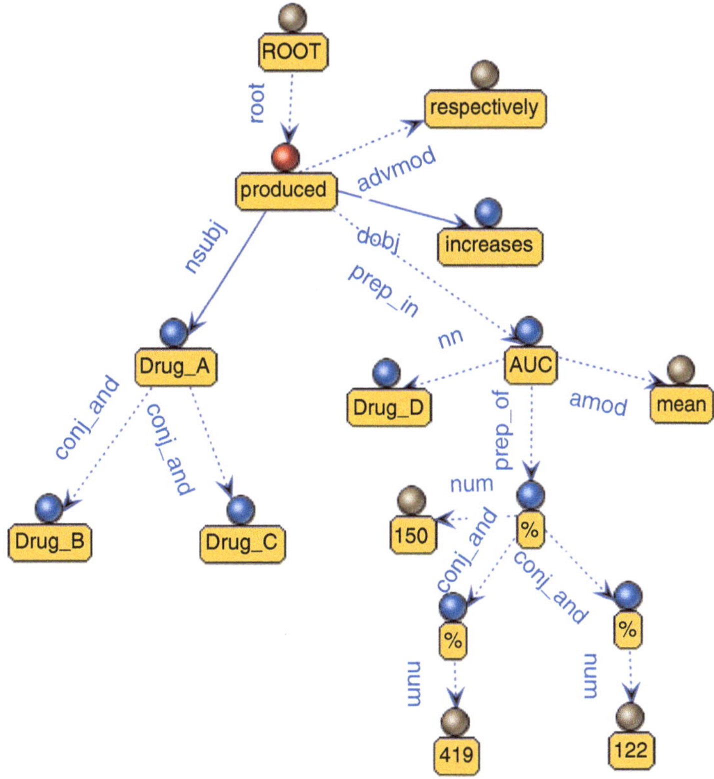

Fig. 4 Dependency graph tree of example 3

Acknowledgment

This work is supported by the US National Institutes of Health grant R01 GM74217 (Lang Li).

References

1. Second Annual Adverse Drug/Biologic Reaction Report (1987) US Food and Drug Administration
2. Becker ML et al (2007) Hospitalisations and emergency department visits due to drug–drug interactions: a literature review. Pharmacoepidemiol Drug Saf 16:641–651
3. Chou TC (2006) Theoretical basis, experimental design, and computerized simulation of synergism and antagonism in drug combination studies. Pharmacol Rev 58(3):621–681
4. Magro L, Moretti U, Leone R (2012) Epidemiology and characteristics of adverse drug reactions caused by drug–drug interactions. Expert Opin Drug Saf 11(1):83–94
5. Juurlink DN et al (2003) Drug–drug interactions among elderly patients hospitalized for drug toxicity. JAMA 289(13): 1652–1658
6. Merle L et al (2005) Predicting and preventing adverse drug reactions in the very old. Drugs Aging 22(5):375–392
7. Johansson I, Ingelman-Sundberg M (2011) Genetic polymorphism and toxicology: with emphasis on cytochrome p450. Toxicol Sci 120(1):1–13

8. Ajayi FO, Sun H, Perry J (2000) Adverse drug reactions: a review of relevant factors. J Clin Pharmacol 40(10):1093–1101
9. DiMasi JA, Grabowski HG (2007) The cost of biopharmaceutical R&D: is biotech different? Manage Decis Econ 28:469–479
10. Pang KS, Rodrigues AD, Peter RM (2010) Enzyme- and transporter-based drug–drug interactions, vol 746. Springer, New York
11. The European Medicines Agency (2012) Guideline on the investigation of drug interactions. The European Medicines Agency, London
12. Jia J et al (2009) Mechanisms of drug combinations: interaction and network perspectives. Nat Rev Drug Discov 8(2):111–128
13. Rowland M, Tozer TN (1995) Clinical pharmacokinetics: concepts and applications. Lippincott Williams & Wilkins, London
14. Wienkers LC, Heath TG (2005) Predicting in vivo drug interactions from in vitro drug discovery data. Nat Rev Drug Discov 4(10): 825–833
15. Rostami-Hodjegan A, Tucker G (2004) 'In silico' simulations to assess the 'in vivo' consequences of 'in vitro' metabolic drug–drug interactions. Drug Discov Today 1(4): 441–448
16. Huang SM et al (2007) Drug interaction studies: study design, data analysis, and implications for dosing and labeling. Clin Pharmacol Ther 81(2):298–304
17. Li L, Yu M, Chin R, Lucksiri A, Flockhart D, Hall S (2007) Drug–drug interaction prediction: a Bayesian meta-analysis approach. Stat Med 26(20):3700–3721
18. Yu M et al (2008) A Bayesian meta-analysis on published sample mean and variance pharmacokinetic data with application to drug–drug interaction prediction. J Biopharm Stat 18(6): 1063–1083
19. Zhou J et al (2009) A new probabilistic rule for drug–drug interaction prediction. J Pharmacokinet Pharmacodyn 36:1–18
20. Zhou J, Qin Z, Kim S, Wang Z, Hall DS, Li L (2009) Drug–drug interaction prediction assessment. J Pharmacokinet Pharmacodyn 19:641–657
21. Wang Z, Kim S, Quinney SK, Zhou J, Li L (2010) Non-compartment model/compartment model transformation. BMC System Biol 4(1):S8
22. Li L (2007) Discussion on parameter estimation for differential equations: a generalized smoothing approach. J Royal Stat Soc B 69: 787–788
23. Chien JY, Lucksiri A, Ernest CS, Gorski JC, Wrighton SA, Hall SD (2006) Stochastic prediction of CYP3A-mediated inhibition of midazolam clearance by ketoconazole. Drug Metab Dispos 34(7):1208–1219
24. Quinney SK, Zhang X, Lucksiri A, Gorski JC, Li L et al (2010) Physiologically based pharmacokinetic model of mechanism-based inhibition of CYP3A by clarithromycin. Drug Metab Dispos 38(2):241–248
25. Hachad H, Ragueneau-Majlessi I, Levy RH (2010) A useful tool for drug interaction evaluation: the University of Washington Metabolism and Transport Drug Interaction Database. Hum Genomics 5(1):61–72
26. Hewett M et al (2002) PharmGKB: the pharmacogenetics knowledge base. Nucleic Acids Res 30(1):163–165
27. Knox C et al (2011) DrugBank 3.0: a comprehensive resource for 'omics' research on drugs. Nucleic Acids Res 39(Database issue): D1035–D1041
28. Gottlieb A et al (2012) INDI: a computational framework for inferring drug interactions and their associated recommendations. Mol Syst Biol 8:592
29. Nadkarni PM, Ohno-Machado L, Chapman WW (2011) Natural language processing: an introduction. J Am Med Inform Assoc 18: 544–551
30. Zweigenbaum P et al (2007) Frontiers of biomedical text mining: current progress. Brief Bioinform 8(5):358–375
31. Kim JD et al (2003) GENIA corpus–semantically annotated corpus for bio-text mining. Bioinformatics 19(Suppl 1):i180–i182
32. Wilbur WJ, Rzhetsky A, Shatkay H (2006) New directions in biomedical text annotation: definitions, guidelines and corpus construction. BMC Bioinformatics 7:356
33. Muller HM, Kenny EE, Sternberg PW (2004) Textpresso: an ontology-based information retrieval and extraction system for biological literature. PLoS Biol 2(11):e309
34. Feldman R et al (2002) Mining biomedical literature using information extraction. Curr Drug Discov 2:19–23
35. Fundel K, Küffner R, Zimmer R (2007) RelEx: relation extraction using dependency parse trees. Bioinformatics 23:365–371
36. Qian L, Zhou G (2012) Tree kernel-based protein–protein interaction extraction from biomedical literature. J Biomed Inform 45(3): 535–543
37. Airola A et al (2008) All-paths graph kernel for protein–protein interaction extraction with evaluation of cross-corpus learning. BMC Bioinformatics 9(Suppl 11):S2
38. Pyysalo S et al (2008) Comparative analysis of five protein–protein interaction corpora. BMC Bioinformatics 9(Suppl 3):S6

39. Tikk D et al (2010) A comprehensive benchmark of kernel methods to extract protein–protein interactions from literature. PLoS Comput Biol 6:e1000837
40. Chen Y, Liu F, Manderick B (2009) Normalizing interactor proteins and extracting interaction protein pairs using support vector machines. In: BioCreative II. 5 Workshop 2009 on Digital Annotations
41. Zhou D, He Y (2008) Extracting interactions between proteins from the literature. J Biomed Inform 41(2):393–407
42. Krallinger M, Leitner F, Valencia A (2009) The BioCreative II.5 challenge overview. In: Proceedings of the BioCreative II. 5 Workshop 2009 on Digital Annotations
43. Segura-Bedmar I, Martinez P, de Pablo-Sanchez C (2011) A linguistic rule-based approach to extract drug–drug interactions from pharmacological documents. BMC Bioinformatics 12(Suppl 2):S1
44. Segura-Bedmar I, Martinez P, Sanchez-Cisneros D (2011) The 1st DDIExtraction-2011 challenge task: extraction of drug–drug interactions from biomedical texts. In: Proceedings of the 1st challenge task on drug–drug interaction extraction 2011, Spain
45. Percha B, Garten Y, Altman RB (2012) Discovery and explanation of drug–drug interactions via text mining. Pac Symp Biocomput 410–421
46. Tari L et al (2010) Discovering drug–drug interactions: a text-mining and reasoning approach based on properties of drug metabolism. Bioinformatics 26(18):i547–i553
47. Segel IH (1975) Enzyme kinetics: behavior and analysis of rapid equilibrium and steady state enzyme systems. Wiley, New York
48. Consortium IT et al (2010) Membrane transporters in drug development. Nat Rev Drug Discov 9(3):215–236
49. Rostami-Hodjegan A, Tucker G (2004) In silico simulations to assess the in vivo consequences of in vitro metabolic drug–drug interactions. Drug Disc Today Technol 1:441–448
50. Lam YW, Alfaro CL, Ereshefsky L, Miller M (2003) Pharmacokinetic and pharmacodynamic interactions of oral midazolam with ketoconazole, fluoxetine, fluvoxamine, and nefazodone. J Clin Pharmacol 43(11):1274–1282
51. Gibaldi M, Perrier D (1982) Pharmacokinetics, 2nd edn. Marcel Dekker, New York
52. Vazquez M et al (2011) Text mining for drugs and chemical compounds: methods, tools and applications. Mol Inform 30:506–519
53. Spasic I et al (2005) Text mining and ontologies in biomedicine: making sense of raw text. Brief Bioinform 6(3):239–251
54. Kim JD, Ohta T, Tsujii J (2008) Corpus annotation for mining biomedical events from literature. BMC Bioinformatics 9:10
55. Brunton LL, Chabner BA, Knollmann BC (2011) Goodman & Gilman's the pharmacological basis of therapeutics, 12th edn. McGraw-Hill, New York
56. Witte R, Kappler T, Baker CJO (2007) Ontology design for biomedical text mining, in semantic Web: revolutionizing knowledge discovery in the life sciences. Springer, USA, pp 281–313
57. Giacomini KM et al (2010) Membrane transporters in drug development. Nat Rev Drug Discov 9(3):215–236
58. Guengerich FP (2008) Cytochrome p450 and chemical toxicology. Chem Res Toxicol 21(1): 70–83
59. Rubin DL, Noy NF, Musen MA (2007) Protege: a tool for managing and using terminology in radiology applications. J Digit Imaging 20(Suppl 1):34–46
60. Wang Z et al (2009) Literature mining on pharmacokinetics numerical data: a feasibility study. J Biomed Inform 42(4):726–735
61. Krippendorff K (2004) Content analysis: an introduction to its methodology. SAGE, Thousand Oaks, CA
62. de Marneffe M-C, MacCartney B, Manning CD (2006) Generating typed dependency parses from phrase structure parses. In LREC
63. Karnik S et al (2011) Extraction of drug–drug interactions using all paths graph kernel. In: The 1st challenge task on drug–drug interaction extraction, Huelva, Spain
64. van Deemter K, Kibble R (2000) On coreferring: coreference in muc and related annotation schemes. Comput Linguist 26(4): 629–637
65. Hobbs J (1986) Resolving pronoun references. Readings in natural language processing. Morgan Kaufmann Publishers Inc., San Francisco, CA, USA, pp 339–352
66. Grosz BJ, Weinstein S, Joshi AK (1995) Centering: a framework for modeling the local coherence of discourse. Comput Linguist 21(2):203–225
67. Yoshikawa K et al (2011) Coreference based event-argument relation extraction on biomedical text. J Biomed Semantics 2(Suppl 5):S6
68. Brennan SE, Friedman MW, Pollard CJ (1987) A centering approach to pronouns. In: Proceedings of the 25th annual meeting on Association for Computational Linguistics, Morristown, NJ, USA
69. Elango P (2005) Coreference resolution: a survey. University of Wisconsin, Madison, WI
70. Lee H et al (2013) Deterministic coreference resolution based on entity-centric, precision-ranked rules. Comput Linguist 34(4):885–916

Chapter 5

Biological Information Extraction and Co-occurrence Analysis

Georgios A. Pavlopoulos, Vasilis J. Promponas, Christos A. Ouzounis, and Ioannis Iliopoulos

Abstract

Nowadays, it is possible to identify terms corresponding to biological entities within passages in biomedical text corpora: critically, their potential relationships then need to be detected. These relationships are typically detected by co-occurrence analysis, revealing associations between bioentities through their coexistence in single sentences and/or entire abstracts. These associations implicitly define networks, whose nodes represent terms/bioentities/concepts being connected by relationship edges; edge weights might represent confidence for these semantic connections.

This chapter provides a review of current methods for co-occurrence analysis, focusing on data storage, analysis, and representation. We highlight scenarios of these approaches implemented by useful tools for information extraction and knowledge inference in the field of systems biology. We illustrate the practical utility of two online resources providing services of this type—namely, STRING and BioTextQuest—concluding with a discussion of current challenges and future perspectives in the field.

Key words Text mining, Co-occurrence analysis, Graph theory, Biomedical literature, Semantic networks, Systems biology

1 Introduction

The new advances in high-throughput technologies and large-scale experiments have led to a "data explosion" in the life and health sciences. Results of such studies are presented in biomedical publications, at an ever-increasing pace. This frantic expansion of the biomedical literature can often be difficult to absorb or manually analyze, as efficient searching about a certain topic can become complicated, tedious, and time consuming. The MedLine database and related repositories currently contain more than 22,000,000 articles (accessed 08/06/2013) and follow an exponential growth over time [1], making it impossible to analyze without specific queries. Therefore, efficient and automated search engines are

Vinod D. Kumar and Hannah Jane Tipney (eds.), *Biomedical Literature Mining*, Methods in Molecular Biology, vol. 1159, DOI 10.1007/978-1-4939-0709-0_5, © Springer Science+Business Media New York 2014

necessary to efficiently explore the biomedical literature using text mining techniques [2–6]. Automated knowledge mining, named entity recognition (NER), extraction of meaningful relationships between articles, concept discovery, identification of biomedical terms in texts, and enhanced and more targeted PubMed queries [7] are some examples of the big challenges in the field. The first evidence that text mining techniques can be used in knowledge extraction, and therefore towards the discovery of new findings, came when Swanson published his studies, where he linked Raynaud's disease with fish oil [8].

Ongoing research in biomedical literature mining [2] involves techniques to:

- *Rank search results* (e.g., miSearch [9], Quertle [10, 11], MedlineRanker [12], ETBLAST [13], MScanner [14]).
- *Cluster them into topics* (e.g., Anne O'Tate [15], GoPubMed [16], XplorMed [17], Caipirini [18]).
- *Identify biomedical terms in a text* such as Gene Ontology (GO) terms [19], genes, proteins, chemicals, drugs, diseases, phenotypes, species, pathways (e.g., WhatizIt [20], Abner [21], Reflect [22], OnTheFly [23], TerMine [24]).
- *Extract and display semantics and relations between them* (e.g., MedEvi [25], EBIMed [26], PubNet [27], PubFocus [28]).
- *Improve search interface and retrieval experience* (e.g., askMedline [29–31], BabelMesh [32], PubGet [33], HubMed [34], PubCrawler [35], PubFinder [36]).

Amongst others, there are many efforts trying to *extract protein–protein interactions (PPIs)* from PubMed abstracts or to create a network of related genes/proteins/biological terms [37]. For example, the information extraction system PubGene is detecting associations between genes using terms from the Medical Subject heading (MeSH) index and terms from the gene ontology (GO) database [19]. CoPub Mapper [38] provides online access to co-occurrence associations extracted from PubMed between genes and biomedical terms. Other approaches extract PPIs from the scientific literature by identification of protein names in text and then sentence processing/semantics analytics (collectively called Natural Language Processing or NLP) [39–42], and by connecting proteins to concept profiles following the assumption that proteins that share one or more concept profiles have increased probability to interact [43]. More complex systems such as iHOP [44], search information through hyperlinks so that literature is clustered according to gene names and text topics leading to the discovery of potential PPIs. All PPIs which are experimentally verified are specifically highlighted [44]. In addition to the above methods, there are other tools that incorporate data from many different sources for describing or even predicting a PPI. One such web resource is the STRING database [45]. STRING uses information from the biomedical literature (i.e., co-occurrences of genes in PubMed abstracts), high-throughput experiments, conserved co-expression, gene neighborhood, gene fusion events, phylogenetic profiling, and association in curated databases and contains protein associations for more than 600 species.

In this chapter we provide a step-by-step overview of how such co-occurrence networks are created, stored, and visualized, and we demonstrate their usefulness by providing two biological examples using the STRING [45] and BioTextQuest [46] services.

2 Materials

To use the web applications which are mentioned in this chapter, the following tools should be pre-installed:

1. A web browser (preferably a latest version of Firefox, Chrome or Safari) to support the latest JavaScript libraries.
2. Java and the relevant browser plugins to run Java applets. JRE version >6 is preferred.
3. Adobe Flash player and the relevant browser plugins.

3 Methods

3.1 Co-occurrence Networks

A co-occurrence network is a graphical representation of relationships between terms that belong to a unit of text. Such a network corresponds to a graph $G=(V, E)$ consisting of vertices (V) and edges (E), where vertices might represent terms and edges the connections between them. Hence, two terms that coexist in the same phrase or paragraph may be linked to each other. Co-occurrence analysis of genes/proteins or other bioentities in published abstracts, entire documents, units of text, or other public databases has been used to derive networks with clear indication that their edges reflect functionally relevant relationships [4, 47–50]. Figure 1 illustrates a representative example of biomedical terms identified to coexist in an abstract (Fig. 1a), their interconnections and the resulting network (Fig. 1b–d). In the first case of *abstract-based* co-occurrences (Fig. 1b), every entity (node) identified in the abstract text is connected to any other entity in the same text, forming a clique [51]. In the second case of *sentence-based* co-occurrences (Fig. 1c), only the entities co-mentioned in the same sentence are connected to each other. A more subtle case is the one of *semantic networks* (Fig. 1d). These networks are usually directed multi-edged networks. Each edge in such networks represents an action between two nodes. If more than one action occurs between the nodes, these are then connected with more than one edge.

a

Abstract

The forkhead box transcription factor FOXM1 is considered to be a promising target for cancer therapy. However, the significance of FOXM1 in tumors harboring mutation in p53, which is very common, is unclear. In this study, we investigated the efficacy of FoxM1-targeting in spontaneous p53-null tumors using genetic ablation as well as using a peptide-inhibitor of FOXM1. We show that conditional deletion of FoxM1 inhibits growth of the p53 null thymic lymphoma and sarcoma cells. In addition, deletion of FoxM1 induces apoptotic cell death of the p53 null tumors, accompanied by reduced expression of the FOXM1 target genes Survivin and Bmi1. An ARF-derived peptide that inhibits the activity of FOXM1, by targeting it to the nucleolus, also induces apoptosis in the p53 null sarcoma and lymphoma, leading to a strong inhibition of their metastatic colonization. Together, our observations suggest that FOXM1 is critical for survival and growth of the p53-null lymphoma and sarcoma, and provide proof-of-principle that FOXM1 is an effective therapeutic target for sarcoma and lymphoma carrying loss of function mutation in p53. *PMID:23427295*

b Abstract-based

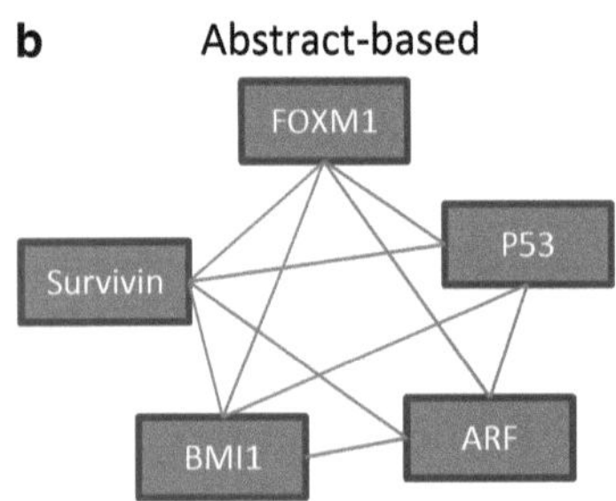

	FOXM1	*P53*	*ARF*	*BMI1*	*Survivin*
FOXM1	1	1	1	1	1
P53	1	1	1	1	1
ARF	1	1	1	1	1
BMI1	1	1	1	1	1
Survivin	1	1	1	1	1

c Sentence-based

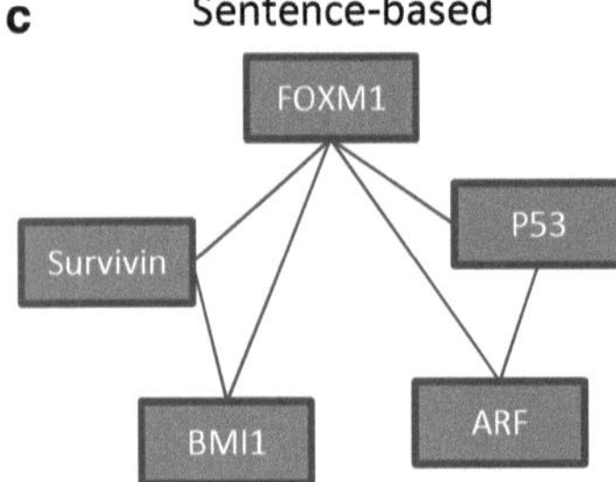

	FOXM1	*P53*	*ARF*	*BMI1*	*Survivin*
FOXM1	1	1	1	1	1
P53	1	1	1	0	0
ARF	1	1	1	0	0
BMI1	1	0	0	1	1
Survivin	1	0	0	1	1

d Directed

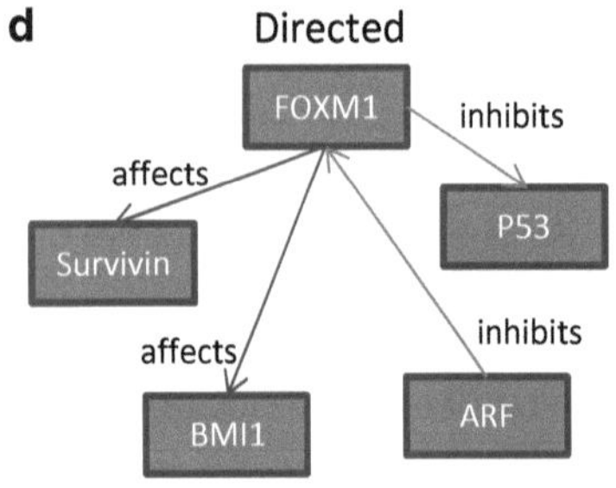

	FOXM1	*P53*	*ARF*	*BMI1*	*Survivin*
FOXM1	1	1	0	1	1
P53	0	1	0	0	0
ARF	1	0	1	0	0
BMI1	0	0	0	1	0
Survivin	0	0	0	0	1

Fig. 1 (**a**) Gene/protein names highlighted in blue for abstract [62]. (**b**) Abstract-based co-occurrence: All genes identified in the abstract are interlinked. (**c**) Sentence-based co-occurrence: Genes that coexist in a sentence are linked. (**d**) Directed network: Genes which coexist in a sentence are linked to each other with edges accompanied by an attribute showing an action between two genes. For example, FOXM1 "*inhibits*" P53. A matrix representation (**b–d**, *right*) for the networks is also available. Notice that the matrix for the semantic (directed) network (**d**) is not necessarily symmetric

For example, within a sentence where it is stated that *A* "*inhibits*" *B*, an arrow starting from node A and ending to B is used. In the case where A and B are related in multiple ways, for example by their coexistence in databases, by being evolutionarily related or found to be co-expressed in an experiment, then multiple lines or arrows of different meaning (colors) are being used; each color representing one action type.

As co-occurrence analysis can be performed for very large corpora, the resulting networks often become dense and incomprehensible. To better understand the topology of the network and quickly identify closely related terms, visualization techniques play a key role. Specialized visualization tools [52, 53] try to tackle this problem by incorporating high-quality clustering algorithms and employing sophisticated layout procedures in order to (1) minimize the crossovers between the edges and (2) simultaneously place closely associated nodes next to each other to form functionally related groups. Visualization tools which worth to me mentioned are the BioLayout [54], Ondex [55], Pajek [56], Cytoscape [57], Arena3D [58, 59], Medusa [60], and VisANT [61].

3.2 Literature Network

Retrieving a representative set of documents related to a topic is not an easy task, as most of the current search engines use keywords to retrieve related records. In most cases, the results are shown as unordered lists which are difficult to manually filter by relevance. Especially in those instances where non-experts query for more generic topics, these lists may contain hundreds or even thousands of documents that are hard to comprehend and difficult to process. Therefore, text retrieval and text categorization still remain, among others, two of the biggest challenges in the field of text mining. A partial solution to the problem is to pre-annotate PubMed abstracts by using the Medical Subject Headings (MeSH) thesaurus "*Mesh term: comprehensive controlled vocabulary for the purpose of indexing journal articles and books in the life sciences*" and the Gene Ontology terms, and subsequently order them by subject, as XplorMed [17], GOPubMed [16], and McSyBi [63] do. Another approach is to apply name entity recognition (NER) techniques in abstracts to identify bioentities (e.g., proteins, drugs, genes, diseases, or pathways) and perform a co-occurrence analysis. Abstracts sharing common annotations or words are likely to be related to each other and to the initial query. Tools such as EBIMed [64] and ReleMed [65], for example, apply co-occurrence analysis at the sentence level to predict relationships between genes. Alternatively, a single abstract or text paragraph can be used as a model to find similar records using the PubMed-related article feature [66].

An illustrative example on how to relate articles in an automated way is presented in Fig. 2, through a four-step process. First, biological terms and entities are identified in an article based on pre-calculated dictionaries. In the second step, each article is represented by a vector consisting of identified terms/words. A similarity measure such as *Pearson*, *Cosine*, or *Tanimoto coefficient* is

a

In this report, we investigated the role and regulation of forkhead box M1 (FOXM1) in breast cancer and epirubicin resistance. We generated epirubicin-resistant MCF-7 breast carcinoma (MCF-7-EPI(R)) cells and found FOXM1 protein levels to be higher in MCF-7-EPI(R) than in MCF-7 cells and that FOXM1 expression is downregulated by epirubicin in MCF-7 but not in MCF-7-EPI(R) cells. We also established that there is a loss of p53 function in MCF-7-EPI(R) cells and that epirubicin represses FOXM1 expression at transcription and gene promoter levels through activation of p53 and repression of E2F activity in MCF-7 cells. Using p53(-/-) mouse embryo fibroblasts, we showed that p53 is important for epirubicin sensitivity. Moreover, transient promoter transfection assays showed that epirubicin and its cellular effectors p53 and E2F1 modulate FOXM1 transcription through an E2F-binding site located within the proximal promoter region. Chromatin immunoprecipitation analysis also revealed that epirubicin treatment increases pRB (retinoblastoma protein) and decreases E2F1 recruitment to the FOXM1 promoter region containing the E2F site. We also found ataxia-telangiectasia mutated (ATM) protein and mRNA to be overexpressed in the resistant MCF-7-EPI(R) cells compared with MCF-7 cells and that epirubicin could activate ATM to promote E2F activity and FOXM1 expression. Furthermore, inhibition of ATM in U2OS cells with caffeine or depletion of ATM in MCF-7-EPI(R) with short interfering RNAs can resensitize these resistant cells to epirubicin, resulting in downregulation of E2F1 and FOXM1 expression and cell death. In summary, our data show that ATM and p53 coordinately regulate FOXM1 via E2F to modulate epirubicin response and resistance in breast cancer. *Abstract PMID: 21518729*

The forkhead box transcription factor FOXM1 is considered to be a promising target for cancer therapy. However, the significance of FOXM1 in tumors harboring mutation in p53, which is very common, is unclear. In this study, we investigated the efficacy of FoxM1-targeting in spontaneous p53-null tumors using genetic ablation as well as using a peptide-inhibitor of FOXM1. We show that conditional deletion of FoxM1 inhibits growth of the p53 null thymic lymphoma and sarcoma cells. In addition, deletion of FoxM1 induces apoptotic cell death of the p53 null tumors, accompanied by reduced expression of the FOXM1 target genes Survivin and Bmi1. An ARF-derived peptide that inhibits the activity of FOXM1, by targeting it to the nucleolus, also induces apoptosis in the p53 null sarcoma and lymphoma, leading to a strong inhibition of their metastatic colonization. Together, our observations suggest that FOXM1 is critical for survival and growth of the p53-null lymphoma and sarcoma, and provide proof-of-principle that FOXM1 is an effective therapeutic target for sarcoma and lymphoma carrying loss of function mutation in p53. *Abstract PMID: 23427295*

The forkhead box m1 (Foxm1) transcription factor is essential for initiation of carcinogen-induced liver tumors; however, whether FoxM1 constitutes a therapeutic target for liver cancer treatment remains unknown. In this study, we used diethylnitrosamine/phenobarbital treatment to induce hepatocellular carcinomas (HCCs) in either WT mice or Arf(-/-)Rosa26-FoxM1b Tg mice, in which forkhead box M1b (FoxM1b) is overexpressed and alternative reading frame (ARF) inhibition of FoxM1 transcriptional activity is eliminated. To pharmacologically reduce FoxM1 activity in HCCs, we subjected these HCC-bearing mice to daily injections of a cell-penetrating ARF(26-44) peptide inhibitor of FoxM1 function. After 4 weeks of this treatment, HCC regions displayed reduced tumor cell proliferation and angiogenesis and a significant increase in apoptosis within the HCC region but not in the adjacent normal liver tissue. ARF peptide treatment also induced apoptosis of several distinct human hepatoma cell lines, which correlated with reduced protein levels of the mitotic regulatory genes encoding polo-like kinase 1, aurora B kinase, and survivin, all of which are transcriptional targets of FoxM1 that are highly expressed in cancer cells and function to prevent apoptosis. These studies indicate that ARF peptide treatment is an effective therapeutic approach to limit proliferation and induce apoptosis of liver cancer cells in vivo. *Abstract PMID: 17173139*

b

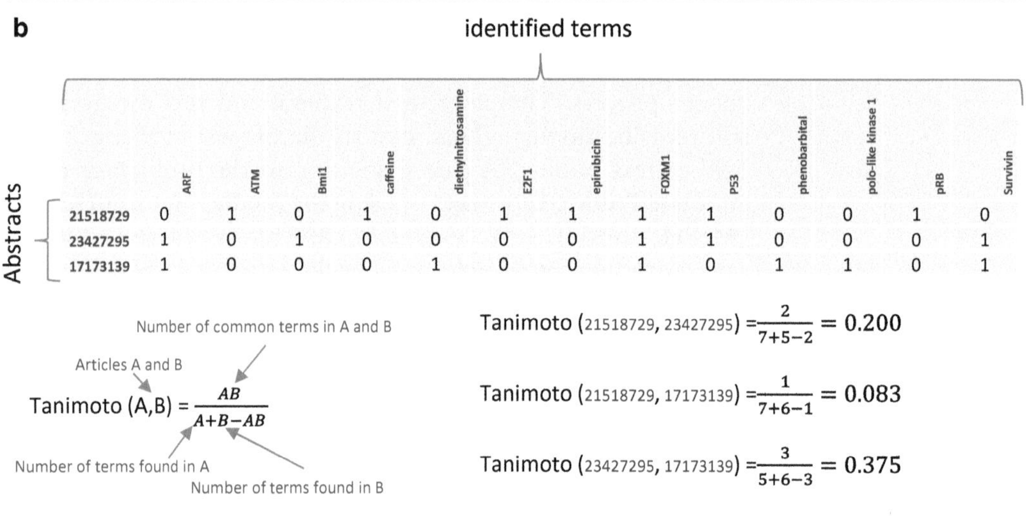

c

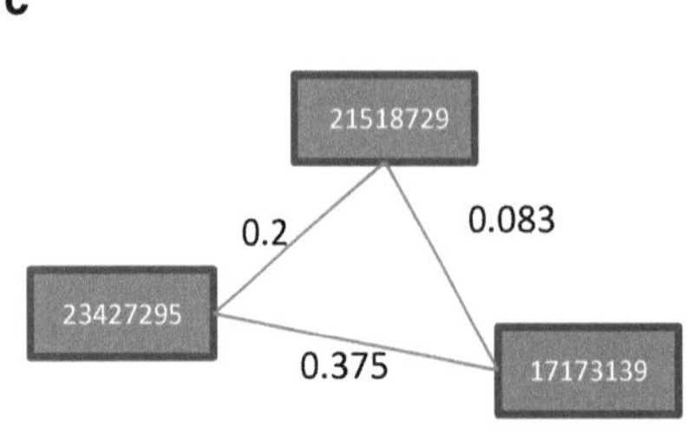

Correlation Matrix (C)

	21518729	23427295	17173139
21518729	1	0.2	0.083
23427295	0.2	1	0.375
17173139	0.083	0.375	1

D=1-C

Distance Matrix (D)

	21518729	23427295	17173139
21518729	0	0.8	0.917
23427295	0.8	0	0.625
17173139	0.917	0.6255	0

Fig. 2 (**a**) Gene and protein names highlighted in blue for certain abstracts [62, 80, 81]. (**b**) Vectors of biomedical terms and text similarities using the Tanimoto coefficient. Synonymous terms (e.g., "forkhead box m1" and Foxm1) are considered as a single entity. (**c**) Graphical and matrix representation of the abstract weighted correlation network

calculated in step 3, to quantify the pairwise similarities between each pair of articles, resulting to a weighted graph. During step 4, clustering algorithms are applied to the network in order to isolate groups of related articles that share common terms or other features. jClust [67], NeAT [68], MCL [69], Affinity propagation [70], MCODE [71], Clique [72], LCMA [73], DPClus [74], CMC [75], SCAN [76], Cfinder [77], GIBA [78], and PCP [79] are graph-theoretic algorithms and platforms for detecting highly connected subnetworks. Notably, for larger amounts of literature data, a good practice is to pre-compute the network clustering, as clustering techniques are often intensive in terms of computer resources, such as processor time and memory.

3.3 Protein–Protein Interaction Networks (PPIs) Using STRING

PPIs play a crucial role in most complex cellular processes. Such processes include cell cycle control, differentiation, protein folding, signaling, transcription, translation, posttranslational modification, and transportation [82]. The field of predicting PPIs is a heavily studied field in systems biology and high-throughput experimental techniques such as pull-down assays [83], tandem affinity purification [84], yeast two hybrid systems—Y2H [85], mass spectrometry [86], and microarrays [87] are widely used to infer such PPIs. Public databases such as the Yeast Proteome Database—YPD [88], MIPS [89], MINT [90], IntAct [91], DIP [92], BIND [93], and BioGRID [94] store such binary interactions of proteins or whole complexes in an organized way along with annotations. Most of these databases are manually curated, freely available to download and for further processing using text mining techniques. Co-occurrence analyses, for example, can prioritize search results by relevance and also suggest new interactions that are not previously known and can be experimentally validated. Assuming that two pairs of proteins A–B and B–C are known to interact and are mentioned in articles, a co-occurrence analysis can reveal indirect connections, i.e., A–C, which might be of importance.

To demonstrate the usefulness of text mining and co-occurrence analysis, we queried the STRING database [45, 95] for the known yeast ARP2/3 complex [96]. The ARP2/3 complex is a known seven-subunit protein complex involved in the regulation of the actin cytoskeleton and is found mostly in actin cytoskeleton-containing eukaryotic cells [97]. Using information only from databases, experiments and co-expression, the STRING database was able to collect six highly connected subunits of the complex (Fig. 3). Similarly, using only the text-mining approach, the STRING database was able to reproduce the same network with high confidence, showing that text mining analysis can stand as key player in biological research to not only predict already known interactions as shown before but also generate new hypotheses.

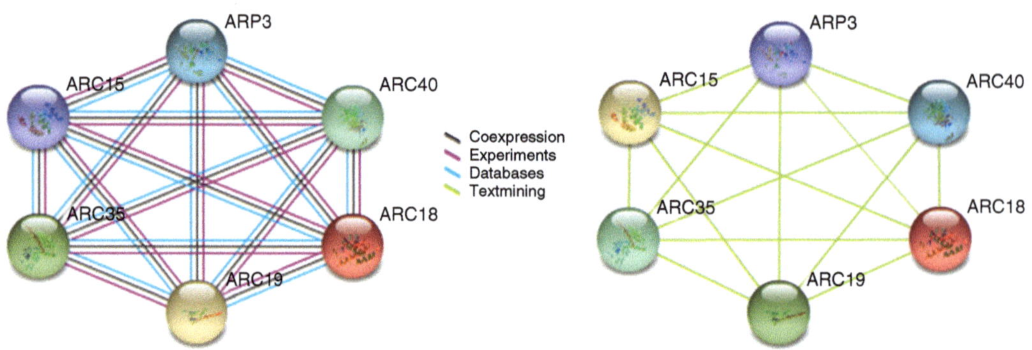

Fig. 3 String query for "*ARP2/3*" complex components (yeast). *Left*: Six components of the ARP2/3 complex based on experiments, databases, and co-expression. *Right*: Six components of the ARP2/3 complex based on text mining

3.4 Concept Discovery Using BioTextQuest

To demonstrate the power of text analysis in concept discovery, we use the BioTextQuest [46] web server. The core component of BioTextQuest was based on the TextQuest algorithm [98]. BioTextQuest is implemented for automated discovery of significant terms in article clusters with structured knowledge annotation, accompanied by NER services and semantic annotation of terms for visualization purposes. BioTextQuest is able to cluster articles based on their common terms and provides a tag-cloud-based visualization of terms mentioned in a document. The size of these terms demonstrates their level of representation in the given set of documents, whereas a color scheme is associated to the type of the bio-entity (e.g., gene, protein). We queried BioTextQuest with a complex PubMed-like query for abstracts containing any of the terms "*anterior-posterior*" or "*dorsal-ventral*" and mentioning the species "*Drosophila*". These terms describe the development of the Drosophila embryo in the two aforementioned axes. Three clusters were obtained (Fig. 4). The first cluster is related to the development of the Drosophila embryo in the dorsal ventral axis (the size of the term "dorsal ventral" dominates the cluster), while characteristic genes of this pathway, such as *cactus* and *sog*, are present. The second cluster is dominated by the term "anterior posterior" and genes such as *hunchback*, *distaless*, and *ubx* which play important role in the formation of the anterior posterior axis during embryogenesis. In the third cluster, the term "oogenesis" is over-represented, together with both "anterior posterior" and "dorsal ventral" terms. Additionally, in this cluster, we observe genes such *gurken* and *oskar* (which are characteristic for the formation of these axes during oogenesis, prior to fertilization). A hypothesis that can be generated from this BioTextQuest analysis is that the two developmental axes are not well defined during the early stages of embryonic development, discussed elsewhere in the literature [99].

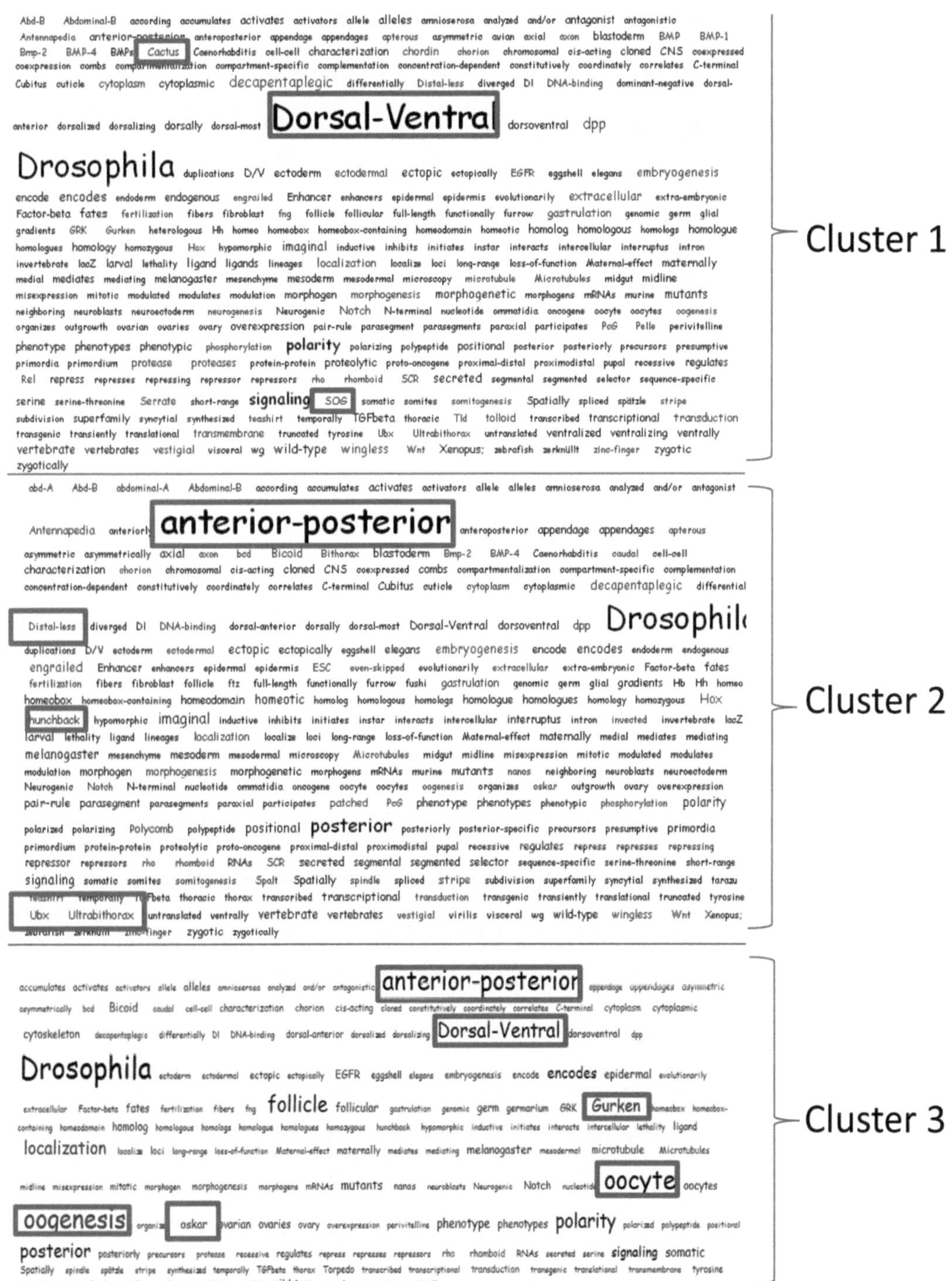

Fig. 4 Query of "*anterior-posterior*" or "*dorsal-ventral*" terms for Drosophila returns three clusters of documents. Cluster 1: documents related to dorsal ventral developmental axis are collected. Cluster 2: collected documents are related to anterio-posterior developmental axis. Cluster 3: terms from both categories are mixed. The term "oogenesis" is overrepresented, thus generating the hypothesis that developmental axes are not well defined in early stages but at later stages in oogenesis [99]

4 Notes

4.1 A Step-by-Step Guide for Generating Protein–Protein Interaction Networks (PPIs) with STRING

To produce the protein interaction networks mentioned in Subheading 3.3, the steps below should be followed:

1. Visit the URL: http://string.embl.de
2. In the field "*organism*", please select: *yeast* or *Saccharomyces cerevisiae*
3. In the field "protein name" please type: *ARP2/3*
4. Press "GO!" button
5. A disambiguation page will appear. To continue, please select the option: *ARC18–ARC18–ARP2/3 complex 21 kDa subunit (p21-ARC); Functions as component of the ARP2/3 complex which is involved in regulation of actin polymerization and together with an activating nucleation-promoting factor (NPF) mediates the formation of branched actin networks (By similarity)*. This option will use the ARC18 protein as a seed to generate a network with its interactors.
6. A new page showing a PPI network will appear. In the "*info and parameters*" section, please select "*High confidence score 0.7*" and type the number five in the field of "*interactions shown*" to reduce the number of interactors. Then hit the "*update parameters*" button and the page will reload showing a new network (smaller).
7. To view the network which is generated by using only TextMining approach, please go to the "*info and parameters*" panel under the "*Active Prediction Methods*" menu and deselect every prediction method except from the "Text Mining" option. Then hit the "*update parameters*" button. The page will reload and show a PPI network consisting of six components of the ARP2/3 complex as they are predicted by the TextMining method only.
8. To view the same network consisting of experimentally validated connections between the components of the Arp2/3 complex, please repeat the procedure of the previous step by deselecting all of the "*Active Prediction Methods*" except from the options: "*databases*", "*experiments*," and "*CoExpression*". After hitting the "*update parameters*" button, STRING will generate the same network as before based on different evidence of information.
9. Connections between the components of ARP2/3 complex are colored in green "Text Mining" for step 7, whereas connections of the multi-edged graph for option 8 are shown in pink (Experiments), black (Coexpression), and blue (Databases).

4.2 A Step-by-Step Guide for Using BioTextQuest

To demonstrate the functionality of BioTextQuest, as mentioned in Subheading 3.4, please follow the steps below:

1. Visit the URL: http://bioinformatics.med.uoc.gr/biotextquest
2. In the "*query PubMed*" text field please type: "(*anterior-posterior AND drosophila*) *OR* (*dorsal-ventral AND drosophila*) *1:2001/03[dp]*"
3. In the "*advanced options*" menu, please select "*K-Means*" for the "*clustering algorithm*" and three for the "*number of clusters*".
4. Hit the button "*Send Data Now*" and be patient for the results to be generated. At this stage, BioTextQuest collects all of the PubMed abstracts that are related to the queried terms for "*Drosophila*" organism and performs a Text Mining analysis.
5. After a while, a new multi-tabbed page will be generated. The "*Tag clouds*" tab will show clusters of terms by highlighting and adjusting the size of the overrepresented terms in a cluster. Bold and words of bigger size are the most representative in the cluster.
6. Similarly, the "*Documents*" tab, will show the PubMed articles, clustered according to their context.

5 Future Perspectives and Challenges

Text mining techniques and co-occurrence analyses are promising methodologies that can assist us to cope with the overload of textual information and extract meaningful conclusions, which are often hidden in literature. Typical text mining tasks include text prioritization, document clustering, concept and entity extraction, sentiment analysis, document summarization, and entity relation modeling. Such techniques can significantly assist large-scale experiments in collaboration with other in silico approaches, such as phylogenetic profiles, fusion events, and gene neighborhood, towards a more complete data integration approach. While text mining techniques can stand as key players in automatic knowledge extraction and targeted literature searching, they are still at their infancy [100] as they are often accompanied by a high false-positive rate. To overcome this challenge in the field and be able to perform accurate co-occurrence analyses, very rich and high-quality dictionaries for NER should become available; despite the current efforts we are far from this point. Term disambiguation still remains a bottleneck as bioentities, such as genes, can be found in multiple forms (e.g., HUGO names, Entrez identifiers, free text). Furthermore, there is no consistency or a unique mapping between a name and a bioentity. The term *C1R* for example represents a cell-line but also a gene name (ENSG00000159403). NER is becoming even more complicated for inter-species entity name disambiguation, as there are no unique identifiers for genes and proteins of a certain species. In addition, full text processing rather than abstract-only annotation is necessary. Efficient and fast converters to transform any type of document such as PDFs, HTMLs, PPTs, XLSs into simple structured XML files are important for full text analysis, especially when there is no need to separately process titles, abstracts, tables, and figure legends. Finally, semantic networks are still very immature as dictionaries that map a verb or a phrase to an action such as "*inhibits*," "*blocks*," or "*regulates*" are currently limited. Natural Language Processing (NLP) approaches should become mature enough to understand the syntax of a sentence and generate rule-based networks to generate new ideas and concepts that are implicitly present but not explicitly detected by readers and researchers alike.

Acknowledgments

The work was supported in part by the European Commission FP7 programme "Translational Potential" (TransPOT; EC contract number 285948).

References

1. Hunter L, Cohen KB (2006) Biomedical language processing: what's beyond PubMed? Mol Cell 21(5):589–594
2. Lu Z (2011) PubMed and beyond: a survey of web tools for searching biomedical literature. Database (Oxford) 2011:baq036
3. Cohen AM, Hersh WR (2005) A survey of current work in biomedical text mining. Brief Bioinform 6(1):57–71
4. Rodriguez-Esteban R (2009) Biomedical text mining and its applications. PLoS Comput Biol 5(12):e1000597
5. Zhu F, Patumcharoenpol P, Zhang C, Yang Y, Chan J, Meechai A, Vongsangnak W, Shen B (2012) Biomedical text mining and its applications in cancer research. J Biomed Inform 46(2):200–211
6. Rebholz-Schuhmann D, Oellrich A, Hoehndorf R (2012) Text-mining solutions for biomedical research: enabling integrative biology. Nat Rev Genet 13(12):829–839
7. Lu Z, Wilbur WJ, McEntyre JR, Iskhakov A, Szilagyi L (2009) Finding query suggestions for PubMed. AMIA Annu Symp Proc 2009: 396–400
8. Swanson DR (1986) Fish oil, Raynaud's syndrome, and undiscovered public knowledge. Perspect Biol Med 30(1):7–18
9. States DJ, Ade AS, Wright ZC, Bookvich AV, Athey BD (2009) MiSearch adaptive pubMed search tool. Bioinformatics 25(7):974–976
10. Giglia E (2011) Quertle and KNALIJ: searching PubMed has never been so easy and effective. Eur J Phys Rehabil Med 47(4):687–690
11. Hymel GM (2011) PubMed central inclusion, quertle indexing, outbound reference linking, and editorial board successions: encouraging developments in the IJTMB's evolution. Int J Ther Massage Bodywork 4(1):1–2
12. Fontaine JF, Barbosa-Silva A, Schaefer M, Huska MR, Muro EM, Andrade-Navarro MA (2009) MedlineRanker: flexible ranking of biomedical literature. Nucleic Acids Res 37(Web Server issue):W141–W146
13. Errami M, Wren JD, Hicks JM, Garner HR (2007) eTBLAST: a web server to identify expert reviewers, appropriate journals and similar publications. Nucleic Acids Res 35(Web Server issue):W12–W15
14. Poulter GL, Rubin DL, Altman RB, Seoighe C (2008) MScanner: a classifier for retrieving Medline citations. BMC Bioinformatics 9:108
15. Smalheiser NR, Zhou W, Torvik VI (2008) Anne O'Tate: a tool to support user-driven summarization, drill-down and browsing of

PubMed search results. J Biomed Discov Collab 3:2

16. Doms A, Schroeder M (2005) GoPubMed: exploring PubMed with the gene ontology. Nucleic Acids Res 33(Web Server issue): W783–W786
17. Perez-Iratxeta C, Bork P, Andrade MA (2001) XplorMed: a tool for exploring MEDLINE abstracts. Trends Biochem Sci 26(9):573–575
18. Soldatos TG, O'Donoghue SI, Satagopam VP, Barbosa-Silva A, Pavlopoulos GA, Wanderley-Nogueira AC, Soares-Cavalcanti NM, Schneider R (2012) Caipirini: using gene sets to rank literature. BioData Min 5(1):1
19. Harris MA, Clark J, Ireland A, Lomax J, Ashburner M, Foulger R, Eilbeck K, Lewis S, Marshall B, Mungall C, Richter J, Rubin GM, Blake JA, Bult C, Dolan M, Drabkin H, Eppig JT, Hill DP, Ni L, Ringwald M, Balakrishnan R, Cherry JM, Christie KR, Costanzo MC, Dwight SS, Engel S, Fisk DG, Hirschman JE, Hong EL, Nash RS, Sethuraman A, Theesfeld CL, Botstein D, Dolinski K, Feierbach B, Berardini T, Mundodi S, Rhee SY, Apweiler R, Barrell D, Camon E, Dimmer E, Lee V, Chisholm R, Gaudet P, Kibbe W, Kishore R, Schwarz EM, Sternberg P, Gwinn M, Hannick L, Wortman J, Berriman M, Wood V, de la Cruz N, Tonellato P, Jaiswal P, Seigfried T, White R (2004) The Gene Ontology (GO) database and informatics resource. Nucleic Acids Res 32(Database issue):D258–D261
20. Rebholz-Schuhmann D, Arregui M, Gaudan S, Kirsch H, Jimeno A (2008) Text processing through Web services: calling Whatizit. Bioinformatics 24(2):296–298
21. Settles B (2005) ABNER: an open source tool for automatically tagging genes, proteins and other entity names in text. Bioinformatics 21(14):3191–3192
22. Pafilis E, O'Donoghue SI, Jensen LJ, Horn H, Kuhn M, Brown NP, Schneider R (2009) Reflect: augmented browsing for the life scientist. Nat Biotechnol 27(6):508–510
23. Pavlopoulos GA, Pafilis E, Kuhn M, Hooper SD, Schneider R (2009) OnTheFly: a tool for automated document-based text annotation, data linking and network generation. Bioinformatics 25(7):977–978
24. Frantzi K, Ananiadou S, Mima H (2000) Automatic recognition of multi-word terms. Int J Digit Libr 3(2):117–132
25. Kim JJ, Pezik P, Rebholz-Schuhmann D (2008) MedEvi: retrieving textual evidence of relations between biomedical concepts from Medline. Bioinformatics 24(11):1410–1412
26. Rebholz-Schuhmann D, Kirsch H, Arregui M, Gaudan S, Riethoven M, Stoehr P (2007) EBIMed—text crunching to gather facts for proteins from Medline. Bioinformatics 23(2): e237–e244
27. Douglas SM, Montelione GT, Gerstein M (2005) PubNet: a flexible system for visualizing literature derived networks. Genome Biol 6(9):R80
28. Plikus MV, Zhang Z, Chuong CM (2006) PubFocus: semantic MEDLINE/PubMed citations analytics through integration of controlled biomedical dictionaries and ranking algorithm. BMC Bioinformatics 7:424
29. Fontelo P, Liu F, Ackerman M, Schardt CM, Keitz SA (2006) askMEDLINE: a report on a year-long experience. AMIA Annu Symp Proc 923
30. Fontelo P, Liu F, Ackerman M (2005) MeSH Speller + askMEDLINE: auto-completes MeSH terms then searches MEDLINE/PubMed via free-text, natural language queries. AMIA Annu Symp Proc 957
31. Fontelo P, Liu F, Ackerman M (2005) askMEDLINE: a free-text, natural language query tool for MEDLINE/PubMed. BMC Med Inform Decis Mak 5:5
32. Liu F, Ackerman M, Fontelo P (2006) BabelMeSH: development of a cross-language tool for MEDLINE/PubMed. AMIA Annu Symp Proc 1012
33. Featherstone R, Hersey D (2010) The quest for full text: an in-depth examination of Pubget for medical searchers. Med Ref Serv Q 29(4):307–319
34. Eaton AD (2006) HubMed: a web-based biomedical literature search interface. Nucleic Acids Res 34(Web Server issue):W745–W747
35. Hokamp K, Wolfe KH (2004) PubCrawler: keeping up comfortably with PubMed and GenBank. Nucleic Acids Res 32(Web Server issue):W16–W19
36. Goetz T, von der Lieth CW (2005) PubFinder: a tool for improving retrieval rate of relevant PubMed abstracts. Nucleic Acids Res 33(Web Server issue):W774–W778
37. Thomas J, Milward D, Ouzounis C, Pulman S, Carroll M (2000) Automatic extraction of protein interactions from scientific abstracts. Pac Symp Biocomput 5:538–549
38. Alako BT, Veldhoven A, van Baal S, Jelier R, Verhoeven S, Rullmann T, Polman J, Jenster G (2005) CoPub Mapper: mining MEDLINE based on search term co-publication. BMC Bioinformatics 6:51
39. Ono T, Hishigaki H, Tanigami A, Takagi T (2001) Automated extraction of information on protein-protein interactions from the biological literature. Bioinformatics 17(2): 155–161

40. Novichkova S, Egorov S, Daraselia N (2003) MedScan, a natural language processing engine for MEDLINE abstracts. Bioinformatics 19(13):1699–1706
41. Rebholz-Schuhmann D, Jimeno-Yepes A, Arregui M, Kirsch H (2010) Measuring prediction capacity of individual verbs for the identification of protein interactions. J Biomed Inform 43(2):200–207
42. Iacucci E, Tranchevent LC, Popovic D, Pavlopoulos GA, De Moor B, Schneider R, Moreau Y (2012) ReLiance: a machine learning and literature-based prioritization of receptor—ligand pairings. Bioinformatics 28(18):i569–i574
43. van Haagen HH, t Hoen PA, Botelho Bovo A, de Morree A, van Mulligen EM, Chichester C, Kors JA, den Dunnen JT, van Ommen GJ, van der Maarel SM, Kern VM, Mons B, Schuemie MJ (2009) Novel protein-protein interactions inferred from literature context. PLoS One 4(11):e7894
44. Hoffmann R, Valencia A (2004) A gene network for navigating the literature. Nat Genet 36(7):664
45. Szklarczyk D, Franceschini A, Kuhn M, Simonovic M, Roth A, Minguez P, Doerks T, Stark M, Muller J, Bork P, Jensen LJ, von Mering C (2011) The STRING database in 2011: functional interaction networks of proteins, globally integrated and scored. Nucleic Acids Res 39(Database issue):D561–D568
46. Papanikolaou N, Pafilis E, Nikolaou S, Ouzounis CA, Iliopoulos I, Promponas VJ (2011) BioTextQuest: a web-based biomedical text mining suite for concept discovery. Bioinformatics 27(23):3327–3328
47. Zhu S, Okuno Y, Tsujimoto G, Mamitsuka H (2006) Application of a new probabilistic model for mining implicit associated cancer genes from OMIM and medline. Cancer Inform 2:361–371
48. Schuemie MJ, Weeber M, Schijvenaars BJ, van Mulligen EM, van der Eijk CC, Jelier R, Mons B, Kors JA (2004) Distribution of information in biomedical abstracts and full-text publications. Bioinformatics 20(16):2597–2604
49. Jenssen TK, Laegreid A, Komorowski J, Hovig E (2001) A literature network of human genes for high-throughput analysis of gene expression. Nat Genet 28(1):21–28
50. Stapley BJ, Benoit G (2000) Biobibliometrics: information retrieval and visualization from co-occurrences of gene names in Medline abstracts. Pac Symp Biocomput 529–540
51. Pavlopoulos GA, Secrier M, Moschopoulos CN, Soldatos TG, Kossida S, Aerts J, Schneider R, Bagos PG (2011) Using graph theory to analyze biological networks. BioData Min 4:10
52. Pavlopoulos GA, Wegener AL, Schneider R (2008) A survey of visualization tools for biological network analysis. BioData Min 1:12
53. Gehlenborg N, O'Donoghue SI, Baliga NS, Goesmann A, Hibbs MA, Kitano H, Kohlbacher O, Neuweger H, Schneider R, Tenenbaum D, Gavin AC (2010) Visualization of omics data for systems biology. Nat Methods 7(3 Suppl):S56–S68
54. Enright AJ, Ouzounis CA (2001) BioLayout—an automatic graph layout algorithm for similarity visualization. Bioinformatics 17(9):853–854
55. Kohler J, Baumbach J, Taubert J, Specht M, Skusa A, Ruegg A, Rawlings C, Verrier P, Philippi S (2006) Graph-based analysis and visualization of experimental results with ONDEX. Bioinformatics 22(11):1383–1390
56. Breitkreutz BJ, Stark C, Tyers M (1998) Pajek—program for large network analysis. Connections 21:47–57
57. Shannon P, Markiel A, Ozier O, Baliga NS, Wang JT, Ramage D, Amin N, Schwikowski B, Ideker T (2003) Cytoscape: a software environment for integrated models of biomolecular interaction networks. Genome Res 13(11):2498–2504
58. Secrier M, Pavlopoulos GA, Aerts J, Schneider R (2012) Arena3D: visualizing time-driven phenotypic differences in biological systems. BMC Bioinformatics 13:45
59. Pavlopoulos GA, O'Donoghue SI, Satagopam VP, Soldatos TG, Pafilis E, Schneider R (2008) Arena3D: visualization of biological networks in 3D. BMC Syst Biol 2:104
60. Pavlopoulos GA, Hooper SD, Sifrim A, Schneider R, Aerts J (2011) Medusa: a tool for exploring and clustering biological networks. BMC Res Notes 4(1):384
61. Hu Z, Hung JH, Wang Y, Chang YC, Huang CL, Huyck M, DeLisi C (2009) VisANT 3.5: multi-scale network visualization, analysis and inference based on the gene ontology. Nucleic Acids Res 37(Web Server issue):W115–W121
62. Wang Z, Zheng Y, Park HJ, Li J, Carr JR, Chen YJ, Kiefer MM, Kopanja D, Bagchi S, Tyner AL, Raychaudhuri P (2013) Targeting FoxM1 effectively retards p53-null lymphoma and sarcoma. Mol Cancer Ther 12(5):759–767
63. Yamamoto Y, Takagi T (2007) Biomedical knowledge navigation by literature clustering. J Biomed Inform 40(2):114–130
64. Rebholz-Schuhmann D, Kirsch H, Arregui M, Gaudan S, Rynbeek M, Stoehr P (2006)

Protein annotation by EBIMed. Nat Biotechnol 24(8):902–903
65. Siadaty MS, Shu J, Knaus WA (2007) Relemed: sentence-level search engine with relevance score for the MEDLINE database of biomedical articles. BMC Med Inform Decis Mak 7:1
66. Lin J, Wilbur WJ (2007) PubMed related articles: a probabilistic topic-based model for content similarity. BMC Bioinformatics 8:423
67. Pavlopoulos GA, Moschopoulos CN, Hooper SD, Schneider R, Kossida S (2009) jClust: a clustering and visualization toolbox. Bioinformatics 25(15):1994–1996
68. Brohee S, Faust K, Lima-Mendez G, Sand O, Janky R, Vanderstocken G, Deville Y, van Helden J (2008) NeAT: a toolbox for the analysis of biological networks, clusters, classes and pathways. Nucleic Acids Res 36(Web Server issue):W444–W451
69. Enright AJ, Van Dongen S, Ouzounis CA (2002) An efficient algorithm for large-scale detection of protein families. Nucleic Acids Res 30(7):1575–1584
70. Frey BJ, Dueck D (2007) Clustering by passing messages between data points. Science 315(5814):972–976
71. Bader GD, Hogue CW (2003) An automated method for finding molecular complexes in large protein interaction networks. BMC Bioinformatics 4:2
72. Spirin V, Mirny LA (2003) Protein complexes and functional modules in molecular networks. Proc Natl Acad Sci U S A 100(21): 12123–12128
73. Li XL, Tan SH, Foo CS, Ng SK (2005) Interaction graph mining for protein complexes using local clique merging. Genome Inform 16(2):260–269
74. Altaf-Ul-Amin M, Shinbo Y, Mihara K, Kurokawa K, Kanaya S (2006) Development and implementation of an algorithm for detection of protein complexes in large interaction networks. BMC Bioinformatics 7:207
75. Liu G, Wong L, Chua HN (2009) Complex discovery from weighted PPI networks. Bioinformatics 25(15):1891–1897
76. Mete M, Tang F, Xu X, Yuruk N (2008) A structural approach for finding functional modules from large biological networks. BMC Bioinformatics 9 Suppl 9:S19
77. Adamcsek B, Palla G, Farkas IJ, Derenyi I, Vicsek T (2006) CFinder: locating cliques and overlapping modules in biological networks. Bioinformatics 22(8):1021–1023
78. Moschopoulos CN, Pavlopoulos GA, Schneider R, Likothanassis SD, Kossida S (2009) GIBA: a clustering tool for detecting protein complexes. BMC Bioinformatics 10 Suppl 6:S11
79. Chua HN, Ning K, Sung WK, Leong HW, Wong L (2008) Using indirect protein-protein interactions for protein complex prediction. J Bioinform Comput Biol 6(3): 435–466
80. Gusarova GA, Wang IC, Major ML, Kalinichenko VV, Ackerson T, Petrovic V, Costa RH (2007) A cell-penetrating ARF peptide inhibitor of FoxM1 in mouse hepatocellular carcinoma treatment. J Clin Invest 117(1):99–111
81. Millour J, de Olano N, Horimoto Y, Monteiro LJ, Langer JK, Aligue R, Hajji N, Lam EW (2011) ATM and p53 regulate FOXM1 expression via E2F in breast cancer epirubicin treatment and resistance. Mol Cancer Ther 10(6):1046–1058
82. Moschopoulos CN, Pavlopoulos GA, Iacucci E, Aerts J, Likothanassis S, Schneider R, Kossida S (2011) Which clustering algorithm is better for predicting protein complexes? BMC Res Notes 4:549
83. Vikis HG, Guan KL (2004) Glutathione-S-transferase-fusion based assays for studying protein-protein interactions. Methods Mol Biol 261:175–186
84. Puig O, Caspary F, Rigaut G, Rutz B, Bouveret E, Bragado-Nilsson E, Wilm M, Seraphin B (2001) The tandem affinity purification (TAP) method: a general procedure of protein complex purification. Methods 24(3): 218–229
85. Ito T, Chiba T, Ozawa R, Yoshida M, Hattori M, Sakaki Y (2001) A comprehensive two-hybrid analysis to explore the yeast protein interactome. Proc Natl Acad Sci U S A 98(8): 4569–4574
86. Gavin AC, Bosche M, Krause R, Grandi P, Marzioch M, Bauer A, Schultz J, Rick JM, Michon AM, Cruciat CM, Remor M, Hofert C, Schelder M, Brajenovic M, Ruffner H, Merino A, Klein K, Hudak M, Dickson D, Rudi T, Gnau V, Bauch A, Bastuck S, Huhse B, Leutwein C, Heurtier MA, Copley RR, Edelmann A, Querfurth E, Rybin V, Drewes G, Raida M, Bouwmeester T, Bork P, Seraphin B, Kuster B, Neubauer G, Superti-Furga G (2002) Functional organization of the yeast proteome by systematic analysis of protein complexes. Nature 415(6868): 141–147
87. Stoll D, Templin MF, Bachmann J, Joos TO (2005) Protein microarrays: applications and future challenges. Curr Opin Drug Discov Devel 8(2):239–252

88. Costanzo MC, Hogan JD, Cusick ME, Davis BP, Fancher AM, Hodges PE, Kondu P, Lengieza C, Lew-Smith JE, Lingner C, Roberg-Perez KJ, Tillberg M, Brooks JE, Garrels JI (2000) The yeast proteome database (YPD) and Caenorhabditis elegans proteome database (WormPD): comprehensive resources for the organization and comparison of model organism protein information. Nucleic Acids Res 28(1):73–76
89. Mewes HW, Frishman D, Mayer KF, Munsterkotter M, Noubibou O, Pagel P, Rattei T, Oesterheld M, Ruepp A, Stumpflen V (2006) MIPS: analysis and annotation of proteins from whole genomes in 2005. Nucleic Acids Res 34(Database issue):D169–D172
90. Licata L, Briganti L, Peluso D, Perfetto L, Iannuccelli M, Galeota E, Sacco F, Palma A, Nardozza AP, Santonico E, Castagnoli L, Cesareni G (2012) MINT, the molecular interaction database: 2012 update. Nucleic Acids Res 40(Database issue):D857–D861
91. Kerrien S, Alam-Faruque Y, Aranda B, Bancarz I, Bridge A, Derow C, Dimmer E, Feuermann M, Friedrichsen A, Huntley R, Kohler C, Khadake J, Leroy C, Liban A, Lieftink C, Montecchi-Palazzi L, Orchard S, Risse J, Robbe K, Roechert B, Thorneycroft D, Zhang Y, Apweiler R, Hermjakob H (2007) IntAct—open source resource for molecular interaction data. Nucleic Acids Res 35(Database issue):D561–D565
92. Xenarios I, Salwinski L, Duan XJ, Higney P, Kim SM, Eisenberg D (2002) DIP, the Database of Interacting Proteins: a research tool for studying cellular networks of protein interactions. Nucleic Acids Res 30(1):303–305
93. Bader GD, Betel D, Hogue CW (2003) BIND: the Biomolecular Interaction Network Database. Nucleic Acids Res 31(1):248–250
94. Stark C, Breitkreutz BJ, Reguly T, Boucher L, Breitkreutz A, Tyers M (2006) BioGRID: a general repository for interaction datasets. Nucleic Acids Res 34(Database issue): D535–D539
95. von Mering C, Huynen M, Jaeggi D, Schmidt S, Bork P, Snel B (2003) STRING: a database of predicted functional associations between proteins. Nucleic Acids Res 31(1):258–261
96. Machesky LM, Gould KL (1999) The Arp2/3 complex: a multifunctional actin organizer. Curr Opin Cell Biol 11(1):117–121
97. Veltman DM, Insall RH (2010) WASP family proteins: their evolution and its physiological implications. Mol Biol Cell 21(16): 2880–2893
98. Iliopoulos I, Enright AJ, Ouzounis CA (2001) Textquest: document clustering of Medline abstracts for concept discovery in molecular biology. Pac Symp Biocomput 384–395
99. Riechmann V, Ephrussi A (2001) Axis formation during Drosophila oogenesis. Curr Opin Genet Dev 11(4):374–383
100. Dai H-J, Chang Y-C, Tzong-Han Tsai R, Hsu W-L (2010) New challenges for biological text-mining in the next decade. J Comput Sci Tech 25(1):169

Part II

Seeking New Biology by Unlocking "Hidden" Information

Chapter 6

Roles for Text Mining in Protein Function Prediction

Karin M. Verspoor

Abstract

The Human Genome Project has provided science with a hugely valuable resource: the blueprints for life; the specification of all of the genes that make up a human. While the genes have all been identified and deciphered, it is proteins that are the workhorses of the human body: they are essential to virtually all cell functions and are the primary mechanism through which biological function is carried out. Hence in order to fully understand what happens at a molecular level in biological organisms, and eventually to enable development of treatments for diseases where some aspect of a biological system goes awry, we must understand the functions of proteins. However, experimental characterization of protein function cannot scale to the vast amount of DNA sequence data now available. Computational protein function prediction has therefore emerged as a problem at the forefront of modern biology (Radivojac et al., Nat Methods 10(13):221–227, 2013).

Within the varied approaches to computational protein function prediction that have been explored, there are several that make use of biomedical literature mining. These methods take advantage of information in the published literature to associate specific proteins with specific protein functions. In this chapter, we introduce two main strategies for doing this: association of function terms, represented as Gene Ontology terms (Ashburner et al., Nat Genet 25(1):25–29, 2000), to proteins based on information in published articles, and a paradigm called LEAP-FS (Literature-Enhanced Automated Prediction of Functional Sites) in which literature mining is used to validate the predictions of an orthogonal computational protein function prediction method.

Key words Protein function prediction, Text mining, Biomedical natural language processing, Concept recognition in text, Gene ontology annotation

1 Introduction

It has been well documented that the biomedical literature is a highly valuable source of biological knowledge, which is most effectively accessed via automated text mining methods given the context of exploding numbers of publications and the extensive human resources required to manually review and catalog (curate) the information therein [3–5]. This is as true for protein function information as for any other biological information. In this chapter, we therefore explore the use of text mining for protein function prediction.

Vinod D. Kumar and Hannah Jane Tipney (eds.), *Biomedical Literature Mining*, Methods in Molecular Biology, vol. 1159, DOI 10.1007/978-1-4939-0709-0_6, © Springer Science+Business Media New York 2014

As highlighted by Friedberg [6], one of the challenges in approaching the task of protein function prediction is obtaining agreement on what is meant by "protein function" and how protein function should be described in a computationally amenable way. Most research in this area, including the recent CAFA (Critical Assessment of Function Annotation) experiment [1] and early text mining work performed in the context of the BioCreAtIvE (Critical Assessment of Information Extraction in Biology) experiment for protein function prediction [7], targets the vocabulary of the Gene Ontology [2], which includes terminology in three "subontologies" describing *molecular function*, *biological processes*, and *cellular components*. However, a more physical interpretation of protein function is also possible, in which a protein function is described in terms of the specific physical interactions the protein participates in, such as catalysis of a specific reaction or binding of a particular molecule (e.g., a small molecule or *ligand*). The two methods we introduce below each address one of these possible functional descriptions.

Still other interpretations of protein function are possible, and several have also been addressed with literature mining methods. Protein sequence similarity, as captured in the Protein Family (Pfam) resource (http://pfam.sanger.ac.uk/), is strongly associated with protein function and classification of proteins into Pfam on the basis of literature terms has been demonstrated to have good performance [8]. Subcellular localization of a protein helps to identify its role in biological processes, function, and potential as a drug target; the SherLoc subcellular localization prediction system integrates protein sequence information with text-based features [9].

2 Materials

The primary material required to address text mining of protein function prediction is the text itself, i.e., the biomedical literature. All abstracts available in the PubMed system can be accessed through the MEDLINE system, either through a lease (http://www.nlm.nih.gov/databases/journal.html) or accessed via a set of programming utilities known as the E-Utilities for interacting with the MEDLINE database (http://www.ncbi.nlm.nih.gov/books/NBK25501/). Full text articles are available through PubMed Central (http://www.ncbi.nlm.nih.gov/pmc/), and an open access collection (called PubMed Central Open Access, or PMC-OA), which computational tools explicitly have the right to access, can be downloaded in bulk (http://www.ncbi.nlm.nih.gov/pmc/tools/openftlist/).

The Gene Ontology (http://www.geneontology.org) vocabulary that serves as the primary target vocabulary for gene/gene product function is publicly available in a variety of formats, including OWL-RDF/XML (http://purl.obolibrary.org/obo/go.owl).

Curated Gene Ontology annotations (GOA) for specific proteins can provide training and test data for text mining systems addressing protein function prediction. There are many annotations available (http://geneontology.org/GO.downloads.annotations.shtml); one particular source of annotations that is commonly considered because many annotations have an associated literature source is from UniProtKB/Swiss-Prot.

It is unfortunately not possible to reproduce the original BioCreAtIvE task addressing protein function prediction (http://www.biocreative.org/tasks/biocreative-i/task-2-functional-annotations/) [7] as it required significant manual effort by domain experts. Recently, the task has been updated and at the time of writing there is an ongoing evaluation addressing Gene Ontology annotation methods (http://www.biocreative.org/tasks/biocreative-iv/track-4-GO/). Data for training machine learning algorithms has been made available through that evaluation and should serve as a resource for direct assessment specifically of GO annotation from the literature.

3 Methods

3.1 Information Extraction of Function Terms from Text

Several strategies for enabling protein function prediction using text-mined features can be identified. One approach involves identification of terms from the Gene Ontology (GO) in texts that are known to be associated with a given protein. This approach is captured schematically in Fig. 1; systems expect a document and a protein of interest as input and identify GO terms in that document that are associated to the protein. Such algorithms are primarily useful in the context of GO annotation curation, i.e., assignment of GO terms to proteins specifically on the basis of published information. While this is not directly cast as a protein function prediction task, this can be thought of as such, especially given that evaluations of protein function prediction often use curated GO annotations as the "gold standard" for evaluation of protein function prediction (*see* **Notes 1** and **2**). The main difference between this task and the more general protein function prediction

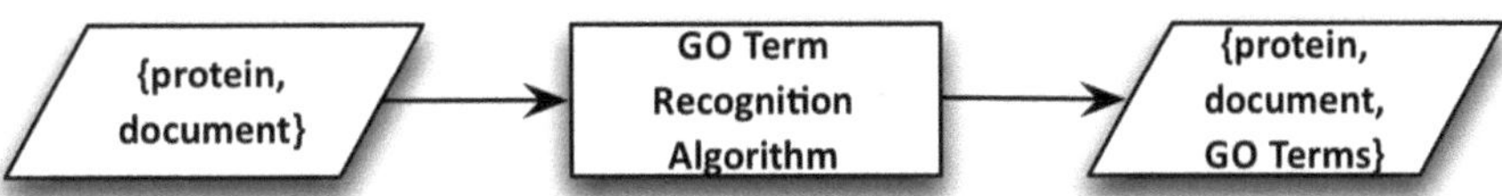

Fig. 1 Schematic architecture for single-document protein function prediction algorithms that center on recognizing Gene Ontology terms in a provided text and associating those terms to a given protein of interest

task is the focus on identifying functional information from a single (provided) document source, rather than a broader set of literature.

The most straightforward approach to recognizing GO terms in text is simple exact matching of the vocabulary terms. However, is known to be inadequate for matching GO terms to natural text [10, 11]. Therefore, various algorithms for GO term recognition have been explored that go beyond exact string matching of the GO terms; these algorithms aim to address the extensive variability in expression of GO concepts in the literature. The two basic paradigms for this are to do direct extraction of gene ontology terms, in which the task is to recognize an explicit mention of a GO concept in a text or to cast the task as a classification problem, in which a classifier makes a decision about whether or not a document is relevant to a GO term [12].

To handle the significant variability in how such terms are expressed, direct extraction algorithms typically accumulate evidence to support the presence of a GO term by considering the contribution of words or substrings from within the term or its definition [13, 14], or through identification of informative or discriminating terms for a GO concept, learned from training examples [12, 15, 16]. The latter work also takes advantage the hierarchical structure of the GO and is therefore also an example of a second general approach to GO term recognition, which involves using the ontology structure to generalize examples to more general concepts in the hierarchy or to concepts in the same "semantic space" [17, 18].

GO term recognition algorithms can form part of a broader strategy for protein function prediction that incorporates information from literature. Figure 2 presents the overview of this sort of strategy, indicating basic steps of selecting relevant documents from the literature, associating a given protein to a subset of those documents, recognizing GO terms in the documents, and associating those GO terms to the input protein. However, many architectures for implementation are possible. The document selection and the protein-document association strategy may in fact be the same, for instance, if document selection occurs by doing a PubMed search on a protein name or if curated links to the literature are utilized (e.g., from UniProt or the Protein Data Bank (PDB)). Similarly, there may not be an explicit GO term recognition step applied to all documents in the protein-related collection, but rather proteins could be associated to GO terms via a more indirect method such as a machine learning algorithm that incorporates general text-based features such as word n-grams.

One effective method in this paradigm has been developed that combines information extraction of GO terms from text with machine learning for general protein function prediction [19]. In this method, protein names and GO terms are directly recognized in texts on a large scale, and then protein-GO term associations are identified on the basis of co-occurrence. These associations are provided, along with a range of other features related to protein

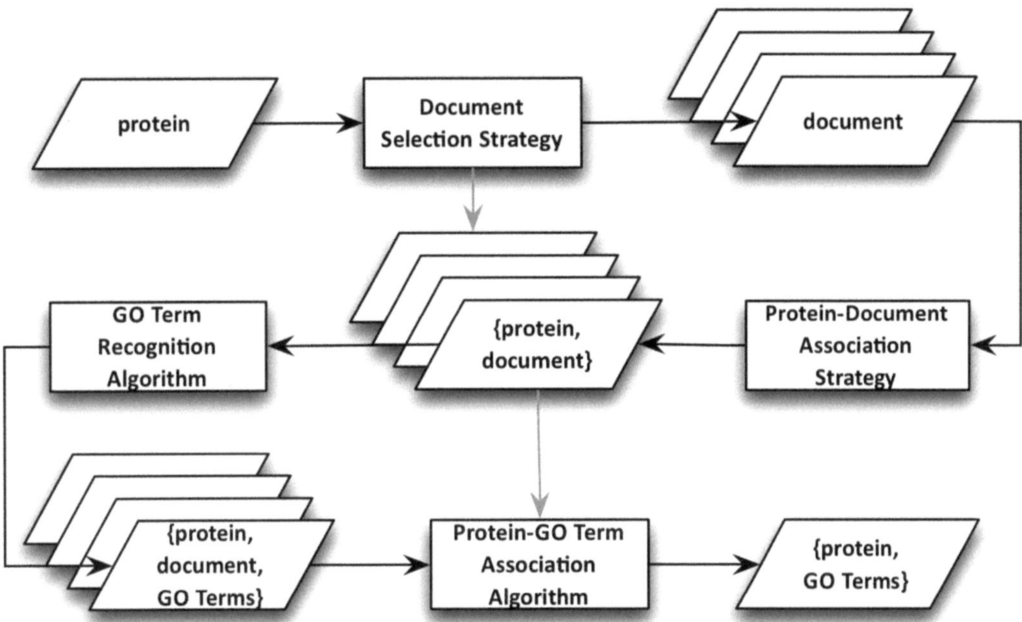

Fig. 2 Schematic architecture for a general strategy for literature-based protein function prediction targeting the Gene Ontology, typically involving information integration from a multiplicity of documents related to a given protein of interest

function, as input to a machine learning algorithm, called GOstruct [20], which respects the hierarchical structure of the target ontology space. The details of this method are beyond the scope of this chapter as the success of the method in the CAFA experiment hinged on integration of a large set of heterogenous data, including features related to the protein sequence itself, gene expression data, and curated interaction data. However, the protein-GO associations derived from text proved critical to the success of the approach.

Other methods that integrate text information into protein function prediction have also been explored. For instance, protein network-based methods that employ graph–theoretic prediction algorithms have been tested, where protein networks may be derived by considering protein co-occurrence or protein interaction information mined from the literature [21].

3.2 The LEAP-FS Paradigm

As introduced above, a physical representation of protein function, in which function is described in terms of the direct interactions a protein participates in and the physical characteristics of the protein, is an alternative to the more qualitative descriptions in resources such as the Gene Ontology. This representation might include information such as the location of active residues on a protein and the biochemical properties of protein–ligand interactions.

To support information extraction of such physical characteristics from the published literature, an approach called

Literature-Enhanced Automated Prediction of Functional Sites (LEAP-FS) was developed [22]. It is a novel paradigm that closely couples literature mining with protein structure modeling; in this paradigm the text mining results are used to *validate* the predictions of an orthogonal structural model rather than make fully independent predictions. The overall architecture of the LEAP-FS approach is presented in Fig. 3.

Prior work on predicting GO annotations exists that also uses information extracted from the literature to validate predictions [23];

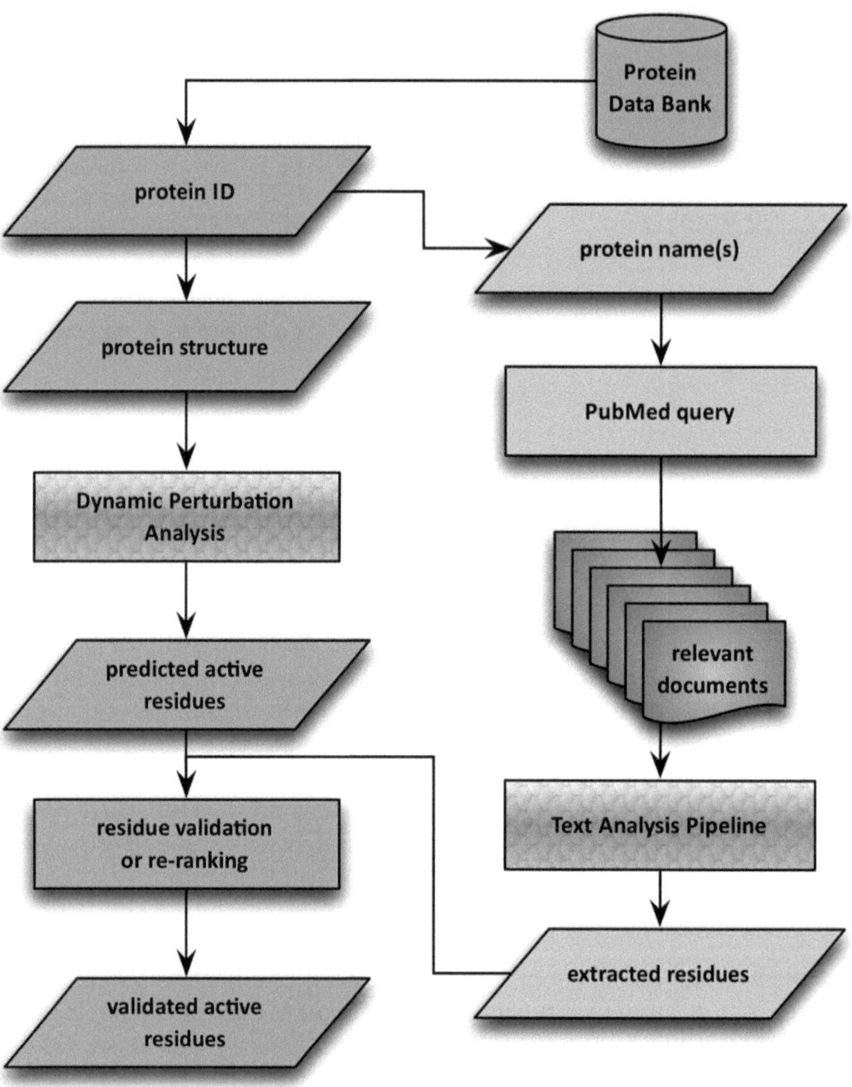

Fig. 3 The Literature-Enhanced Automated Prediction of Functional Sites (LEAP-FS) architecture utilizing text mined position-localized protein residues to validate the outputs of a structure-based prediction model targeting a physical characterization of protein function, specifically prediction of active sites on the surface of a protein

in that work sentence-level co-occurrences of protein mentions and GO term mentions were used for validation, and a large proportion of predictions could not be validated (approximately 60 % were missing from the literature set). The LEAP-FS approach, in contrast, focuses on extraction of much more specific, easier to identify, information about proteins in text, namely localizable protein residues, i.e., a protein residue at a specific position in a protein sequence. In both studies, predictions that can be validated through the literature are found to be highly accurate.

The LEAP-FS approach capitalizes on the fact that the functional site predictions themselves provide unexploited, valuable context for the literature search. In particular, providing a literature search with a specific protein of interest and an associated protein sequence simultaneously enables a more targeted search for information (focusing on the given protein) and a more relaxed search for site-specific information about that protein, because false-positive sites can be filtered by validation with respect to the given protein sequence. This is in contrast to more open-ended text mining that does not make use of external, structured information to improve precision.

The integration of protein functional site prediction and text-based functional site predictions in this way has recently been demonstrated [22]. A structure analysis approach developed for locating small-molecule binding sites on proteins, called Dynamics Perturbation Analysis (DPA), was applied to a comprehensive set of ~100,000 domains in the SCOP (Structural Classification of Proteins) database (http://scop.mrc-lmb.cam.ac.uk/scop/) [24]. The predictions were found to recapitulate much of the information about functional sites in existing curated databases, specifically the Catalytic Site Atlas (CSA) [25] and the Binding MOAD database [26], but many of the predictions of this structural method were left unvalidated. In parallel, text mining was used to automatically extract residue mentions from abstracts of papers about the protein structure. The analysis provided evidence that residues mentioned in abstracts are often functionally important and showed that DPA sites containing residues mentioned in text are more likely to have validating information in the databases.

The text mining strategy was to extract from the text corpus all mentions of a specific residue, i.e., an amino acid at a specific position in the sequence. A simple mention of, e.g., "*Glycine*" was not sufficient to support extraction; a localizable mention such as "*Glycine 23*" was required. Residues mentioned in the context of a mutation at a specific site, e.g., "*Gly23Ala*" were included. The implementation used a set of regular expressions representing patterns corresponding to such localizable mentions of protein residues and mutations. The performance of this extraction algorithm

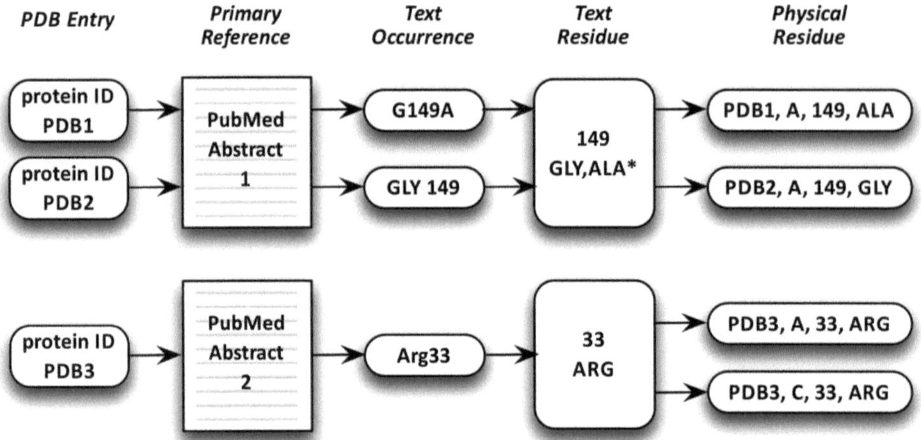

Fig. 4 High-level view of the text mining process in LEAP-FS, with validation against the protein sequence enabling mapping of residues mentioned in text to physical residues, potentially to multiple protein structures (*top*) or to multiple chains of the same protein (*bottom*)

was characterized on three different corpora (http://bionlp-corpora.sourceforge.net/proteinresidue/). The performance was good and warranted its use in a validation context.

A high level view of the process that builds on the extraction of residue mentions is shown in Fig. 4. The process aims to map residues mentioned in the text to physical residues in the relevant protein domains. By numerous measures, it was found that text residues do indeed provide supporting evidence for structure-based predictions of functional sites [22]. In addition, text residues were often found for sites for which annotations are not yet available. Taken together, the data showed that the functional importance of many residues not yet documented in the databases can be inferred from the identification of physically verified text residues. These results illustrate the potential for text analysis to make a substantial impact in providing supporting evidence for predictions and identifying new annotations. Further work has applied machine learning to more precisely characterize the specific activity at a given site based on the context of text mentions with promising results [27].

Figure 5 illustrates an example of literature-enhanced function prediction applied to a putative antibiotic resistance protein from Mycobacterium tuberculosis, Protein Data Bank (PDB) entry 1YK3 [28]. Structure-based prediction using DPA predicted a large functional site. Automated analysis of the abstract text highlighted the functional importance *His130* and *Asp168*. There was no evidence for the importance of *His130* found in any of the databases examined. The *Asp168* residue contacts a biologically relevant ligand in another structure in the same SCOP family.

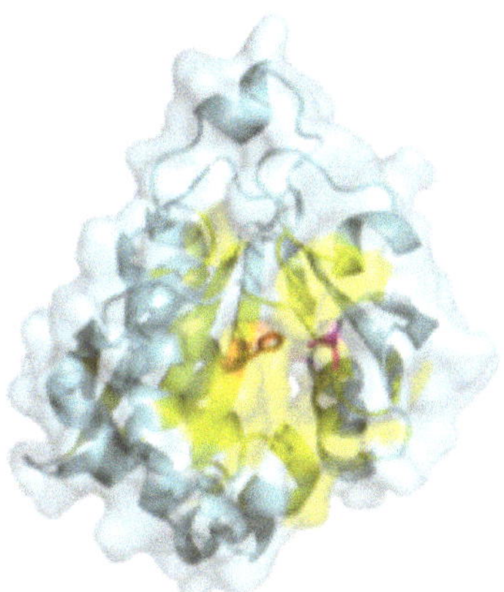

"Modeling the postulated substrate the N(epsilon)-hydroxylysine side chain of mycobactin, into the acceptor substrate binding groove identifies two residues at the active site, **His130** and **Asp168**, that have putative roles in substrate binding and catalysis"

Fig. 5 Example of LEAP-FS applied to a putative antibiotic resistance protein from Mycobacterium tuberculosis, PDB 1YK3. The protein is displayed as a *light blue ribbon* with a semitransparent surface. The predicted functional site from structure-based analysis is colored *yellow*, and the predicted residues mentioned in the abstract of the primary reference are rendered as sticks colored *orange* (*His130*) and *magenta* (*Asp168*). Text from the crystal structure publication abstract is shown on the *right*, with residue mentions in *bold font*

The full text of the primary reference, which was not automatically analyzed due to restrictions on access to full text, but could be if such access were made possible, states that *His130* and *Asp168* are likely catalytic residues in the putative active site. The active site also includes many other DPA-predicted residues. Overall the integrated analysis highlighted a putative active site that might be worthy of annotation and suggested the possibility of a previously unappreciated functional role of several protein residues, perhaps as an allosteric site.

4 Notes

Computational protein function prediction remains a very active area of research, and the best strategies for accurate protein function annotation remain far from clear. Indeed, even the best measures for evaluating the performance of different strategies remains unclear.

1. *Evaluation of method performance*: To assess the performance of any predictive method, it is important to have test sets that can be used to evaluate the method. Protein function prediction methods have been evaluated in the past by considering the ability of prediction methods to reproduce curated GO annotations of proteins (e.g., using the GOA annotations

introduced above [29]). Typically these evaluations use a protein sequence or a protein name as the starting point for prediction.

The CAFA challenge used quite an interesting strategy for evaluating predictions [1]. Rather than using existed curated annotations, the organizers selected sequences in the Swiss-Prot database that did not have functional annotations, under the assumption that a reasonable proportion of those sequences would acquire curated annotations during an evaluation period. This prevented any "gaming" of the predictive systems, as the annotations of the test data were entirely blind to the computational systems. The second CAFA challenge is now underway utilizing the same basic strategies, but targeting the Human Phenotype ontology in addition to the GO, and should provide insight into improvements in the state of the art methods.

Evaluation of methods specifically aiming at GO term extraction from text has been hampered by lack of structured test sets. In 2003, the first BioCreative included a task that addressed protein function prediction from a given text [7]. However, systems were required to produce evidence derived from an article for a given prediction, in addition to the prediction itself, and the evaluation methods hinged on curators judging the relevance of that text. Since there was no gold standard of expected evidence text, that evaluation was done entirely manually. The assessment of that task is therefore not reproducible. In BioCreative IV, being held at the time of this writing, there is a new task on GO term recognition, for which training and evaluation data has been provided (http://www.biocreative.org/tasks/biocreative-iv/track-4-GO/). That task follows the paradigm in Fig. 1. Participants were provided with input full text articles and associated gene information; teams were asked to return a list of relevant GO terms for each of the input genes in a paper. Since this task focuses on predictions derived from a single paper, it is not completely analogous to the broader protein function prediction problem, but rather aims to stimulate development of text mining tools that can support curators who perform GO annotation. However, such tools clearly could form part of broader-scope protein function prediction system, as suggested in Fig. 2.

A recently released corpus provides additional opportunities to investigate methods for recognizing GO terms in text. The Colorado Richly Annotated Full Text Corpus (CRAFT) is a full-text corpus that has been annotated with terms from several ontologies, including the GO [30, 31]. Indeed, several publications have already targeted this resource for investigation of GO term recognition tools [11, 32, 33].

2. *Evaluation measures for hierarchical prediction*: An important consideration in any protein function prediction methodology that targets a complex structure such as the Gene Ontology is how to measure the performance of the predictive system within that complex structure. While this issue is not unique to text mining-based systems, the problem of evaluating classification performance where classes are hierarchically organized is a problem that has been studied mostly in the context of text classification [34, 35]. The standard evaluation measures used for classification, namely Precision, Recall, and F-score, do not allow for "near misses" in the ontology space. For instance, if a system predicts a class $G(x) = \{C\}$ for input x while the gold standard associates x to $F(x) = \{D\}$, where D subsumes C in the hierarchy (e.g., D is the parent or grandparent of C), the standard measures would count this prediction as an error. Intuitively, however, the prediction is *close* to the correct answer and indeed given the hierarchical semantics, C implies D (i.e., C is a D), the correct answer is implied by the system prediction.

   To capture this semantics, the standard evaluation measures can be extended to the hierarchical context. Hierarchical Precision, Hierarchical Recall, and Hierarchical F-score can be defined as follows, following [29]. Here, $T(x)$ is the set of known (true) class assignments for an input x, $P(x)$ is the set of predictions for input x, and $\uparrow q$ indicates the set of ancestors of a node $q \in Q$ in the hierarchical structure, that is, all of the nodes in the subgraph from node q up to the root of the tree.

$$\mathrm{HP} = \frac{1}{|P(x)|} \sum_{p \in P(x)} \max_{t \in T(x)} \frac{\uparrow t \cap \uparrow p}{|\uparrow p|}$$

$$\mathrm{HR} = \frac{1}{|T(x)|} \sum_{t \in T(x)} \max_{p \in P(x)} \frac{\uparrow t \cap \uparrow p}{|\uparrow t|} \qquad \mathrm{HF} = \frac{2 \times \mathrm{HP} \times \mathrm{HR}}{|\mathrm{HP} + \mathrm{HR}|}$$

   These measures calculate the relative sizes of the intersection sets (overlapping subgraphs) of the true answers as compared to the predicted answers. The more ancestors two elements of the hierarchy share, the "closer" they are judged to be. These measures thereby capture two important intuitions of measuring correctness of an answer in hierarchical space: (1) correctness is scored based on how "close" the predicted answer(s) are from the gold answer(s) such that predicted answers can receive "partial credit" for being "close", and (2) the more specific the predictions (the further down the hierarchy), the less penalty they receive for being "close" but not exact. The latter follows from the measures using the size of the subgraphs and reflects the intuition that errors at the higher, more general levels on the hierarchy are more severe.

   In recent work, an information-theoretic framework for evaluation in hierarchies has been proposed [36], specifically in the context of evaluation of computational protein function prediction into the GO. This framework models the prior probability of a protein's function in terms of a Bayesian network reflecting the hierarchical structure, and introduces the measures *misinformation* and *remaining uncertainty* as information-theoretic analogs of hierarchical precision and hierarchical recall, respectively. However, rather than focusing on the intersection of the two subgraphs, these measures look at the *differences* between the two subgraphs—i.e., how many nodes not in one as compared to the other—and consider the information content of those differences. Interested readers are referred to the original publication [36] for more details.

5 Conclusions

The published literature is a rich source of the most up-to-date information about protein function; a source that has typically been under-utilized in computational protein function prediction algorithms. However, there is growing research on topics that are useful as components of a broader protein function prediction strategy, including recognition of Gene Ontology terms in text and machine learning algorithms that take advantage of the structural properties of hierarchically structured output spaces, such as the Gene Ontology vocabulary. Given initial results from the first CAFA experiment that demonstrate significant value from integrating text mined-features into protein function prediction, coupled with the promise of using text mining to validate "noisy" predictions from orthogonal data sources, including protein structure, we can expect to see more research that explores novel algorithms for drawing on the rich context that text provides in protein function prediction.

References

1. Radivojac P, Clark WT, Oron TR, Schnoes AM, Wittkop T, Sokolov A, Graim K, Funk C, Verspoor K, Ben-Hur A et al (2013) A large-scale evaluation of computational protein function prediction. Nat Methods 10(13):221–227
2. Ashburner M, Ball CA, Blake JA, Botstein D, Butler H, Cherry JM, Davis AP, Dolinski K, Dwight SS, Eppig JT et al (2000) Gene ontology: tool for the unification of biology. Nat Genet 25(1):25–29
3. Blaschke C, Valencia A (2013) The Functional Genomics Network in the evolution of biological text mining over the past decade. N Biotechnol 30(3):278–285
4. Valencia A (2005) Automatic annotation of protein function. Curr Opin Struct Biol 15(3):267–274
5. Baumgartner WA Jr, Cohen KB, Fox L, Acquaah-Mensah GK, Hunter L (2007) Manual curation is not sufficient for annotation of genomic databases. Bioinformatics 23:i41–i48
6. Friedberg I (2006) Automated protein function prediction—the genomic challenge. Brief Bioinform 7(3):225–242
7. Blaschke C, Leon E, Krallinger M, Valencia A (2005) Evaluation of BioCreAtIvE assessment of task 2. BMC Bioinformatics 6(Suppl 1):S16

8. Maguitman AG, Rechtsteiner A, Verspoor K, Strauss CE, Rocha LM (2006) Large-scale testing of bibliome informatics using Pfam protein families. Pac Symp Biocomput 76–87
9. Shatkay H, Hoglund A, Brady S, Blum T, Donnes P, Kohlbacher O (2007) SherLoc: high-accuracy prediction of protein subcellular localization by integrating text and protein sequence data. Bioinformatics 23(11): 1410–1417
10. Verspoor CM, Joslyn C, Papcun GJ (2003) The gene ontology as a source of lexical semantic knowledge for a biological natural language processing application. In: SIGIR workshop on Text Analysis and Search for Bioinformatics, 51–56
11. Funk C, Baumgartner W, Garcia B, Roeder C, Bada M, Cohen KB, Hunter LE, Verspoor K (2014) Large-scale biomedical concept recognition: an evaluation of current automatic annotators and their parameters. BMC Bioinformatics 15:59. doi: 10.1186/1471-2105-15-59
12. Wong A, Shatkay H (2013) Protein function prediction using text-based features extracted from the biomedical literature: the CAFA challenge. BMC Bioinformatics 14(Suppl 3):S14
13. Krallinger M, Padron M, Valencia A (2005) A sentence sliding window approach to extract protein annotations from biomedical articles. BMC Bioinformatics 6 Suppl 1
14. Couto FM, Silva MJ, Coutinho PM (2005) Finding genomic ontology terms in text using evidence content. BMC Bioinformatics 6 Suppl 1
15. Ray S, Craven M (2005) Learning statistical models for annotating proteins with function information using biomedical text. BMC Bioinformatics 6(Suppl 1):S18
16. Verspoor K, Cohn J, Joslyn C, Mniszewski S, Rechtsteiner A, Rocha LM, Simas T (2005) Protein annotation as term categorization in the gene ontology using word proximity networks. BMC Bioinformatics 6 Suppl 1
17. Martin D, Berriman M, Barton G (2004) GOtcha: a new method for prediction of protein function assessed by the annotation of seven genomes. BMC Bioinformatics 5:178
18. Conesa A, Gotz S, Garcia-Gome J, Terol J, Talon M, Robles M (2005) Blast2GO: a universal tool for annotation, visualization and analysis in functional genomics research. Bioinformatics 21(18):3674–3676
19. Sokolov A, Funk C, Graim K, Verspoor K, Ben-Hur A (2013) Combining heterogeneous data sources for accurate functional annotation of proteins. BMC Bioinformatics 14(Suppl 3):S10
20. Sokolov A and Ben-Hur A (2010) Hierarchical classification of Gene Ontology terms using the GOstruct method. Journal of Bioinformatics and Computational Biology 8(2):357–376
21. Gabow AP, Leach SM, Baumgartner WA Jr, Hunter L, Goldberg DS (2008) Improving protein function prediction methods with integrated literature data. BMC Bioinformatics 9:198
22. Verspoor KM, Cohn JD, Ravikumar KE, Wall ME (2012) Text mining improves prediction of protein functional sites. PLoS One 7(2):e32171
23. Jaeger S, Gaudan S, Leser U, Rebholz-Schuhmann D (2008) Integrating protein-protein interactions and text mining for protein function prediction. BMC Bioinformatics 9(Suppl 8):S2
24. Andreeva A, Howorth D, Chandonia J-M, Brenner SE, Hubbard TJP, Chothia C, Murzin AG (2008) Data growth and its impact on the SCOP database: new developments. Nucleic Acids Res 36(Database issue):D419–25
25. Porter CT, Bartlett GJ, Thornton JM (2004) The Catalytic Site Atlas: a resource of catalytic sites and residues identified in enzymes using structural data. Nucleic Acids Res 32(Database issue):D129–133
26. Benson ML, Smith RD, Khazanov NA, Dimcheff B, Beaver J, Dresslar P, Nerothin J, Carlson HA (2008) Binding MOAD, a high-quality protein–ligand database. Nucleic Acids Res 36(suppl 1):D674–D678
27. Verspoor K, MacKinlay A, Cohn JA, Wall ME (2013) Detection of protein catalytic sites in the biomedical literature. Pac Symp Biocomput 18:433–444
28. Card GL, Peterson NA, Smith CA, Rupp B, Schick BM, Baker EN (2005) The crystal structure of Rv1347c, a putative antibiotic resistance protein from Mycobacterium tuberculosis, reveals a GCN5-related fold and suggests an alternative function in siderophore biosynthesis. J Biol Chem 280(14): 13978–13986
29. Verspoor K, Cohn J, Mniszewski S, Joslyn C (2006) A categorization approach to automated ontological function annotation. Protein Sci 15(6):1544–1549
30. Verspoor K, Cohen KB, Lanfranchi A, Warner C, Johnson HL, Roeder C, Choi JD, Funk C, Malenkiy Y, Eckert M et al (2012) A corpus of full-text journal articles is a robust evaluation tool for revealing differences in performance

of biomedical natural language processing tools. BMC Bioinformatics 13:207

31. Bada M, Eckert M, Evans D, Garcia K, Shipley K, Sitnikov D, Baumgartner WA Jr, Cohen KB, Verspoor K, Blake JA et al (2012) Concept annotation in the CRAFT corpus. BMC Bioinformatics 13:161
32. Campos D, Matos S, Oliveira JL (2013) Neji: a tool for heterogeneous biomedical concept identification. In: Proceedings of BioLINK SIG 2013; ISMB/ECCB 2013, Berlin, Germany, pp 28–31, See: http://biolinksig.org/past-meetings/biolink-2013/
33. Jacob C, Thomas P, Leser U (2013) Comprehensive Benchmark of gene ontology concept recognition tools. In: Proceedings of BioLINK SIG 2013; ISMB/ECCB 2013, Berlin, Germany, pp 20–26, See: http://biolinksig.org/past-meetings/biolink-2013/
34. Li X, Ling C, Wang H (2013) Effective top-down active learning for hierarchical text classification. In: Pei J, Tseng V, Cao L, Motoda H, Xu G (eds) Advances in knowledge discovery and data mining, vol 7819. Springer, Berlin, pp 233–244
35. Silla CN, Freitas AA (2011) A survey of hierarchical classification across different application domains. Data Min Knowl Discov 22(1):31–72
36. Clark WT, Radivojac P (2013) Information-theoretic evaluation of predicted ontological annotations. Bioinformatics 29(13):i53–i61

Chapter 7

Functional Molecular Units for Guiding Biomarker Panel Design

Andreas Heinzel, Irmgard Mühlberger, Raul Fechete, Bernd Mayer, and Paul Perco

Abstract

The field of biomarker research has experienced a major boost in recent years, and the number of publications on biomarker studies evaluating given, but also proposing novel biomarker candidates is increasing rapidly for numerous clinically relevant disease areas. However, individual markers often lack sensitivity and specificity in the clinical context, resting essentially on the intra-individual phenotype variability hampering sensitivity, or on assessing more general processes downstream of the causative molecular events characterizing a disease term, in consequence impairing disease specificity. The trend to circumvent these shortcomings goes towards utilizing multimarker panels, thus combining the strength of individual markers to further enhance performance regarding both sensitivity and specificity. A way of identifying the optimal composition of individual markers in a panel approach is to pick each marker as representative for a specific pathophysiological (mechanistic) process relevant for the disease under investigation, hence resulting in a multimarker panel for covering the set of pathophysiological processes underlying the frequently multifactorial composition of a clinical phenotype.

Here we outline a procedure of identifying such sets of disease-specific pathophysiological processes (units) delineated on the basis of disease-associated molecular feature lists derived from literature mining as well as aggregated, publicly available Omics profiling experiments. With such molecular units in hand, providing an improved reflection of a specific clinical phenotype, biomarker candidates can then be assigned to or novel candidates are to be selected from these units, subsequently resulting in a multimarker panel promising improved accuracy in disease diagnosis as well as prognosis.

Key words Gene sets, Omics profiles, Literature mining, Protein–protein interaction networks, Network modules, Biomarker candidates, Biomarker panels

1 Introduction

The mere data generation nurtured by high throughput "Omics" approaches has outstripped the ability to interpret the data space with respect to biological relevance. In addition, the number of biomarker candidates proposed in the literature is exploding independent of the disease under study [1]. These biomarker

Vinod D. Kumar and Hannah Jane Tipney (eds.), *Biomedical Literature Mining*, Methods in Molecular Biology, vol. 1159, DOI 10.1007/978-1-4939-0709-0_7, © Springer Science+Business Media New York 2014

candidates are often the result of explorative analysis in small patient cohorts in a pilot study setting, and the conclusion of these publications frequently culminates in "*these results need to be validated in a larger prospective study*", but then lack successful follow-up reports. Establishing appropriate sample cohorts for such validation studies is both time and labor intense, and in many cases cannot even be performed by a single research institution but ends up to be a complex, multicenter effort [2]. Once the appropriate biobanks are available and the relevant clinical data for characterizing phenotypes are collected it comes to the critical part of defining which specific markers to measure in such precious samples, as the number of candidates amenable for measurement is naturally limited (next to assay availability) by the available sample volume. Beyond these practical issues certainly the optimal rationale for selecting biomarker candidates for validation is essential, where such candidates ideally are afflicted with molecular disease mechanisms and not only with disease-associated molecular processes, in their totality serving as proxy for the various clinical presentations of a given phenotype.

We in the following discuss a workflow for linking Omics datasets and literature-derived feature sets associated with a disease under study towards identifying molecular models of disease in order to evaluate proposed biomarker candidates, and, if needed, to complement given with novel candidates, ultimately forming multimarker panels.

The following steps compose this workflow, as described in more detail in the materials section, and exemplarily executed for laying the basis for a multimarker panel for cardiovascular disease (CVD) phenotypes (coronary artery disease and atherosclerosis): (1) derive a comprehensive set of molecular features being identified as associated with disease molecular pathophysiology; (2) derive the associated set of molecular processes (units) and their connectivity (molecular model) reflecting the molecular pathophysiology of the disease; (3) select a representative biomarker for each such molecular process, thus generating a multimarker panel capable of quantitatively assessing each individual molecular process of disease relevance, and hence gaining a more complete representation of the entire molecular model characterizing the disease.

The proposed workflow is applicable independent of specific use of such panel, be it in the diagnostic or prognostic setting, as a more detailed differential diagnostic (promised by multimarker panels) in turn relates to improved prognostic readout.

2 Materials

2.1 Omics Studies for Delineating Disease-Associated Molecular Feature Sets

Along with the methodological improvements and standardization of experimental Omics profiling procedures, and efforts of making analysis results as well as raw datasets available, meta-analysis of individual profiling experiments became feasible. Next to publications reporting the results of a specific Omics experiment, publicly available databases holding Omics raw data, as well as result signatures have been established. Controlled vocabularies can be used to retrieve relevant datasets for a disease term of interest.

The most popular biomedical literature database is PubMed (http://www.ncbi.nlm.nih.gov/pubmed), being maintained by the National Center for Biotechnology Information (NCBI). The majority of the currently around 23 million publications indexed in PubMed are annotated with Medical Subject Headings (MeSH), thus facilitating the identification of specific studies. MeSH terms are arranged in a hierarchy with 16 top-level terms, including "Diseases", "Organisms", or "Analytical, Diagnostic and Therapeutic Techniques and Equipment."

As an example, a query for heart failure associated human transcriptomics studies may be formulated as:

"heart failure"[majr] AND ("microarray analysis"[mh] OR "gene expression profiling"[mh]) AND "humans"[mh] NOT review[ptyp]

The field modifier [mh] specifies publications being indexed with the corresponding MeSH term, whereas [majr] further limits the search to publications for which the corresponding MeSH term—as in our example heart failure—is designated as "major" (i.e., of particular relevance in the respective study). Depending on the particular disease of interest, "heart failure" needs to be replaced by the specific disease term, where valid terms for the given ontology can be identified at the MeSH Browser available at http://www.nlm.nih.gov/mesh/MBrowser.html.

The following search terms may be used for other Omics types next to transcriptomics (for which we suggest to use *"microarray analysis"[mh] OR "gene expression profiling"[mh]*): For microRNA studies the following query string should be used: *("microarray analysis"[mh] OR "gene expression profiling"[mh]) AND "microRNAs"[mh]*. For Genome-Wide Association Studies (GWAS), proteomics and metabolomics studies the respective MeSH terms are *"Genome-Wide Association Study"[mh], "proteomics"[mh], and "metabolomics"[mh].*

Adding the terms *"humans"[mh]* and *NOT review* ensures that the resulting studies focus on human samples and are all original articles.

After execution of the query a manual curation step of search results is mandatory for evaluating in detail the study characteristics, specifically including study hypothesis, number of samples used, sample characteristics, sample material, assay platform used, etc., otherwise comparability of individual studies and hence suitability for integration is hampered.

For retrieving relevant molecular features of an Omics study various procedures are to be followed depending on the information provided in the respective publication. The entire list of relevant features (defined as being found as significantly associated with a phenotype in statistical testing) may be provided in the publication itself or as supplementary data. In other cases only an excerpt of relevant feature sets is presented in the paper, but the complete raw—as well as normalized and annotated data sets are provided in one of the public repositories as, e.g., in the Gene Expression Omnibus (GEO) in the case of transcriptomics studies [3]. The obvious advantage of accessing raw data is the possibility of an independent analysis without being restricted to the original setting of the authors (e.g., application of alternative preprocessing and statistical analysis strategies). Another repository for high-throughput transcriptomics data is the ArrayExpress database hosted by the European Bioinformatics Institute (EBI) [4]. Details on available preprocessing methods and a navigation through the statistical and bioinformatics analysis process can be found in Mühlberger et al. [5].

Databases holding findings from genetic association studies include the GWAS Catalog hosted by the National Human Genome Research Institute [6], GWAS Central (http://www.gwascentral.org/), or the GWASdb [7].

Comprehensive databases on tissue-specific proteins in the context of specific diseases are comparably rare. One example is the Human Urinary Proteome database which holds information about protein abundance of roughly 4,000 human urine samples collected from subjects covering a wide spectrum of pathophysiological conditions [8].

One of the most complete and comprehensive curated collection of human metabolite data (although not organized in the context of human diseases) is the Human Metabolome Database (HMDB) with 40,283 entries and partly also disease associations [9]. Table 1 provides an overview of such databases together with the corresponding web links.

2.2 Literature Mining Approaches for Delineating Disease-Associated Molecular Feature Sets

Next to utilizing Omics profiling data for assembling disease-associated features scientific text provides another valuable source. The importance of extracting information out of unstructured text, as for example from all titles and abstracts of scientific publications available at NCBI's Medline, in an automatic way is gaining importance with increasing number of publications [10].

Table 1
Public repositories for Omics data

Database	Web link
Transcriptomics	
NCBI Gene Expression Omnibus	http://www.ncbi.nlm.nih.gov/geo/
EMBL Array Express	http://www.ebi.ac.uk/arrayexpress/
GWAS	
GWAS catalog	http://www.genome.gov/gwastudies/
GWAS central	http://www.gwascentral.org/
GWASdb	http://jjwanglab.org/gwasdb/
Proteomics	
Human Urinary Proteome database	http://www.mosaiques-diagnostics.de/
Metabolomics	
Human Metabolome Database	http://www.hmdb.ca/

One approach is via the semiautomatically curated MeSH information provided in PubMed in conjunction with gene to publication links (gene2pubmed) also provided by NCBI. Publications being annotated with MeSH terms being also linked to genes imply evidence for association between these concepts. The advantage of linking genes to diseases via this approach, thus making use of the manual annotation, is the presumably lower number of false-positive assignments, with the limitation on the other hand of potentially missing a number of weak (but still relevant) evidence links and generating false-positive gene to disease associations due to coarse co-mentioning.

Alternatively, or in addition, text mining methods may be employed to create gene to disease links from paper abstracts or full text articles. Text mining comprises of many technicalities with one being information extraction, being the task of identifying and extracting entities and relations between them.

One information extraction solution publicly available is EBI's Whatizit service [11], offering pipelines for named entity recognition, the task of identifying text passages holding a certain real world entity (e.g., a gene name), and entity normalization, the task of assigning such entity to an actual record in a reference database, be it for genes, proteins, diseases, and other biomedical entities. The complexity of natural language renders identification of relations between entities complicated. Even though pattern recognition and machine learning-based approaches have been developed, the much simpler approach of deriving associations between entities from co-mentioning is often used for inference of molecular feature-to-disease association.

A biomedical resource utilizing text mining for extraction of molecular entities from free text containing keywords of interest is the Fast Automated Biomedical Literature Extraction (FABLE) resource (http://fable.chop.edu). The information hyperlinked over protein (iHop) service as further option uses text mining methods for building a navigable gene interaction network from scientific literature in PubMed [12]. Gene2MeSH employs a statistical approach utilizing MeSH indexing information from PubMed and gene publication links as reported in NCBI gene for annotating genes with concepts available in MeSH.

2.3 Namespace Consolidation (Annotation)

One major challenge in molecular feature consolidation utilizing Omics profiling and literature mining results is overcoming name space fragmentation with the primary aim of linking gene and protein name spaces. Different organizations use different identifiers for the same biological entities (genes, proteins, disease terms, etc.) with the larger institutions setting the standards. Noteworthy organizations providing relevant namespaces are, among others, the NCBI, the European Molecular Biology Laboratory (EMBL), and the Universal Protein Resource (UniProt) [13]. Additionally to individual namespaces, each such institution also provides a cross-reference attempting to match foreign identifiers to its own name space conventions. Such mappings are also plagued by technical shortcomings, with the most relevant ones being incompleteness and inconsistency, i.e., mappings provided by different institutions are not bijective. Reasons are twofold, the first one being of technical nature as the individual update cycles are asynchronous, and the second one being of conceptual nature due to diverging interpretations of, e.g., genome assembly. In an attempt to overcome these shortcomings, initiatives to provide a single standardized namespace exist, such as the HUGO Gene Nomenclature Committee (HGNC) (http://www.genenames.org), which has assigned unique gene symbols and names to over 34,000 human loci so far, of which around 19,000 are protein coding.

Aside from inconsistencies between different namespaces, information aggregation faces additional challenges related to the type of biological entity included in consolidation. Integrative approaches spanning different Omics types (cross-Omics) are particularly vulnerable. To exemplify, an approach attempting to aggregate transcriptomics and proteomics data needs to address the aspect of mRNA splicing, i.e., the RNA of a gene being translated into protein (splice) variants. This situation is further hampered by data incompleteness, e.g., proteomics data inherently addressing a certain splice variant being frequently reported only with the respective gene name. Linking proteomics information directly to the gene ignores the aspect of splicing, while linking transcriptomics information to each of a gene's splice variant generates redundancy and imprecise assignment.

To support integration efforts in overcoming mapping difficulties, several bioinformatics tools have been developed and made publicly available. The ENSEMBL BioMart (http://www.ensembl.org/biomart/martview/) is a comprehensive resource for mapping and retrieving molecular data including exons/introns, homology and sequences for various organisms [14]. The Database for Annotation, Visualization and Integrated Discovery (DAVID) Gene ID Converter (http://david.abcc.ncifcrf.gov/conversion.jsp) aims at mapping identifiers while working in conjunction with additional tools of the DAVID Bioinformatics Resources to provide functional gene set analysis services [15]. The Clone|Gene ID converter (http://idconverter.bioinfo.cnio.es/) is part of the Asterias suite for analyzing genomic and proteomic data, serving from normalization to development of predictive models. Finally, the UniProt ID Mapper (http://www.uniprot.org/mapping/) aims not only at linking identifiers from different namespaces but also at identifying biological entities via sequence alignment [13].

A particular effort is necessary when integrating Omics data beyond genes and proteins. This is the case, among other instances, for bringing metabolites and microRNA (miRNA) information to the gene and protein name space. For metabolites one way is to extract the enzymes being involved in their metabolism, potentially together with transporters. A comprehensive resource for metabolite-protein/enzyme mapping is the previously mentioned HMDB.

miRNAs are short, nonprotein coding RNA fragments that regulate gene expression. As such, when investigating miRNAs determining their target genes is essential. Cascione et al. provide a review of routines for miRNA expression profile analysis together with tools for target prediction [16]. Additionally, the miRTarBase (http://mirtarbase.mbc.nctu.edu.tw/) provides high-quality information on experimentally verified miRNA–target relationships [17].

2.4 Protein–Protein Interaction Data

After extraction of disease term-associated molecular feature sets, followed by name space consolidation, the resulting feature list needs to be mapped on molecular interaction networks. A number of experimental as well as computational methods for identifying protein–protein interactions, with high-throughput methods like yeast-two-hybrid or affinity purification have rendered large-scale characterization of the human protein interactome feasible. Since the first creation of databases holding protein–protein interaction data not only the number of databases but also coverage of the specific types of interactions substantially increased [18]. Next to physical protein–protein interactions data have also become available on genetic interactions or drug–target interactions. Adoption of the PSI-MI data format developed by the Human Proteome Organization Proteomics Standards Initiative (HUPO-PSI) molecular interaction workgroup (MI), together with the use of controlled vocabularies, raised consistency among the various databases, and significantly reduced the effort to search, filter and combine data from different sources [19]. The use of common data formats was also the basis for the development of the Proteomics Standard Initiative Common QUery InterfaCe (PSICQUIC) providing a standardized way for programmatic access to molecular interaction databases [20]. A comprehensive overview on protein–protein interaction databases was recently published by Orchard [18]. In the following we focus in more detail on the Biological General Repository for Interaction Datasets (BioGRID) [21], IntAct [22], and Reactome [23].

BioGRID is a publicly accessible database holding next to human genetic and protein–protein interactions also datasets for selected model organisms. Interaction data in BioGRID are manually curated from published literature, or come from interaction datasets directly deposited by the respective authors. While BioGRID provides extensive coverage of literature for certain selected model organisms, curation of human molecular interactions is limited to particular areas of interest. As of March 2013 BioGRID holds more than 640,000 genetic and protein–protein interactions in total, with more than 180,000 being interactions for human. The interaction data can be accessed via the search portal provided at http://thebiogrid.org or may be directly imported into Cytoscape, a prominent visualization tools for interaction networks [24]. Data are also available for download in one of multiple xml or text-based formats including PSI-MI.

IntAct is an open source database holding next to protein–protein interaction data from various species also interactions of proteins and small molecules, as well as protein–gene interactions. IntAct interaction data are manually curated either from published literature or direct data depositions and are not released before approval by a senior curator. As of April 2013 IntAct in total provides more than 300,000 interactions, with more than 93,000 interactions for human. Besides the IntAct search portal at http://www.ebi.ac.uk/intact, from where data can be directly imported and visualized using CytoscapeWeb [25], data can also be downloaded in the PSI-MI data format.

REACTOME is an open source, publicly accessible, peer reviewed molecular pathway database with primary focus on human pathways. Along with pathway information molecular interactions derived from reactions and complexes available in REACTOME are provided. As of April 2013 REACTOME offers more than 2,000,000 interactions, of which 127,934 interactions are for human. Reactome data can either be explored using the Reactome Pathway Browser available at http://www.reactome.org or downloaded in various file formats also supporting the representation in categories of biological pathways. In addition, molecular interaction data are available for download in PSI-MI data format.

2.5 Network Visualization Tools

Next to quantitative analysis of networks also their qualitative (visual) inspection supports identification of disease-specific network areas. A number of biological data visualization tools have become available with Cytoscape, Osprey [26], or VisANT [27] as representatives with a specific focus on interaction networks, as further discussed in more detail in Suderman et al. [28] as well as in Gehlenborg et al. [29]. The user community, as well as the number of available plugins developed from users all over the world for the open-source software framework Cytoscape is constantly growing [24]. Cytoscape in the meantime has evolved from a tool focusing essentially on visualization of large networks to a graph analysis framework, offering more than 40 plugins available at http://www.cytoscape.org/. Functionalities range from (1) calculating graph properties as node degree, betweenness, or characteristic path length, (2) coloring nodes based on gene expression profile data in order to identify up- or downregulated network areas, or (3) graph layouting based on information regarding subcellular location of individual proteins helping in understanding cross-compartment signaling cascades. The ability to segment large networks into smaller subgraphs holding molecules being part of joint biological processes is another feature of relevance implemented in a number of graph visualization tools which will be discussed in more detail in the next section.

2.6 Segmentation Algorithms and Cluster Procedures

Populating interaction networks with disease-associated molecular feature sets usually results in subgraphs. For approximating distinct molecular processes (units) in such subgraph clustering algorithms may be applied. Topological network clustering algorithms employ different approaches for cluster identification and can be classified for example based on their support for weighted networks, or participation of nodes in final clusters, and information required. Global clustering algorithms, in contrast to local methods, require all nodes to be assigned to at least one cluster [30]. Furthermore, methods allowing overlapping clusters with single nodes belonging to multiple clusters can be discriminated from procedures where a node is deemed being uniquely represented in a single cluster. Local search algorithms allow addressing cases where loading the entire network at once is not computationally feasible, or information is initially only available for some local segments of the network. Currently there is no single best solution for clustering molecular interaction networks, and the different approaches employ different criteria for the generation of clusters, like optimizing betweenness values or number of cuts necessary for cluster formation.

In the following we introduce two popular algorithms for biological network analysis, namely the Molecular Complex Detection (MCODE) algorithm [31], as well as the Markov Clustering Algorithm (MCL) [32]. The reader is referred to [30] and [33] for more details on clustering approaches and comprehensive lists of available methods.

MCODE identifies clusters of densely interconnected nodes in molecular networks following a three-step procedure: Initially all nodes in the graph are assigned weights based on the density of their direct neighborhood. Subsequently, outwards traversals starting from the highest weighted node available adding all nodes above a certain weight threshold are performed for identifying densely connected clusters. Nodes being already assigned to a cluster will neither be considered as start node nor can be added to a growing cluster during outwards traversal. Finally, in a postprocessing step based on the setting of configuration parameters, MCODE removes clusters not meeting minimum topological requirements. For the initial node weighting step information about the entire network is required, whereas for subsequent steps local information is sufficient. MCODE allows nodes to be assigned to more than one cluster if the "fluffing" option is enabled, but does not demand a node to be assigned to a specific cluster. Besides the original command line version a MCODE Cytoscape plugin is available.

MCL divides a network into densely connected regions (clusters) using a random walk-based simulation of flow. The basic principle behind MCL is that random walks are unlikely to leave dense network regions, and as such will only infrequently go from one (high density) cluster to another. Simulated random walks using the two alternating operations of expansion and inflation are used to estimate transition probabilities between nodes. While MCL may be executed on unweighted networks it can also make use of edge weights (if available for a given interaction network) and requires global information about the network. A MCL result holds all nodes in the final clustering and does only allow unique assignment of nodes to clusters.

Utilizing molecular feature sets associated to a disease term mapped on an interaction network, subsequently interpreted on the level of identified clusters can now serve as basis for functional interpretation of such disease-associated molecular clusters (units), in turn serving evaluation of biomarker candidates.

2.7 Functional Interpretation of Gene Clusters

Besides the need to evaluate individual molecular features with respect to relevance in a specific disease (biological) context, a further issue is to functionally characterize an entire set of such features as, e.g., of clusters identified on disease-associated subgraphs.

Given a set of genes being presumably functionally linked, a standard analysis approach used to reveal more information on the biological context of a feature set is Gene Set Enrichment Analysis (GSEA) [34]. GSEA compares the biological roles (e.g., on the level of pathway assignment) of members of a given feature set to the background frequency of feature assignments to such biological roles and utilizes statistical testing for unraveling significant enrichment or depletion of such roles in a given feature set (*see* **Note 1**). Statistical significance is usually computed using the Chi-Squared test or Fisher's Exact test. The pathways of the Kyoto Encyclopedia of Genes and Genomes (KEGG) [35] or Gene Ontology (GO) terms [36] may be used as reference gene sets. The previously mentioned DAVID Functional Annotation Bioinformatics Microarray Analysis tool may be used for automated GSEA on the basis of several resources such as KEGG, GO, or PANTHER.

2.8 Biomarker Candidates in Gene Clusters

Based on the National Institutes of Health (NIH) definition a biomarker is "a characteristic that is objectively measured and evaluated as an indicator of (1) normal biological processes, (2) pathogenic processes, or (3) pharmacologic response to a therapeutic intervention" [37]. Biomarkers can either be classified (among other schemes) based on the type of biomolecule or regarding their use. Next to clinical biomarkers like age or blood pressure, molecular markers on the level of DNA, RNA, proteins, or metabolites may be used for characterizing phenotypes. Regarding use there are (1) screening markers to detect disease at an early stage, (2) diagnostic markers to establish the presence of disease, (3) prognostic markers to allow predicting the course of disease progression, (4) predictive markers to predict outcome with regard to therapeutic intervention, and (5) monitoring markers to measure response to a specific treatment.

For example, identifying biomarker candidates reported in the public domain the MeSH term "Biological Markers" may be used to identify publications with a focus on biomarkers, which covers a wide range of terms like "Biochemical Marker," "Clinical Marker," "Immunological Marker," or "Surrogate Marker," among others. There is a specific MeSH term for tumor markers entitled "Tumor Markers, Biological" as well as for genetic markers entitled "Genetic Markers," both subterms of the MeSH term "Biological Markers." If specifically interested in drug monitoring markers, the MeSH term "Biomarkers, Pharmacological" in conjunction with the MeSH term "Drug Monitoring" may be used.

For numerous disease terms and use scenarios molecular biomarkers have been reported over the last decade. Utilizing the definition of "measuring and evaluating a biological process" such biomarkers may now be assigned to molecular clusters/processes as identified via segmentation of disease term associated subgraphs, as discussed above. Following the assumption that such functional clusters resemble core elements of the molecular pathophysiology of a specific disease, selecting biomarker candidates being members of such process sets promise improved monitoring of the disease as such, be it on the diagnostic or prognostic level. Accordingly, holding the set of clusters (disease-associated processes) allows to (1) specifically select biomarker candidates out of the (frequently extensive) set of reported candidates, or if needed (2) allows selection of novel candidates in case specific processes of relevance are not yet covered by already reported candidates.

The result of this procedure is a multimarker panel, where each marker serves as proxy for an individual process (molecular unit) of relevance, and the panel promises better coverage of the entire molecular disease pathophysiology.

3 Methods

In the following we exemplify the workflow elements discussed above focusing on aspects of cardiovascular disease (CVD, with the specific pathophysiologies of coronary artery disease and atherosclerosis) as an example case. Next to deriving a molecular feature set associated with CVD, followed by populating an interaction network for deriving a CVD-specific subgraph and segmenting this CVD subgraph for deriving CVD-associated molecular units, we evaluate biomarker candidates reported in the context of CVD, thus laying the ground for forming multimarker panels covering CVD pathophysiology.

3.1 Deriving a CVD Molecular Feature Set from Omics Studies

Relevant CVD Omics studies are identified from PubMed and, specifically for transcriptomics, in the public Omics repositories GEO and ArrayExpress.

1. The following query is used at PubMed to identify transcriptomics studies on coronary artery disease and atherosclerosis: ("atherosclerosis"[majr] OR coronary artery disease[majr]) AND ("microarray analysis"[mh] OR "gene expression profiling"[mh]) NOT "review"[ptyp] (*see* **Note 2**). Applying this query results in 234 publications (as of March 2013).
2. Paper title and abstract of the resulting 234 publications are manually checked for the level of disclosed Omics profiling results using arteries as sample material, and appropriateness of the specific disease phenotype included in the individual studies. The following four studies are considered appropriate and are forwarded to feature extraction (Table 2).
3. Lists of differentially expressed genes (DEGs) comparing CVD and control samples are obtained from the publications itself or the supplementary data files for three of the four studies, namely from Archacki et al., Cagnin et al., as well as Volger and colleagues.

Table 2
List of selected transcriptomics studies in the CVD context

First author	Study title	Citation
Archacki et al.	Identification of new genes differentially expressed in coronary artery disease by expression profiling	[46]
Cagnin et al.	Reconstruction and functional analysis of altered molecular pathways in human atherosclerotic arteries	[47]
Volger et al.	Distinctive expression of chemokines and transforming growth factor-beta signaling in human arterial endothelium during atherosclerosis	[48]
Haegg et al.	Multi-organ expression profiling uncovers a gene module in coronary artery disease involving transendothelial migration of leukocytes and LIM domain binding 2: the Stockholm Atherosclerosis Gene Expression (STAGE) study	[49]

Archacki et al. lists differentially expressed genes as explicit table in the publication. The authors provide for each gene next to the GenBank ID, the description and their molecular activity together with the direction of regulation (up/down) in CAD samples. Cagnin et al. provide lists of differentially expressed genes in the associated GEO record GSE11138. Volger et al. provide the list of differentially expressed genes in the supplemental material of the respective publication as found on the Journal of Pathology webpage (http://www.journals.elsevierhealth.com/periodicals/ajpa/article/S0002-9440%2810%2961966-9/addOns).

4. Reanalysis based on the raw expression files is necessary for the study published by Haegg et al., as the complete list of differentially expressed genes associated with CAD is neither given in the paper nor in the supplementary data files. The microarray expression raw data of the study are deposited in NCBI's GEO database with the GEO series ID GSE40231. A link-out to the GEO dataset is provided in the section "Related Information" at the PubMed entry of the publication (http://www.ncbi.nlm.nih.gov/pubmed/19997623). Besides information about the study design the raw data files as well as processed data files are provided. For analysis of the data provided we used CARMAweb for data preprocessing including background correction, summarization of probe set values, as well as MAS5 normalization [38]. Significance analysis of microarrays (SAM) [39] as included in the open source MultiExperiment Viewer (MeV) software [40] is further applied to the normalized expression matrix for identification of significantly differentially expressed genes comparing samples from the atherosclerotic and the non-atherosclerotic arterial wall, setting the false discovery rate to <5 % and the fold change cutoff to >2.

Table 3
Number of identified molecular features associated with CVD in the selected Omics studies as well as following the automatic literature search

Study	Number of molecular features
Archacki 2003 Omics study	90
Cagnin 2009 Omics study	1,070
Volger 2007 Omics study	978
Haegg 2009 Omics study	625
Literature mining approach	1,505
Total number of unique features	*3,781*

The number of identified genes showing deregulation on the mRNA level in the set of four transcriptomics datasets is given in Table 3, being complemented by the set of literature derived features as outlined in the next section.

3.2 Complementing the CVD Omics Feature Set via Literature Mining

In addition to the differentially expressed genes on CVD we also generate a set of relevant molecular features (genes and proteins, respectively) following an automatic literature mining procedure.

1. The following query is used at NCBI PubMed in order to identify publications on coronary artery disease and atherosclerosis: *"atherosclerosis[majr] OR coronary artery disease[majr]"* resulting in a set of 145,208 publications (as of March 2013).
2. Genes of relevance in this set of publications are identified using the gene2pubmed file as provided by NCBI (ftp://ftp.ncbi.nih.gov/gene/DATA/gene2pubmed.gz). The list of PubMed IDs (PMIDs) can be downloaded via the "*Send to file*" mechanism and is subsequently used for identifying genes in the gene2pubmed file being reported in the context of the extracted set of publications. As we are only interested in human genes, we limit the extraction to those entries with the human taxonomy ID 9606, by this deriving a set of 1,505 genes associated to CAD and atherosclerosis.

3.3 Namespace Consolidation of Feature Sets from Omics and Literature Mining

For integrating data from the different sources (transcriptomics profiles as well as literature), it is necessary to map the obtained data to a common namespace. The result of the automatic literature mining process is a list of NCBI ENTREZ Gene IDs. Volger et al. provide Gene Symbols. Cagnin et al. as well as Archacki et al. provide NCBI GenBank identifiers. Reanalysis of the data set from Haegg et al. led to a list of deregulated transcripts with the corresponding Affymetrix probe identifiers. DAVID offers a convenient tool for mapping various identifiers to a common namespace, being in our case the Ensembl Gene IDs [41].

1. At the DAVID webpage (http://david.abcc.ncifcrf.gov/) select the DAVID Gene ID Conversion tool.
2. Upload the list of identifiers for annotation either by pasting the list into the respective form or via file upload.
3. Select the type of identifiers provided from the drop-down menu (for GenBank use GENBANK_ACCESSION, for ENTREZ Gene ID use ENTREZ_GENE_ID, for Gene Symbols use OFFICIAL_GENE_SYMBOL, and for Affymetrix probe IDs derived from Haegg et al. use AFFYMETRIX_3PRIME_IVT_ID).
4. Select the "*gene list*" radio button and submit the list.
5. Once all lists are uploaded proceed with mapping to Ensembl Gene IDs by selecting the respective lists in the List Manager Tab and press the "Use" button.
6. Select "Gene ID Conversion tool" from the "Shortcut to DAVID Tools menu" in the top menu bar, select "ENSEMBL_GENE_ID" as target namespace, and click the "Submit to Conversion" Tool button.
7. The resulting table with aligned identifiers can be downloaded as tab-delimited text file.

3.4 Populating a CVD Protein–Protein Interaction Network

The set of consolidated and annotated molecular entities derived for CVD serve as input to extract protein–protein interaction data from three publicly available databases, thus generating a protein–protein interaction network specifically covering a CVD interactome. For this, molecular protein–protein interaction data from BioGRID, IntAct and Reactome are used.

1. BioGRID interaction data for various organisms are downloaded from http://thebiogrid.org/download.php in Tab 2.0 delimitated text file format (BIOGRID-ORGANISM-3.2.98.tab2.zip as of March 2013).

   IntAct data are accessible at ftp://ftp.ebi.ac.uk/pub/databases/intact/current in the psimitab format.

   Reactome Human protein–protein interaction pairs in tab-delimited format are downloaded from http://www.reactome.org/download/index.html (Fig. 1).
2. Interactions in the human BioGRID interaction file (BIOGRID-ORGANISM-Homo_sapiens-3.2.98.tab2.txt) with annotation in the phenotype field are excluded from further analyses, as are interactions holding an interactor from species other than human. The remaining interactions are mapped to Ensembl Gene IDs.

a

B	C	P	Q	U
Entrez Gene Interactor A	Entrez Gene Interactor B	Organism Interactor A	Organism Interactor B	Phenotypes
6416	2318	9606	9606	-
84665	88	9606	9606	-
90	2339	9606	9606	-
2624	5371	9606	9606	-
6118	6774	9606	9606	-
375	23163	9606	9606	-
377	23647	9606	9606	-
377	27236	9606	9606	-
10327	54512	9606	9606	-

b

A	B	J	K
#ID(s) interactor A	ID(s) interactor B	Taxid interactor A	Taxid interactor B
uniprotkb:Q92918	uniprotkb:P46108	taxid:9606(human)|ta	taxid:9606(human)|tax
uniprotkb:Q92918	uniprotkb:P46109	taxid:9606(human)|ta	taxid:9606(human)|tax
uniprotkb:Q9Y4K4	uniprotkb:P46108	taxid:9606(human)|ta	taxid:9606(human)|tax
uniprotkb:Q9Y4K4	uniprotkb:P46109	taxid:9606(human)|ta	taxid:9606(human)|tax
uniprotkb:P62993	uniprotkb:Q9Y4K4	taxid:9606(human)|ta	taxid:9606(human)|tax
uniprotkb:Q9Y4K4	uniprotkb:P46108	taxid:9606(human)|ta	taxid:9606(human)|tax
uniprotkb:P46109	uniprotkb:Q92918	taxid:9606(human)|ta	taxid:9606(human)|tax
uniprotkb:P46109	uniprotkb:Q92918	taxid:9606(human)|ta	taxid:9606(human)|tax
uniprotkb:Q9Y4K4	uniprotkb:P46108	taxid:9606(human)|ta	taxid:9606(human)|tax
uniprotkb:Q92918	uniprotkb:P46109	taxid:9606(human)|ta	taxid:9606(human)|tax

c

A	D
# 127934unique interactions	
UniProt:A0JLT2	UniProt:O15379
UniProt:A0JLT2	UniProt:O15379
UniProt:A0JLT2	UniProt:O15379
UniProt:A0JLT2	UniProt:O75376
UniProt:A0JLT2	UniProt:O75376
UniProt:A0JLT2	UniProt:O75376
UniProt:A0JLT2	UniProt:O95402
UniProt:A0JLT2	UniProt:P37231
UniProt:A0JLT2	UniProt:P37231

Fig. 1 Protein–protein interaction file formats. Clippings of the different protein–protein interaction files derived from BioGRID (**a**), IntAct (**b**), and Reactome (**c**) are depicted. Only columns with information subsequently used in the analysis are shown

3. The interaction file from the IntAct database (intact.txt) holds interactions from all organisms in one file, and therefore only interactions where both interactors are of human origin are kept for further analyses leaving all rows in the file with the human taxonomy id 9606 in the columns entitled Taxid interactor A and Taxid interactor B. Molecular entity identifiers provided in the fields ID(s) interactor A and ID(s) interactor B are mapped to Ensembl Gene IDs.
4. Interactions holding one nonhuman interactor are also removed in the Reactome file (homo_sapiens.interactions.txt). Remaining interactions are mapped to the Ensembl Gene IDs.

5. Protein–protein interactions from the three different files are combined into one tab-delimited text file holding one interaction per row with the Ensembl IDs as common identifiers of the interactors. Duplicate entries are removed to ensure that each interaction is present only once in the datafile. In addition interactions from one protein with itself (homodimers) are also discarded.
6. Based on the consolidated interaction file the disease-specific network is generated, thus focusing only on interactions holding two members of the consolidated CVD feature set. We in total identify 27,963 protein–protein interactions involving 3,396 proteins reported as associated with CVD.

3.5 Visualization of the CVD Network

The consolidated, CVD-specific protein–protein interaction data file serves as input for the network visualization tool Cytoscape V2.8.3 for graphical representation of the interaction network.

1. The protein–protein interaction data file is imported into Cytoscape using the "Network from Table" wizard (File → Import → Network from Table (Text/MS Excel)).
2. Source and Target interactions are set to the first and second field (column) of the tab-delimited file, respectively, and "default interaction" is used as interaction type.
3. After import of the network the number of sources (Omics studies/features identified through literature mining) reporting each individual gene are imported using the "Attribute from Table (Text/MS Excel)" wizard, node color settings are changed in VizMapper to reflect the source data information, and the network was layouted (*see* **Note 3** for information on layout algorithms). A rendering of the CVD network is provided in Fig. 2.

3.6 Segmenting the CVD Network

The resulting CVD-associated network is further segmented into clusters of highly connected nodes, as we are ultimately interested in sets of nodes (genes/proteins) involved in specific molecular processes.

1. The Molecular Complex Detection (MCODE) algorithm, available as Cytoscape plugin, is used for identifying highly connected subnetworks (*see* **Note 4** on how to install Cytoscape plugins). The MCODE plugin is started via Plugins → MCODE → Start MCODE and executed using the default parameters, namely "include loops no," "degree cutoff 2," "haircut yes," "fluff no," "node score cutoff 0.2, K-Core 2," and "max depth 100."
2. Identified clusters are inspected by selecting one after the other in the Cluster Browser provided in the Cytoscape result panel, and subsequently exported using the "export" functionality

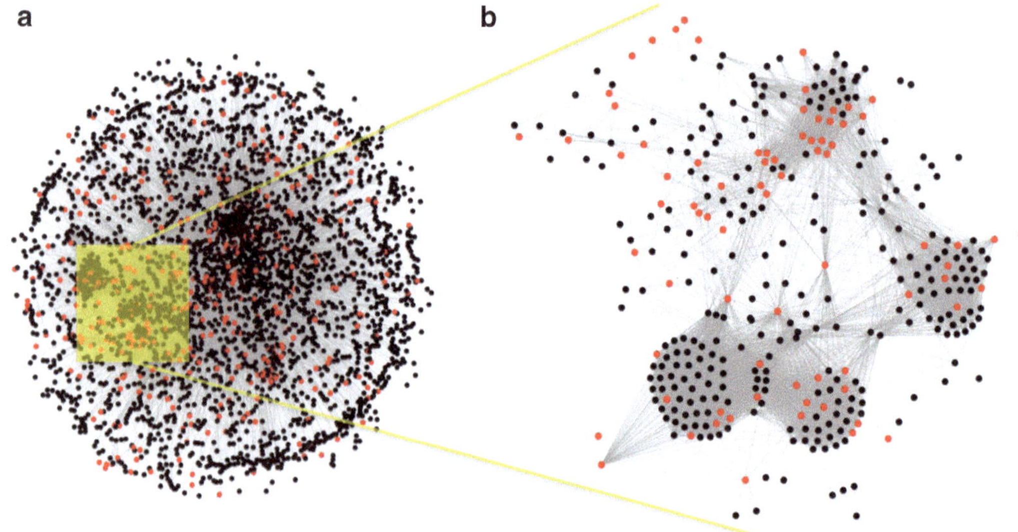

Fig. 2 Giant component of the disease-specific interaction network. Nodes represent proteins associated to CVD, and edges encode protein–protein interactions. Nodes colored in *red* indicate genes being identified as relevant in multiple sources (i.e., the four Omics studies or the literature derived list). (**a**) Holds the entire giant component, whereas (**b**) shows a detailed representation of the area highlighted in (**a**)

offered in the result panel. Executing this procedure we in total identify 50 clusters, which we in the following consider being molecular units, ranging in size (number of molecular features) from 3 to 98.

3. One representative cluster and its location in the overall network is exemplarily given in Fig. 3.
4. In addition, a unit–unit graph resembling a molecular model for CVD is compiled from the clustering result: In such model nodes represent individual units, and edges encode relations between units. The strength of relation between each pair of units is determined by the ratio of relations connecting members from different units with respect to maximum relations possible (being $m \times n$, where m and n are the respective sizes of the two units). A rendering of such molecular model is given in Fig. 4a.

3.7 Functional Interpretation of CVD Units

We next apply gene set enrichment analysis for each of the identified CVD units using the gene ontology reflecting biological processes.

1. Molecular features of each unit are combined into a single text file in DAVID's multi-list file format. This file is subsequently used as input for DAVID's functional annotation tool accessible at http://david.abcc.ncifcrf.gov/ by selecting "Functional Annotation."
2. Input files can be uploaded to DAVID using the same upload form as previously used for uploading gene lists for identifier

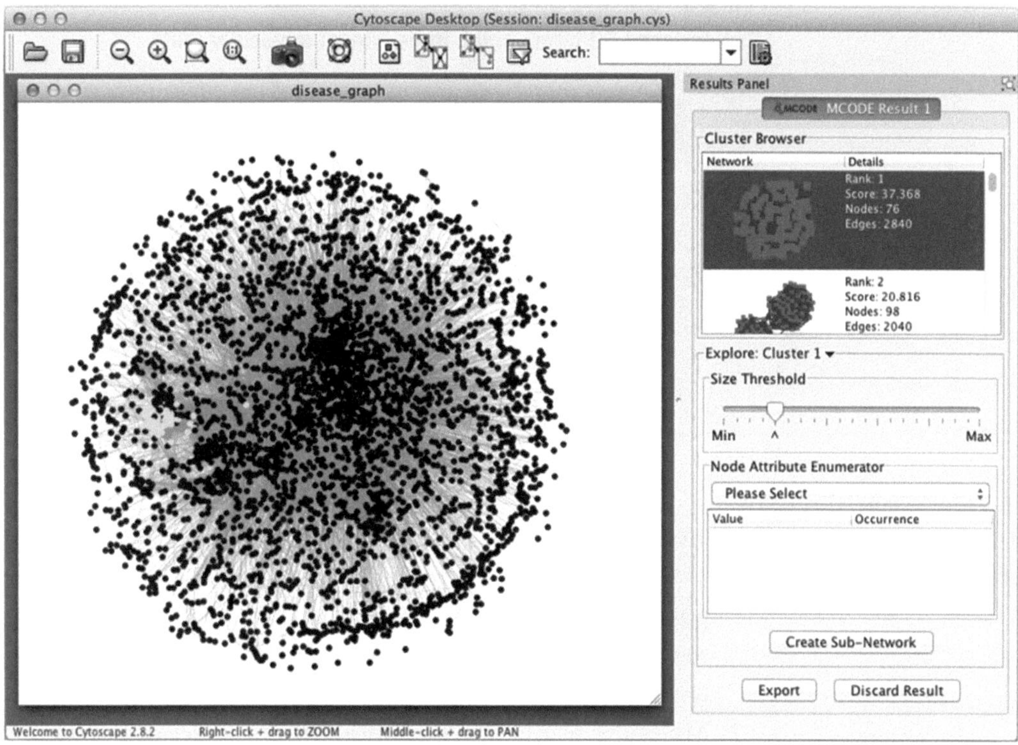

Fig. 3 MCODE output. The Cytoscape screenshot shows the disease-specific network as well as clusters identified by MCODE visible in the MCODE result panel. Nodes belonging to the currently selected cluster (cluster 1) are highlighted in *yellow*

mapping. Attention needs to be paid to activate the checkbox next to "Multi-List File."

3. Once the file is uploaded and the list manager box on the list tab is populated with the individual lists from the supplied multi-list file, the various annotations for an individual list can be retrieved by selecting the respective list in the list manager, then clicking the "use" button, and finally selecting the respective annotation of interest on the right hand side of the page. For "gene ontology process" expand the gene ontology category, and select the "Chart" button next to "GOTERM_BP_FAT" for retrieving the list of process terms being enriched. The result holds the category a specific term is assigned to, the actual term itself, information about the number of genes from the input list assigned to a term, as well as the raw and corrected *p*-value of the Fisher's Exact test expressing statistics of enrichment. The "Download File" link above the upper right corner of the results table allows downloading the entire result as tab-delimited file. The most enriched GO categories for the three largest subnetworks are listed in Table 4.

Table 4
Top enriched GO terms of the three largest molecular units

UNIT	UNIT size	GO term
UNIT 2	98	Complement activation, classical pathway
UNIT 2	98	Humoral immune response mediated by circulating immunoglobulin
UNIT 7	83	Regulation of Rho protein signal transduction
UNIT 7	83	Cytoskeleton organization
UNIT 1	76	G-protein-coupled receptor protein signal pathway
UNIT 1	76	Cell surface receptor-linked signal transduction

3.8 Evaluation of CVD Biomarker Candidates

A set of CVD biomarker candidates is retrieved following an automatic literature search.

1. 4,768 publications (as of March 2013) are identified using the following PubMed query: "*(atherosclerosis[majr] OR coronary artery disease[majr]) AND biological markers[mh:noexp]*" (*see* **Note 5**).
2. Molecular features being annotated in this set of publications in the context of biomarkers discussed are identified using the NCBI gene2pubmed file in the same way as described above for literature mining approaches. We in total derive a set of 333 marker candidates.
3. Identified marker candidates are subsequently mapped on the 50 CVD units. 27 of the 50 units hold at least one marker candidate, addressing 16 out of the 17 largest units (*see* Fig. 4a). This given assembly of biomarker candidates forms a selected panel for CVD, promising improved evidence for involvement in CVD pathophysiology. Such set may be further complemented for also including further CVD clusters not yet covered, needing the selection of novel candidates for such additional clusters. An approach is to screen further clusters for molecular features being reported as biomarker candidates in the context of some other disease phenotype or by screening for candidates with suitable subcellular location for allowing assaying.

Next to selection of marker candidates also drugs and their respective targets may be investigated in the molecular units context. Drugs and their respective drug targets may be retrieved from DrugBank and subsequently used to annotate individual unit members [42].

1. The list of drugs and their respective targets can be conveniently retrieved using the batch exporter "Data Extractor" in DrugBank's Search menu available at http://www.drugbank.ca.

2. The list on the left-hand side of the "Data Extractor" allows adding individual drug details to the search result. The fields "Generic Name" and "DrugBank ID" are automatically included in the search result. Including the "Target Field" "UniProt ID" into the search result allows retrieving respective drug targets. Adding the "Drug Field" and "Drug Group" in addition to the search result and restricting its value to "approved" by entering the string "Approved" in the respective text field allows restricting the search result to approved drugs only.
3. After all required fields are added to the search result the output type may be switched to CSV Excel by choosing the respective option from the drop-down menu provided at the bottom of the page. Finally the search can be executed by clicking on the "*Search*" button and the resulting CSV file can be saved.
4. The list of drugs and their respective targets retrieved from DrugBank comes with Uniprot IDs. With the help of DAVID's Gene Id Conversion tool those IDs can again be mapped to the Ensembl gene IDs.

Utilizing this procedure we in total identify 208 drug targets of 611 different drugs in 35 distinct molecular units of the CVD set of processes (drugbank data status as of March 2013). Unit 13, holding 33 proteins, is exemplarily shown in Fig. 4b, with interactions between individual proteins arranged according to subcellular location of proteins. In addition, the top 5 enriched gene ontology terms are given together with the drugs targeting proteins also reported as biomarker candidates in the context of CVD.

Transforming disease term-associated molecular feature sets into a molecular model as a composition of disease term-specific functional units promises a better reflection of a diseases molecular pathophysiology. In consequence, more informed retrieval of biomarker candidate panels also in conjunction with target/drug information becomes amenable [43].

4 Notes

1. GSEA has certain pitfalls, as the selection of the background, i.e., the total set of genes considered to underlie the investigated lists, naturally impacts statistics. For example, when comparing a set of genes found to have a significant differential regulation in a transcriptomics case–control study to a curated pathway, the question arising is which background to use for the contingency table. Neither the number of features included on the array nor the total size of the pathway database serves as an ideal reference.
2. MeSH query "Coronary artery disease" versus Coronary artery disease:

   Unquoted search terms tagged as MeSH term or MeSH Major Topic (using [mh] or [majr], respectively) are automatically

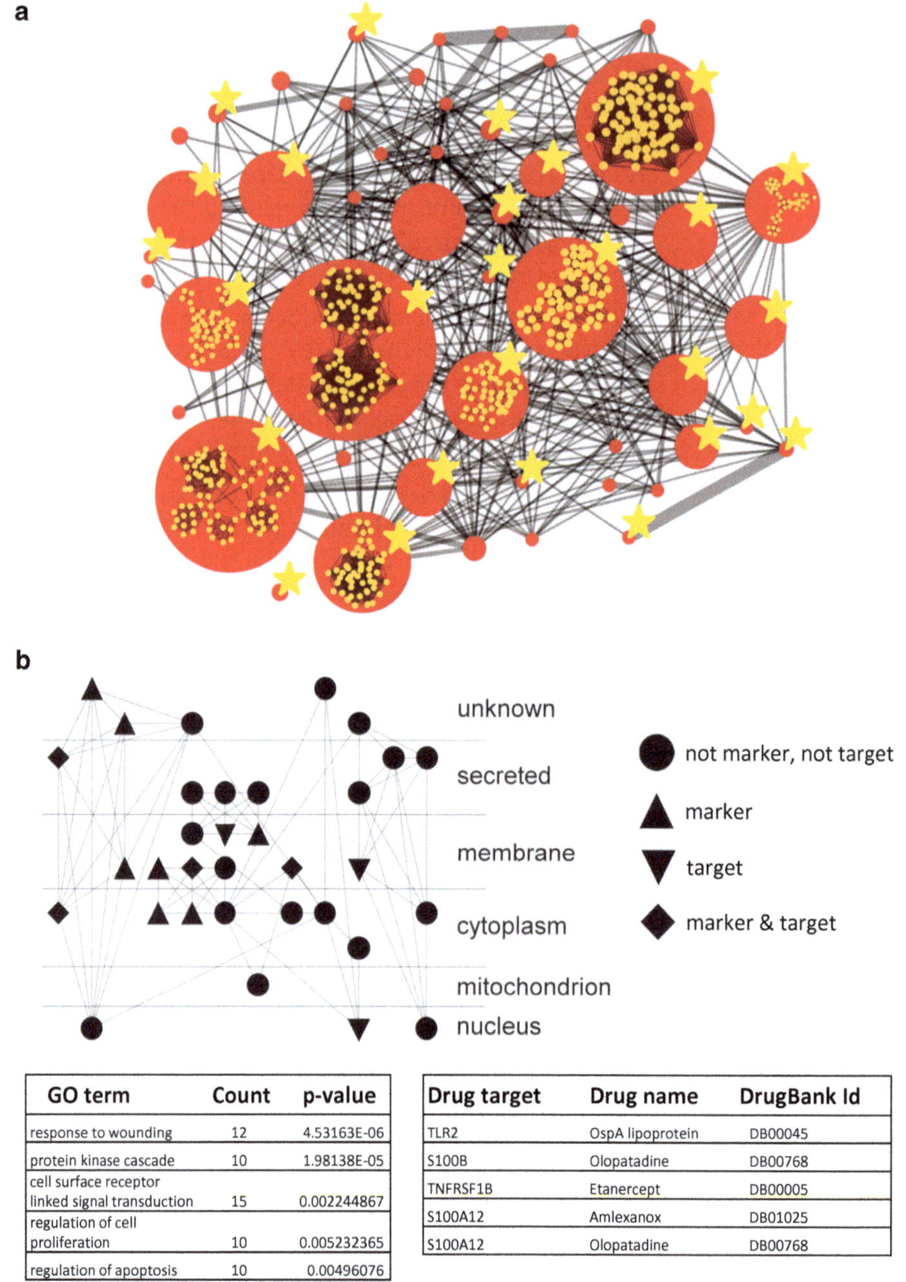

GO term	Count	p-value
response to wounding	12	4.53163E-06
protein kinase cascade	10	1.98138E-05
cell surface receptor linked signal transduction	15	0.002244867
regulation of cell proliferation	10	0.005232365
regulation of apoptosis	10	0.00496076

Drug target	Drug name	DrugBank Id
TLR2	OspA lipoprotein	DB00045
S100B	Olopatadine	DB00768
TNFRSF1B	Etanercept	DB00005
S100A12	Amlexanox	DB01025
S100A12	Olopatadine	DB00768

Fig. 4 Unit–unit graph and detailed representation of unit 13. A schematic representation of the unit–unit graph (molecular model), where individual units are represented by nodes and relations between units based on interaction counts between individual unit members are encoded by edges. (**a**) Protein–protein interactions between unit members are schematically illustrated for the eight largest units. Units holding at least one biomarker are marked with an *asterisk*. (**b**) Exemplarily depicts the network resembling unit 13, where proteins are arranged in different layers according to their subcellular location. Biomarkers, drug targets, and proteins being discussed in both, biomarker and drug target context, are represented by *triangles*, *inverted triangles*, or *diamonds*, respectively. The top five enriched GO biological process terms as well as selected drugs addressing molecules being reported as biomarker in the context of CVD are listed in addition in (**b**). For further details on the construction of (**b**) *see* **Note 6**

mapped to respective (potentially multiple) MeSH terms. In the specific case of the term "Coronary artery disease" the term is mapped to the following two MeSH terms: Coronary Disease and Coronary Artery Disease. For further details on the PubMed search the reader is referred to http://www.ncbi.nlm.nih.gov/books/NBK3827/.

3. Graph layout options: By default Cytoscape arranges nodes in a grid-like manner for network visualization. Layout algorithms available from the "Layout" menu allow for automatic arrangement of nodes and their edges according to topological or user-defined properties.
4. Installation of Cytoscape plugins: Cytoscape plugins can be installed via the plugin manager (Plugins → Manage Plugins). The string "mcode" is used as key word for the search and the version of the MCODE plugin compatible with the used Cytoscape version (indicated by a small checkmark) is selected from the search result (available for Install → Clustering → MCODE v.X.XX) and installed. A guide to Cytoscape plugins is available in [44].
5. The suffix "noexp" has to be used in order to turn off the automatic explosion of MeSH headings thus only searching for the specified MeSH term and none of its child terms.
6. Graph layout according to subcellular location information: The Cerebral plugin for Cytoscape was used to layout the network according to subcellular location information of individual genes [45]. Since Cytoscape does not offer node shapes in the form of inverted triangles those were added in a subsequent step.

Acknowledgements

The research leading to these results has received funding from the European Community's Seventh Framework Programme under the grant agreement no. 2782494 (EU-MASCARA).

References

1. Ptolemy AS, Rifai N (2010) What is a biomarker? Research investments and lack of clinical integration necessitate a review of biomarker terminology and validation schema. Scand J Clin Lab Invest Suppl 242:6–14. doi:10.3109/00365513.2010.493354
2. Ziegler A, Koch A, Krockenberger K, Grosshennig A (2012) Personalized medicine using DNA biomarkers: a review. Hum Genet 131:1627–1638. doi:10.1007/s00439-012-1188-9
3. Barrett T, Troup DB, Wilhite SE et al (2007) NCBI GEO: mining tens of millions of expression profiles—database and tools update. Nucleic Acids Res 35:D760–D765. doi:10.1093/nar/gkl887
4. Rustici G, Kolesnikov N, Brandizi M et al (2013) ArrayExpress update—trends in database

growth and links to data analysis tools. Nucleic Acids Res 41:D987–D990. doi:10.1093/nar/gks1174

5. Mühlberger I, Wilflingseder J, Bernthaler A et al (2011) Computational analysis workflows for omics data interpretation. Methods Mol Biol 719:379–397. doi:10.1007/978-1-61779-027-0_17
6. Hindorff LA, Sethupathy P, Junkins HA et al (2009) Potential etiologic and functional implications of genome-wide association loci for human diseases and traits. Proc Natl Acad Sci U S A 106:9362–9367. doi:10.1073/pnas.0903103106
7. Li MJ, Wang P, Liu X et al (2012) GWASdb: a database for human genetic variants identified by genome-wide association studies. Nucleic Acids Res 40:D1047–D1054. doi:10.1093/nar/gkr1182
8. Coon JJ, Zürbig P, Dakna M et al (2008) CE-MS analysis of the human urinary proteome for biomarker discovery and disease diagnostics. Proteomics Clin Appl 2:964. doi:10.1002/prca.200800024
9. Wishart DS, Jewison T, Guo AC et al (2013) HMDB 3.0—the Human Metabolome Database in 2013. Nucleic Acids Res 41:D801–D807. doi:10.1093/nar/gks1065
10. Rebholz-Schuhmann D, Oellrich A, Hoehndorf R (2012) Text-mining solutions for biomedical research: enabling integrative biology. Nat Rev Genet 13:829–839. doi:10.1038/nrg3337
11. Rebholz-Schuhmann D, Arregui M, Gaudan S et al (2008) Text processing through Web services: calling Whatizit. Bioinformatics 24:296–298. doi:10.1093/bioinformatics/btm557
12. Hoffmann R, Valencia A (2005) Implementing the iHOP concept for navigation of biomedical literature. Bioinformatics 21 Suppl 2:ii252–ii258. doi: 10.1093/bioinformatics/bti1142
13. UniProt-Consortium (2012) Reorganizing the protein space at the Universal Protein Resource (UniProt). Nucleic Acids Res 40:D71–D75. doi:10.1093/nar/gkr981
14. Flicek P, Ahmed I, Amode MR et al (2013) Ensembl 2013. Nucleic Acids Res 41:D48–D55. doi:10.1093/nar/gks1236
15. Huang DW, Sherman BT, Stephens R et al (2008) DAVID gene ID conversion tool. Bioinformation 2:428–430
16. Cascione L, Ferro A, Giugno R et al (2013) Elucidating the role of microRNAs in cancer through data mining techniques. Adv Exp Med Biol 774:291–315. doi:10.1007/978-94-007-5590-1_15
17. Hsu S-D, Lin F-M, Wu W-Y et al (2011) miRTarBase: a database curates experimentally validated microRNA-target interactions. Nucleic Acids Res 39:D163–D169. doi:10.1093/nar/gkq1107
18. Orchard S (2012) Molecular interaction databases. Proteomics 12:1656–1662. doi:10.1002/pmic.201100484
19. Kerrien S, Orchard S, Montecchi-Palazzi L et al (2007) Broadening the horizon—level 2.5 of the HUPO-PSI format for molecular interactions. BMC Biol 5:44
20. Aranda B, Blankenburg H, Kerrien S et al (2011) PSICQUIC and PSISCORE: accessing and scoring molecular interactions. Nat Methods 8:528–529. doi:10.1038/nmeth.1637
21. Chatr-Aryamontri A, Breitkreutz B-J, Heinicke S et al (2013) The BioGRID interaction database: 2013 update. Nucleic Acids Res 41:D816–D823. doi:10.1093/nar/gks1158
22. Kerrien S, Aranda B, Breuza L et al (2012) The IntAct molecular interaction database in 2012. Nucleic Acids Res 40:D841–D846. doi:10.1093/nar/gkr1088
23. Croft D, O'Kelly G, Wu G et al (2011) Reactome: a database of reactions, pathways and biological processes. Nucleic Acids Res 39:D691–D697. doi:10.1093/nar/gkq1018
24. Smoot ME, Ono K, Ruscheinski J et al (2010) Cytoscape 2.8: new features for data integration and network visualization. Bioinformatics 27:431–432. doi:10.1093/bioinformatics/btq675
25. Lopes CT, Franz M, Kazi F et al (2010) Cytoscape web: an interactive web-based network browser. Bioinformatics 26:2347–2348. doi:10.1093/bioinformatics/btq430
26. Breitkreutz B-J, Stark C, Tyers M (2003) Osprey: a network visualization system. Genome Biol 4:R22
27. Hu Z, Hung J-H, Wang Y et al (2009) VisANT 3.5: multi-scale network visualization, analysis and inference based on the gene ontology. Nucleic Acids Res 37:W115–W121. doi:10.1093/nar/gkp406
28. Suderman M, Hallett M (2007) Tools for visually exploring biological networks. Bioinformatics 23:2651–2659. doi:10.1093/bioinformatics/btm401
29. Gehlenborg N, O'Donoghue SI, Baliga NS et al (2010) Visualization of omics data for systems biology. Nat Methods 7:S56–S68. doi:10.1038/nmeth.1436
30. Schaeffer SE (2007) Graph clustering. Comput Sci Rev 1:27–64. doi:10.1016/j.cosrev.2007.05.001

31. Bader GD, Hogue CWV (2003) An automated method for finding molecular complexes in large protein interaction networks. BMC Bioinformatics 4:2
32. Van Dongen S (2000) Graph clustering by flow simulation. PhD thesis, University of Utrecht
33. Wang J, Li M, Deng Y, Pan Y (2010) Recent advances in clustering methods for protein interaction networks. BMC Genomics 11 Suppl 3:S10. doi: 10.1186/1471-2164-11-S3-S10
34. Subramanian A, Tamayo P, Mootha VK et al (2005) Gene set enrichment analysis: a knowledge-based approach for interpreting genome-wide expression profiles. Proc Natl Acad Sci U S A 102:15545–15550. doi:10.1073/pnas.0506580102
35. Kanehisa M (2013) Molecular network analysis of diseases and drugs in KEGG. Methods Mol Biol 939:263–275. doi:10.1007/978-1-62703-107-3_17
36. Ashburner M, Ball CA, Blake JA et al (2000) Gene ontology: tool for the unification of biology. The Gene Ontology Consortium. Nat Genet 25:25–29. doi:10.1038/75556
37. Biomarkers-Definitions-Working-Group (2001) Biomarkers and surrogate endpoints: preferred definitions and conceptual framework. Clin Pharmacol Ther 69:89–95. doi:10.1067/mcp.2001.113989
38. Rainer J, Sanchez-Cabo F, Stocker G et al (2006) CARMAweb: comprehensive R- and bioconductor-based web service for microarray data analysis. Nucleic Acids Res 34:W498–W503
39. Tusher VG, Tibshirani R, Chu G (2001) Significance analysis of microarrays applied to the ionizing radiation response. Proc Natl Acad Sci U S A 98:5116–5121
40. Saeed AI, Sharov V, White J et al (2003) TM4: a free, open-source system for microarray data management and analysis. Biotechniques 34:374–378
41. Huang DW, Sherman BT, Tan Q et al (2007) DAVID Bioinformatics Resources: expanded annotation database and novel algorithms to better extract biology from large gene lists. Nucleic Acids Res 35:W169–W175. doi:10.1093/nar/gkm415
42. Knox C, Law V, Jewison T et al (2011) DrugBank 3.0: a comprehensive resource for "omics" research on drugs. Nucleic Acids Res 39:D1035–D1041. doi:10.1093/nar/gkq1126
43. Heinzel A, Fechete R, Mühlberger I et al (2013) Molecular models of the cardiorenal syndrome. Electrophoresis 34:NA. doi:10.1002/elps.201370101
44. Saito R, Smoot ME, Ono K et al (2012) A travel guide to Cytoscape plugins. Nat Methods 9:1069–1076. doi:10.1038/nmeth.2212
45. Barsky A, Gardy JL, Hancock REW, Munzner T (2007) Cerebral: a Cytoscape plugin for layout of and interaction with biological networks using subcellular localization annotation. Bioinformatics 23:1040–1042. doi:10.1093/bioinformatics/btm057
46. Archacki SR, Angheloiu G, Tian X-L et al (2003) Identification of new genes differentially expressed in coronary artery disease by expression profiling. Physiol Genomics 15:65–74. doi:10.1152/physiolgenomics.00181.2002
47. Cagnin S, Biscuola M, Patuzzo C et al (2009) Reconstruction and functional analysis of altered molecular pathways in human atherosclerotic arteries. BMC Genomics 10:13. doi:10.1186/1471-2164-10-13
48. Volger OL, Fledderus JO, Kisters N et al (2007) Distinctive expression of chemokines and transforming growth factor-beta signaling in human arterial endothelium during atherosclerosis. Am J Pathol 171:326–337
49. Hägg S, Skogsberg J, Lundström J et al (2009) Multi-organ expression profiling uncovers a gene module in coronary artery disease involving transendothelial migration of leukocytes and LIM domain binding 2: the Stockholm Atherosclerosis Gene Expression (STAGE) study. PLoS Genet 5:e1000754. doi:10.1371/journal.pgen.1000754

Chapter 8

Mining Biological Networks from Full-Text Articles

Jan Czarnecki and Adrian J. Shepherd

Abstract

The study of biological networks is playing an increasingly important role in the life sciences. Many different kinds of biological system can be modelled as networks; perhaps the most important examples are protein–protein interaction (PPI) networks, metabolic pathways, gene regulatory networks, and signalling networks. Although much useful information is easily accessible in publicly databases, a lot of extra relevant data lies scattered in numerous published papers. Hence there is a pressing need for automated text-mining methods capable of extracting such information from full-text articles. Here we present practical guidelines for constructing a text-mining pipeline from existing code and software components capable of extracting PPI networks from full-text articles. This approach can be adapted to tackle other types of biological network.

Key words Named entity recognition, Relationship extraction, Biological networks, Protein–protein interactions

1 Introduction

Mapping the various interaction networks within cells is a central task of modern biology [1]. A critical aspect of this process is the need to integrate data from disparate sources. In practice, a lot of relevant information is only available in scientific papers; given the vast and increasing volume of papers being published, the development of computational pipelines capable of automating the extraction of this information is a priority for the text-mining community.

Here we demonstrate that computational pipelines for extracting protein–protein interaction (PPI) networks from full-text articles can be built using existing code and software components. This approach can be adapted to tackle other types of biological network; for example, an algorithm for extracting metabolic reactions from full-text articles has been developed using the same computational framework used here [2]. Such a pipeline consists of several discrete stages:

Vinod D. Kumar and Hannah Jane Tipney (eds.), *Biomedical Literature Mining*, Methods in Molecular Biology, vol. 1159, DOI 10.1007/978-1-4939-0709-0_8, © Springer Science+Business Media New York 2014

(a) *The retrieval of relevant documents.* Given an appropriate user query, such as the name of a species and the name of a protein, this step involves the automated retrieval of relevant full-text documents.

(b) *Extracting text from documents.* Documents come in many formats, whereas text-mining tools generally require plain text (with characters encoded in a standard way). Hence it is typically necessary to extract text from documents using a text-extraction tool or library.

(c) *Named entity recognition.* In each of the retrieved documents, the entities relevant to the network—whether proteins, genes, species, enzymes, and/or metabolites—need to be identified. In many tasks, it is necessary to identify several entities. In the case of PPI network extraction, it is necessary to identify the names of proteins and the names of species.

(d) *Relationship extraction.* Given two or more entities, the pipeline needs to decide whether they are involved in a relationship. For example, the sentence "protein A interacts with protein B" constitutes evidence for there being a relationship between proteins A and B in a PPI network.

(e) *Synonym resolution.* The same entity may be referred to by different names in different documents (or, potentially, within the same document). If this task is not performed successfully for network entities, the network will not have the proper connectivity because the same entity will appear multiple times with different names.

At each of these stages different choices are available regarding what tools or program code to deploy. Given that new, potentially better, components are being developed all the time, it is sensible to develop a text-mining pipeline in a modular way within a widely supported development framework. Here we use the Unstructured Information Management Architecture (UIMA) framework [4], which is widely used by the natural language processing community.

Arguably the key stage within this pipeline is the relationship extraction stage, as it is a very challenging task and a wide range of approaches of contrasting complexity can be deployed to tackle it. The simplest approach is co-occurrence [5]. If the names of two entities occur in close proximity to each other (e.g., in the same sentence), they are deemed to have a relationship with each other. This approach typically produces high coverage (few false negatives) but low precision (many false positives). At the other end of the spectrum are sophisticated tools that analyze the structure of a given sentence using a natural language parser. A detailed analysis of PPI relationship extraction methods [3] showed that simple heuristic rules (e.g., using lists of interaction keywords) were often as effective as these sophisticated methods and much easier to use in practice, but the best component to use will depend on the precise application and the state of the art at the time the pipeline is being constructed (*see* **Note 1**).

2 Materials

The computational pipeline described here assumes that it will be developed using Java, but equivalent libraries exist for other programming languages. Here we assume that the full-text documents are available in PDF format.

2.1 Development Framework

1. Download and install the UIMA development framework (from http://uima.apache.org/). Conveniently, UIMA can be used as a plug-in within the widely used Eclipse Integrated Development Environment (which can be downloaded from http://www.eclipse.org/). Henceforth we assume that UIMA is used within the Eclipse IDE.
2. Create a new skeleton UIMA Type System Descriptor file within Eclipse. This will subsequently be used to store various type-related information associated with specific tools discussed below.

2.2 General Language Processing Tools

1. Download and install the OpenNLP toolkit (from http://opennlp.apache.org/). The toolkit provides Java code that supports a range of standard natural language processing tasks. In the present context, the tools to use are the SentenceDetector (for identifying sentences within a document) and the Tokenizer (for splitting a sentence into its component words or other tokens). OpenNLP has built-in UIMA support.
2. Import the OpenNLP Type System Descriptor file into the UIMA Type System Descriptor file.
3. In the UIMA Analysis Engine Descriptor file for Sentence Detector, set the Sentence type parameter to opennlp.uima.Sentence and *use the relevant pre-trained English language model, en-sent.bin, under the Resources tab* (*see* **Note 2**).
4. In the UIMA Analysis Engine Descriptor file for Tokenizer, set the Token type parameter to opennlp.uima.Token and *use the relevant pre-trained English language model, en-token.bin, under the Resources tab* (*see* **Note 2**).
5. Download and install PDFBox (from http://pdfbox.apache.org/), a Java PDF library that includes the PDFTextStripper utility for extracting text from a PDF.

2.3 Named Entity Recognition Tools

1. Download and install the BANNER Named Entity Recognition System (from http://banner.svn.sourceforge.net/viewvc/banner/trunk/) (*see* **Note 3**). BANNER [6] is used to identify the names of proteins in PPI network extraction (*see* **Note 4**).
2. BANNER provides a UIMA Analysis Engine Descriptor file and a BANNER configuration file called banner_UIMA_TEST.xml. Place them in a subdirectory within your project resources directory. The Descriptor file requires a single parameter: configFile. Set this parameter to the path of the configuration file. The configuration file requires the paths of a number of models and dictionaries (namely, modelFilename, lemmatiserDataDirectory, posTaggerDataDirectory, and dictionaryFile) (*see* **Note 5**).
3. BANNER also provides its own UIMA Type System Descriptor file. Import the contents of this file into the project's UIMA Type System Descriptor file created in Subheading 2.1.
4. Download and install the LINNAEUS species name recognition library (from http://linnaeus.sourceforge.net/). LINNAEUS does not have built-in UIMA support, but it does provide a UIMA wrapper, which can be downloaded from the same webpage.
5. The LINNAEUS wrapper provides a UIMA Analysis Engine Descriptor file, and the LINNAEUS library provides a configuration file called javaProperties.conf. Place them in a subdirectory within your project resources directory. The Descriptor file requires a single parameter: ConfigFile. Set this parameter to the path of the configuration file. The configuration file requires the paths of a number of models and dictionaries (*see* **Note 6**).
6. The LINNAEUS wrapper also provides its own UIMA Type System Descriptor file. Import the contents of this file into the project's UIMA Type System Descriptor file created in Subheading 2.1.
7. Create a protein synonym database. Download the UniProtKB/TrEMBL database text file (from http://www.uniprot.org/downloads). In your SQL database management system of choice, create a table mapping a given protein UniProt ID to its organism's NCBI taxonomy ID and a table mapping a given protein UniProt ID to its one or more synonymous protein names. Populate these tables from the database text file (*see* **Note** 7).

3 Methods

In cases where a collection of PDFs from which data is to be extracted is already available (e.g., stored in a local directory), the only part of Subheading 3.1 that needs to be undertaken is the extraction of plain text from PDFs using the PDFBox library.

3.1 Retrieval of Free Text from Relevant Documents

1. The pipeline presented here accepts a PubMed query as input. The form of the query is down to the user but would typically incorporate the name of an organism and the name of a protein of interest, for example (*Homo sapiens*) AND (alcohol dehydrogenase) (*see* **Note 8**).
2. PubMed queries are incorporated into the computational pipeline using one of the Entrez Programming Utilities (E-utilities) available from the NCBI. This is achieved using the following URL:

   http://eutils.ncbi.nlm.nih.gov/entrez/eutils/esearch.fcgi?db=pubmed&term=query&retmax=numOfArticles

   where *query* is the desired PubMed query. The utility returns a list of PubMed unique identifiers (UIDs). By default, the maximum number of UIDs retrieved is 20, but this can be modified using the *numOfArticles* field up to a maximum of 100,000 (*see* **Note 9**). The URL can be integrated into the UIMA framework using a UIMA collection reader (an object that extends CollectionReader_ImplBase).
3. For each PubMed UID, the following additional information needs to be retrieved: the article title, article abstract, and URL to the relevant publisher's webpage (from which we will attempt to retrieve the PDF of the full-text article). An XML document containing the title and abstract of the article can be retrieved using the following URL:

   http://eutils.ncbi.nlm.nih.gov/entrez/eutils/efetch.fcgi?db=pubmed&retmode=xml&id=pubmedUID

   An XML document containing a URL to a relevant webpage, or (where unavailable) to the article's page on PubMed, can be retrieved using the following URL:

   http://eutils.ncbi.nlm.nih.gov/entrez/eutils/elink.fcgi?dbfrom=pubmed&retmode=ref&cmd=prlinks&id=pubmedUID
4. Sometimes the publisher's URL retrieved from PubMed provides a direct link to the article PDF or points to a webpage that contains such a link (e.g., within an HTML href attribute). In these cases, the PDFBox library can be used to extract plain text from the PDF (*see* **Note 10**). When access to the PDF is not available, the natural option is to use the article's abstract instead.

3.2 Document Annotation

1. Create an InteractionKeyword type extending the UIMA Annotation type in the UIMA Type System Descriptor file. This type requires no new fields.
2. Create an Interaction type extending the UIMA Annotation type. This type needs to add the following new fields (*see* **Note 11**):

Field name	Type
gene1	banner.types.uima.Gene
gene2	banner.types.uima.Gene
interactionKeyword	*yourTypesPackage*.InteractionKeyword
score	uima.cas.Integer
organism	uima.cas.FSList
sentence	opennlp.uima.Sentence

3. Generate the classes for all the types within the type system by clicking the UIMA JCasGen button.
4. Write an InteractionKeyword annotator that will be used to match keywords in text that are believed to help identify whether a given pair of protein names are involved in a PPI. A published list of interaction keywords [6] is available at

   http://www.biomedcentral.com/content/supplementary/1471-2105-10-233-S1.txt (*see* **Note 12**).

   Read the keywords into a HashSet. Write code such that the InteractionKeyword annotator tests whether a given token matches any of the keywords in the HashSet (*see* **Note 13**). If a match is found, the annotator should create a new InteractionKeyword annotation at the location of the token within the document.
5. Write an Interaction annotator that creates a new Interaction annotation for every pair of proteins (Gene annotations) that occur within the same sentence and with an InteractionKeyword arising between the two protein names. To obtain a simple score for the interaction, count the number of tokens between the first protein name and the InteractionKeyword and the number of tokens between the second protein name and the InteractionKeyword; the score is the greater of the two numbers.
6. Identify the most likely organism(s) that the interaction is present in (*see* **Note 14**). If any organism names (annotated by LINNAEUS) are present in the same sentence as the interaction, add the organism to the Interaction annotation's organism variable (*see* **Note 15**). If no organism is mentioned in the sentence, assign the interaction to the first organism mentioned in the title and abstract.
7. Once these interaction annotators have been created, the entire annotation pipeline can be run by applying each of the following annotators to the text extracted from a given PDF in the following order (given dependencies between several of these

tools, this ordering is important): the OpenNLP Sentence-Detector; the OpenNLP Tokenizer; BANNER; LINNAEUS; the InteractionKeyword annotator; and the Interaction annotator (*see* **Note 16**).

3.3 Network Construction

1. Construct a global list of all pairs of interacting proteins, together with their interaction scores, from the full set of annotated documents.
2. Use the synonyms and taxonomy IDs in the UniProt database to map as many proteins as possible from the global list to a UniProt ID (*see* **Note 17**).
3. Combine interactions that involve the same interactors. Score each PPI by taking into account the potential multiple interaction scores for the same PPI extracted from multiple sentences/documents (*see* **Note 18**).
4. The interactions can be output in different formats. For the purpose of visualizing in Cytoscape, a comma-separated value (CSV) file is recommended (*see* **Note 19**). Interactions will be automatically joined together to form a network using the protein IDs. However, using a standard format for handling network data (such as SBML) may be more appropriate in many circumstances (*see* **Note 20**).

4 Notes

1. An example of a sophisticated PPI extraction method is AkanePPI [7], which combines the deep syntactic parser Enju [8] with a support vector machine [9] for extracting rules from training corpora. Although AkanePPI performed better than some other tools, it did not outperform the simple method implemented here in a rigorous multi-corpus evaluation and was deemed nontrivial to install and use [3].
2. A selection of pre-trained models can be found at http://opennlp.sourceforge.net/models-1.5/. As the tool only requires the relatively simple functionality of sentence and token detection, the standard English language models are adequate. More complex functionality (such as POS tagging and sentence parsing) would require models trained on scientific text.
3. At the time of writing, the version of BANNER available from the URL given above is regularly updated and has the added advantage of being UIMA compatible, so can be easily integrated into the computational pipeline we present here. On the other hand, the version of the tool available from the BANNER homepage (http://cbioc.eas.asu.edu/banner/) is several years out-of-date.

4. BANNER, like other comparable NER tools, makes no attempt to distinguish between the names of proteins and genes.
5. While the BANNER configuration file should be stored in the project resources, the files referenced in the configuration file itself cannot be in the project resources. These files must be stored outside the compiled jar file and relative paths used to access them.
6. Unlike BANNER, all the LINNAEUS files can be stored in the project resources. Any referenced paths in the UIMA Type System Descriptor file or the LINNAEUS configuration file must be prefixed with internal: however.
7. Each row in the database text file has a two-letter identifier:
   (a) AC—UniProt ID
   (b) OX—NCBI taxonomy ID
   (c) GN—gene names

   The gene names row is in the form
   GN Name = RecommendedName; Synonyms = Synonym1, Synonym2;

   Import the recommended name and all synonyms into the database. This method of synonym resolution requires that a queried name matches a synonym in the database exactly. Various steps can be taken to maximize the chance of matching protein name from the literature to a UniProt synonym, including normalization (e.g., converting all characters to lowercase, removal of punctuation) and automated variant generation (e.g., conversion between Roman and Arabic numerals, Greek and Latin lettering) [10]. A more sophisticated solution is to implement a full search index (Lucene being a popular choice for the Java language).
8. Whereas PubMed searches automatically incorporate synonyms for the names of an organism (e.g., a search for *Homo sapiens* will additionally return matches for the word human), synonyms for the names of other entities are not incorporated automatically. In Subheading 2.3 we explain how to find the synonyms of key network entities. Using these approaches, it is possible to incorporate the synonymous names of entities such as proteins in a query to ensure that the complete set of relevant documents is retrieved. For example, the original query "(*Homo sapiens*) AND (alcohol dehydrogenase)" can be expanded to incorporate synonymous names as follows: "(*Homo sapiens*) AND ((alcohol dehydrogenase) OR adh1b OR adh2)".
9. Typically it will be appropriate to retrieve more articles than the default number of 20, although the precise number is up to the user and will depend on the context. Factors that may

sensibly be taken into account include the frequency with which articles are published about the species and proteins of interest (the more articles published, the larger the number it is likely to be appropriate to retrieve) and the availability of time and computational resources (the more articles retrieved, the longer the pipeline will take to run).

10. The URL provided by PubMed could point directly to a PDF, a webpage with the PDF embedded in an iframe element, or a webpage containing a link to the PDF. As a URL pointing to a PDF does not necessarily end with the file extension .pdf, to determine if a given URL points to a PDF, simply attempt to load the URL with the PDFBox library; if the URL does not point to a PDF, an IOException will be thrown. If neither the page URL nor any src attributes within an iframe element point to PDFs, check whether any links on the webpage (between the opening and closing tags of an HTML a element) contain the text "pdf" or "PDF." Similarly, determine if the href attribute of a given element, or the src attribute of any iframe element within the linked page, points to a PDF.

    Note that there may be errors in the way PDFBox extracts plain text from a PDF (the same is true of other PDF-to-text conversion methods). One common problem is the inability to distinguish between a soft hyphen (i.e., a discretionary hyphen that permits a word to be split when it occurs at the end of a line) and a hard hyphen. In cases where a soft hyphen is treated as if it is a hard hyphen, a single word may be split into two. In such cases, it may be worth using an electronic dictionary to identify word fragments that need to be joined together.

11. As noted already (*see* **Note 4**), BANNER makes no distinction between gene and protein names. The BANNER UIMA type banner.types.uima.Gene uses the term "gene" but applies equally to proteins.

12. The file containing the list of interaction keywords should be stored in the project resources and the path to the file passed to the Analysis Engine as a parameter in the UIMA Analysis Engine Descriptor file.

13. To ensure that all appropriate matches are found, it is worth ensuring that there is no whitespace at the start or the end of a keyword token and that all tokens are lowercase.

14. This simple algorithm assumes that the papers mined typically involve research on a single organism but may refer to interactions present in other organisms. A more sophisticated algorithm would be necessary to reliably assign interactions to organisms in papers dealing with multiple species (such as virus–host interactions).

15. Adding an organism to an Interaction annotation's organisms variable is not an entirely trivial operation, because UIMA-type

objects cannot have arbitrary objects assigned to their fields. The solution is to either use an FSList object (a type of tree list, an object that many Java programmers will not be familiar with) or simply store the taxonomy IDs within a single string (i.e., with the multiple taxonomy IDs separated by a delimiter).

16. To join annotators together create a new aggregate annotator. Individual annotators can be imported and arranged in any order. Types, parameters, and resources specified in the individual annotators will automatically be imported into the aggregate annotator.
17. As described in **Note 7**, many variants of the same protein name may occur in the literature that do not appear in UniProt synonym lists. If normalized variants of protein names have been pregenerated and stored in the database, putative protein names extracted from the literature should be normalized before searching for matches within the database.
18. Scoring should take into account both the score of each putative interaction extracted and the number of times that putative interaction has been extracted. The optimal scoring scheme is context dependent, and a discussion of the options (which vary greatly in their statistical sophistication) lies outside the scope of this chapter.
19. Cytoscape does not require a particular CSV format. It only requires two columns for interactors and a column for the interaction type. The interactors should be specified by their UniProt ID. This allows Cytoscape to automatically link interactions together to form networks. Arbitrary columns (for the text-mining score, for instance) can also be included allowing them to be visualized. To visualize scores, use the Cytoscape Vizmapper to make edge widths dependent on scores.
20. Although Cytoscape will draw SBML files, by default it cannot handle information stored within SBML annotations, which is where the PPI scores need to be stored. However, Cytoscape can handle PPI scores stored in a CSV file without needing a plug-in.

References

1. Barabási AL, Oltvai ZN (2004) Network biology: understanding the cell's functional organization. Nat Rev Genet 5:101–113
2. Czarnecki J, Nobeli I, Smith AM, Shepherd AJ (2012) A text-mining system for extracting metabolic reactions from full-text articles. BMC Bioinformatics 13:172
3. Kabiljo R, Clegg AB, Shepherd AJ (2009) A realistic assessment of methods for extracting gene/protein interactions from free text. BMC Bioinformatics 10:233
4. Ferrucci D, Lally A, Gruhl D, Epstein E, Schor M, Murdock JW, Frenkiel A, Brown EW, Hampp T, Doganata Y, Welty C, Amini L, Kofman G, Kozakov L, Mass Y (2006) Towards an interoperability standard for text and multi-modal analytics. IBM research report

5. Manning CD, Schütze H (1999) Foundations of statistical natural language processing. MIT Press, Cambridge, MA
6. Leaman R, Gonzalez G (2008) BANNER: an executable survey of advances in biomedical named entity recognition. Pac Symp Biocomput 2008:652–63
7. Sætre R, Kenji S, Tsujii J (2008) Syntactic features for protein-protein interaction extraction. In: Short paper proceedings of the 2nd international symposium on languages in biology and medicine (LBM 2007). ISSN 1613-0073319. Singapore, pp 6.1–6.14, CEUR workshop proceedings (CEUR-WS.org)
8. Hara T, Miyao Y, Tsujii J (2007) Evaluating impact of re-training a lexical disambiguation model on domain adaptation of an HPSG parser. In: Proceedings of IWPT 2007 Prague, Czech Republic
9. Moschitti A (2004) A study on convolution kernels for shallow semantic parsing. In: Proceedings of the 42nd conference on association for computational linguistic (ACL-2004), Barcelona, Spain
10. Clegg AB, Shepherd AJ (2008) Text mining. In: Keith JM (ed) Bioinformatics volume II: structure, function and applications, vol 453, Methods in molecular biology. Humana Press, New York, pp 471–491

Chapter 9

Scientific Collaboration Networks Using Biomedical Text

Siddhartha R. Jonnalagadda, Philip S. Topham, Edward J. Silverman, and Ryan G. Peeler

Abstract

The combination of scientific knowledge and experience is the key success for biomedical research. This chapter demonstrates some of the strategies used to help in identifying key opinion leaders with the expertise you need, thus enabling an effort to increase collaborative biomedical research.

Key words Natural language processing, Biomedical literature, Health news, Social network analysis (SNA), Collaboration networks

1 Introduction

Biomedical research is a team sport. The many terms—collaboration, competition, cooperation, partnership, co-authorship, symposia, department, school, or simply fellows—highlight that we humans are social creatures. We organize and cooperate together to accomplish more than our own two hands can do. However, with millions of researchers, spread all around the globe working across different disciplines, finding and connecting to the right people can be challenging and often left to a chance meeting. Beyond the research itself, translating new discoveries from bench top to market is also a team sport requiring the right mix of scientists, policy makers, and businessmen.

Balas and Boren [1] estimated that translating biomedical discoveries into practical treatments takes around 17 years, and 86 % of research knowledge is lost during this transition through peer-review process, bibliographic indexing, and meta-analysis. At the commercial end, pharmaceutical companies on average spend 24 % of their total marketing budgets on opinion leader activities [2].

By effectively connecting those who produce knowledge with those who apply it, we can get discoveries to market quicker, produce better products, and ultimately improve quality of life. An important

Vinod D. Kumar and Hannah Jane Tipney (eds.), *Biomedical Literature Mining*, Methods in Molecular Biology, vol. 1159, DOI 10.1007/978-1-4939-0709-0_9, © Springer Science+Business Media New York 2014

step in this direction is the large-scale discovery of subject experts and key opinion leaders involved in specific areas of research based on their mentions in literature and news articles.

Having a ready product is not sufficient to ensure adoption. Public health programs are often focused on changing behaviors and social norms [3]. It is increasingly common in the domain of medical informatics to use social network analysis (SNA) [4, 5] to study interaction patterns of scientists in relation to a research area or a department. Although there are systems that assign topics of expertise to the identified persons [6, 7], there are no systems that identify the opinion leaders themselves.

Although in public health programs the focus is often on changing an individual's behavior, recent work by Christakis and Fowler [8] has shown that behaviors are contagious. They have shown that your propensity for obesity is impacted by your friends, your friends' friends, and—even up to three degrees of separation—your friends' friends' friends. In a more practical sense, people are influenced by people they see and by messages from people they do not directly see.

As social creatures, we intuitively know that we are impacted by the world we cannot see. By tapping into biomedical literature and other data sources we can now create strategies to engage key leaders across all stakeholders involved in scientific discovery and translational science.

This chapter covers techniques and strategies for using biomedical literature and related data sources, involving both structured and unstructured data, to discover the relationships inherent in groups.

2 Materials

While most of the methods being described are executable in any modern-day computer, some of the methods require access to a server or a cluster of servers (such as hadoop [9]). Please refer to our previous publications on these topics [10–12] for more details. These include two journal papers on SNA using text-mining methods and three white papers on importance and advantages of using social network methods.

2.1 Data Cleaning

This is the process of identifying which distinct names refer to the same person and which similar names refer to different persons. It is accomplished in two steps—using a program that disambiguates names and then a graphical mapping program equipped with fuzzy logic to find more loosely matched candidate names.

2.2 Mapping and Visual Displays

Visual mapping programs such as Gephi [13] or Cytoscape [14]. Gephi is an open-source program that refers to itself as a Graph Visualization Engine. Gephi can network up to 50,000 nodes (people) and 1,000,000 edges (links between nodes). There are features that enable clustering as well as SNA and numerous plug-ins that are available via marketplace.gephi.org. One plug-in, *Force Atlas 3D*, allows for a 3D layout. Another, *Noverlap*, can make the graph appear more cleanly. *Geolayout* can incorporate geomapped data. An extensive wiki that includes Manuals for Gephi can be found at http://wiki.gephi.org/index.php/Main_Page.

Cytoscape is also an open-source software package. Cytoscape can be altered to handle more nodes and edges but typically is limited to 70,000–150,000 combined nodes and edges (http://cytoscape.org/manual/Cytoscape2_6Manual.html) Like Gephi, many apps are available in order to customize and enhance the product for an individual's needs. The website, http://www.cytoscape.org, contains extensive tutorials. The software was originally designed to deal with molecular and genetic interaction datasets but can be used to map any relational data. Plug-ins like *jActiveModules* and *MCODE* can be used in clustering analysis. *Centiscape* can be used to calculate the centrality measures used in SNA. There are also apps like *Genemania or PSICQUICUniversal Client*, which import interaction information from public databases and would be of considerable use to a life scientist engaging in network analysis. A full list of Apps for Cytoscape can be found here: http://apps.cytoscape.org/.

2.3 Statistical Software

R suite of statistical software [15].

3 Methods

3.1 Named Entity Recognition

This is the task of extracting relevant named entities or concepts (such as persons, organizations, and locations) from unstructured or semi-structured text. We will describe how to extract the names of scientists and organizations from PubMED® and names of scientists and organizations from webpages. The organization names are used at the next stage for disambiguating the names of the scientists and to create social networks and key organizations in a medical topic too.

3.1.1 Extracting the Names of Scientists from PubMED®

1. Retrieve PubMed XML files through PubMed's web service [16]. PubMed supplies a complete description of each article in XML format.
2. Parse PubMed XML files into individual fields after they have been retrieved from PubMed.
3. Load PubMed XML files, and parse out a logical record for each author who contributed to the article.

(a) Articles where individual authors are not listed (group authorship), but investigators are listed, should use the investigators in place of authors.

(b) Articles without any author or investigators are not processed.

3.1.2 Extracting the Names of Organizations [11] from PubMED®

1. Use steps similar to the above to obtain affiliation strings.
2. Split the affiliation strings into phrases based on a regular expression that takes into account various punctuation marks. One such example is (?<!(\\.))[^\\,\\;\\:\\@]+((\\,|\\;|\\:|$)(?=(\\p{Z}|[A-Z]|$))|(\\.)(?=(\\p{Z}|$))).
3. Find all the abbreviations within the individual phrases using a regular expression such as (?<!(\\.))[^\\,\\;\\:\\@]+((\\,|\\;|\\:|$)(?=(\\p{Z}|[A-Z]|$))|(\\.)(?=(\\p{Z}|$))).
4. Expand the abbreviations into their full form using available dictionaries.
5. Detect which phrases correspond to the country using dictionaries for country name, region name, city name, languages, e-mail, country top-level domain, and zip codes.
6. Detect the e-mail addresses using a regular expression such as [a-zA-Z0-9!#$%&''*+,/=?^_`{|}~-])+(\\.[a-zA-Z0-9!#$%&''*+/=?^_`{|}~-]+)*@([A-Za-z0-9]([A-Za-z0-9-]*[A-Za-z0-9])?\\.)+[A-Za-z0-9]([A-Za-z0-9-]*[A-Za-z0-9])?.
7. Detect the URLs using a regular expression such as (?<=(\\s|^))(https?://|www\\.)[-\\w]+(\\.\\w[-\\w]*)+(?![\\w\\.-]*@).
8. In the remaining phrases, find which phrases correspond to street addresses using regular expressions that also take into account keywords for addresses, directions, etc.
9. Then determine the cities and states present using dictionaries for cities, states, and zip codes.
10. Now, the remaining phrases are likely to be organization named entities. Depending on the number of phrases remaining and presence of keywords for organizations, they are tagged as organizations.
11. If it is needed to disambiguate the organization names, techniques such as clustering and string similarity described in detail in our journal paper [11] are used.

3.1.3 Extracting the Names of Persons and Organization Names [10] from Webpages

1. Use a web crawler such as E-FFC [17] or UbiCrawler [18] to extract the webpages relevant to the topic.
2. If a rule-based system would be used, build lexicons for persons and organizations from resources such as those mentioned in the previous section.

3. If a pre-built machine-learning system is a preference, the Stanford NER tool is available at http://nlp.stanford.edu/software/CRF-NER.shtml. This is a well-supported tool and has state-of-the-art performance of close to 90 % for extracting person names, organizations, and locations.
4. If there is a need to train a machine-learning system for a particular corpus (set of documents) or a particular set of features (lexical, syntax, semantic, and pragmatic), one might use the conditional random field (CRF) algorithm—a state-of-the-art algorithm—for most named entity recognition tasks. Mallet [19] provides a widely used available implementation—http://mallet.cs.umass.edu/sequences.php.
5. Organization and person names not in the list found in the previous steps are filtered out, as they are less likely to represent scientific organizations and scientists.
6. A list of words such as Dr, MD, and PhD in the proximity of the person names are also used, optionally, to further increase the sensitivity of extracting the names of scientists.

3.2 Data Cleaning

Before analyzing any network, it is essential to make sure that the nodes in the network are represented accurately. Just as a scientist would carefully validate a protein–protein interaction by making sure that both proteins were correctly cloned, expressed, and active, a network analyst has to make sure that the people in the network are correctly identified.

1. Clinically relevant scientific information composed of grant abstracts, clinical trials, symposia information, and publications is harvested.
2. Every piece of scientific information harvested will have the names and often the organizational affiliation of the authors.
3. Data cleaning starts with taking the harvested information and determining whether two similar names refer to the same person or not. If there is a James Smith at Beth Israel Deaconess, it is necessary to determine, for example, if he is the same person as J. Smith from Harvard Medical School.
4. A software program similar to Authority [20] is used to assist with the process of pre-matching names that might be matches based on attributes like affiliation, coauthors in common, and name.
5. An analyst then makes the determination (in our example) if James Smith and J. Smith are actually the same person by looking at the scientific information itself as well as publicly available biographical information available on the Web.
6. After this process, we import the names and connections into graphical mapping that allows the use of fuzzy logic to find names that might be similar. This helps identify researchers

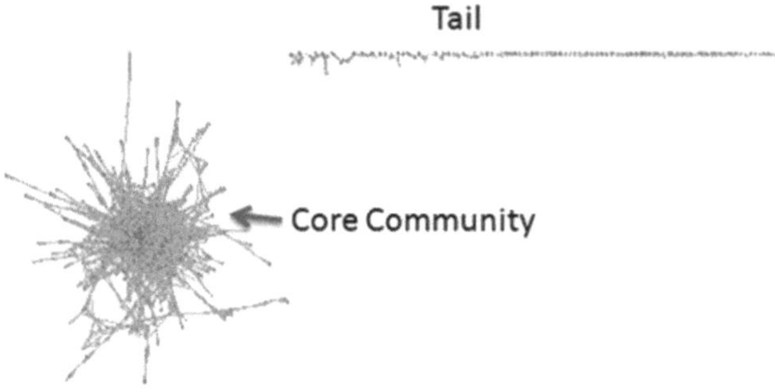

Fig. 1 Defining the community

who may have gotten married and changed their last name as well as individuals who have typographical errors or multiple forms of their name on scientific information. For example, it is then possible to determine if Maria Gomez is the same as Maria Gomez-Hidalgo.

3.3 Generating Meaningful Visual Displays

1. Once the data has been validated and cleaned, it is uploaded into a mapping and visual display program such as i2, Gephi, Cytoscape, and CartoDB.
2. The people and connections between the scientists are then drawn as a map in order to determine who is part of the core collaborative network (Core Community) and who is not (Tail) (Fig. 1).

3.4 SNA Algorithms

The percent of the people in the community helps to determine the success of the SNA; any network with 30 % or more of the network in the core yields actionable results. Once the core community has been identified, SNA algorithms using select centrality measures are used in order to identify the KOLs within a therapeutic area.

Centrality measures:

1. *Degree* centrality, also called *first-degree centrality*, counts the number of nodes with a single intervening connection and describes the number of connections a node has in a network. For example, if a researcher collaborates with 20 other researchers in three different instances (publications, grants, etc.) that researcher's degree is 20.
2. *Closeness* centrality is the measure of the total distance of one researcher from all other researchers in a network. The smaller that distance, the larger the closeness score. The closeness value is a good predictor of the speed of transmitting information or influence through a network. High closeness value for a particular researcher means that the network will learn about that researcher's work sooner than a researcher with a lower closeness score.

3. *Betweenness* centrality quantifies the number of times a researcher is the shortest path between two other researchers. A researcher with a high betweenness score is able to exert significant control over communications between researchers in a network.
4. *Eigenvector* centrality measures the connections a researcher has to high-scoring researchers using a relative degree connection scoring system. This is similar to the algorithm search engines used to rank webpages. This is commonly referred to as a researcher's prestigious connection score.
5. *Katz* centrality provides a combination of degree centrality and betweenness centrality by accounting for attenuation of influence of researchers whose closeness scores are relatively lower. Researchers who collaborate with others who have low closeness scores do not influence the research network as dramatically as groups of researchers due to a low Katz centrality.
6. *Alpha* centrality closely resembles eigenvector centrality with an important exception. Attributes about a researcher that are external to the network are applied to the analysis. For example, if a researcher was the chairperson of a prestigious professional society, or if the researcher was a well-known public speaker in the popular media, and only occasionally published in peer-reviewed journals, alpha centrality measures would affect the researcher's importance in the network.

3.5 Reach, Reach Curves, and Inflection Points

Understanding how all of the researchers in a community are connected allows an analyst to quantify reach or maximum spread of an idea in a first-degree network. This is generally plotted on a curve with the inflection point on the power law distribution curve giving a generalized idea of the diminishing margin of returns for percent reach into a community by including additional researchers.

1. *Identify the first-degree connections (FDCs) of the KOLs*—The database will have relational data. An analyst can use this to determine the quantity and identity of all the people to whom the KOLs are connected.
2. *Calculate the reach*—The amount of people and scientific information the KOLs and the FDCs influence can then be calculated. As shown in Fig. 2, 7 % of the people reach 79 % of the core community and access and impact 97 % of all scientific information.
3. *Generate a reach curve*—The amount of KOLs can be adjusted up or down, according to the needs of a client, and the community and scientific research reach then calculated to generate a curve (*see* Fig. 3). From this analysis an inflection point, the point at which the derivative of the slope of the curve changes from positive to negative, can be calculated. This inflection point determines the number of people in the

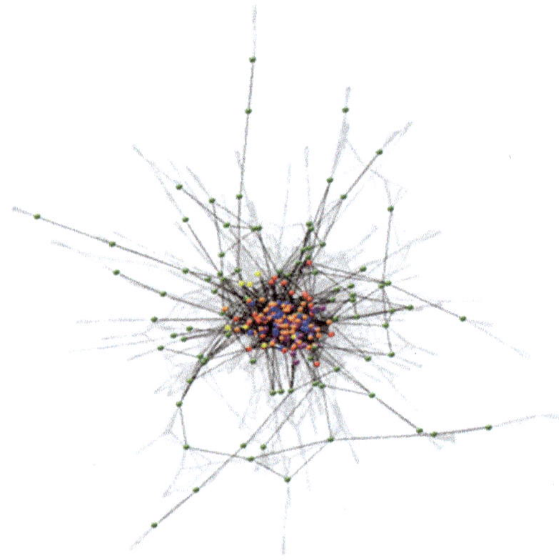

Fig. 2 Identifying the KOLs—The *colored circles* in figure represent KOLs identified for this therapeutic area. The connections between the KOLs have also been *darkened*. After the algorithms identify the KOLs in the community, we can then measure the impact of those important individuals, reach

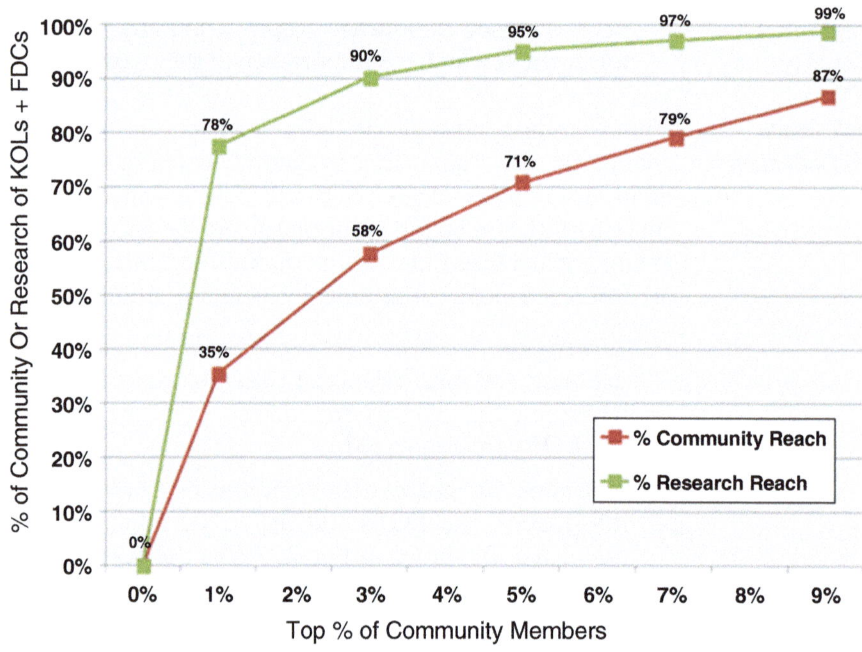

Fig. 3 Reach curve. The *X*-axis shows the top *X*% of community members (KOLs) referenced against the *Y*-axis, which represents the % of community members reached by the KOLs (*red*) or the amount of research reached by the KOLs and their first-degree connections (*green*)

network who reach the most people non-redundantly and how few people one needs to identify to understand who produces the bulk of the scientific information.

3.6 Clustering

Beyond being able to determine what person in a particular network is important, it is also desirable to know which groups are important.

1. *Cluster identification*—Determining which groups are most influential is typically done through connectivity-based clustering or hierarchy clustering [21]. In research networks, clustering manifests itself in groups of people who serially collaborate. The most common clustering algorithms use distance measures, which receive weights based on linkage criteria specified by the analyst.
2. *Label the cluster*—The reasons for serial collaborations are varied, and we have also observed that geographic proximity, committee appointments, grants, and highly specific interests all influence preferential serial collaboration.
3. *Group the clusters by theme*—An analyst will then group the clusters by theme [22]. Themes can include genetics, clinical trials, drug treatments, and treatment guidelines. *See* Fig. 4. Below is an example of a clustered KOL map identified by theme.

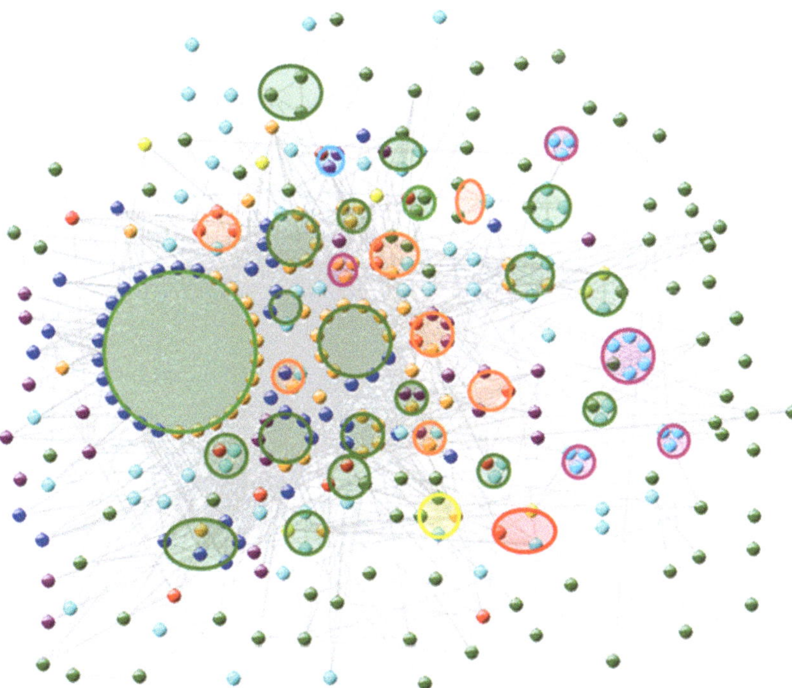

Fig. 4 Clustered map of KOLs identified by theme (*green*—clinical treatment guidelines, *pink*—drug safety, *orange*—genetics, *red*—pathogenesis, *blue*—drug safety)

4 Conclusions

1. *Not all high publishers are important*: Social network analytics serves as a tool for identifying important researchers. Researchers who publish frequently ("high publishers") are generally considered to have greater expertise. "Low publishers," researchers who publish less frequently, are considered to be new or less well-known individuals. How can a researcher who is a "low publisher" be a highly respected expert? There are many reasons and situations that could create such a pattern, such as the following: (1) Rising stars: individuals early in their careers who have published very impactful papers but have not yet generated a volume of work; (2) individuals on sabbatical or semi-retired who have established respect through previous works but are now publishing at a lower volume; (3) individuals who join or leave the industry and thus have a hiatus in their academic research frequency; or (4) highly respected KOLs with prestige in another—typically related—field, yet few publications in the field being studied.
2. *Brokers between themes and groups*: Social network analytics also values researchers who are brokers of information. Researchers who form bridges between communities are commonly high-value and high-impact individuals according to network scores regardless of the frequency of their publications.
3. *Community density*: Depending on community size, density, and other properties the inflection point on a reach curve can be highly movable. Anecdotally, the breast cancer research community is mature and the graph of the largest component of the community is over 65 % of the researchers in the community. Contrast that with researchers in women's health who study a greater variety of topics (bone density, early-onset menopause, delayed-onset menarche, etc.) in which the largest network component is only 30 % of the researchers in a community. *See* Fig. 5.

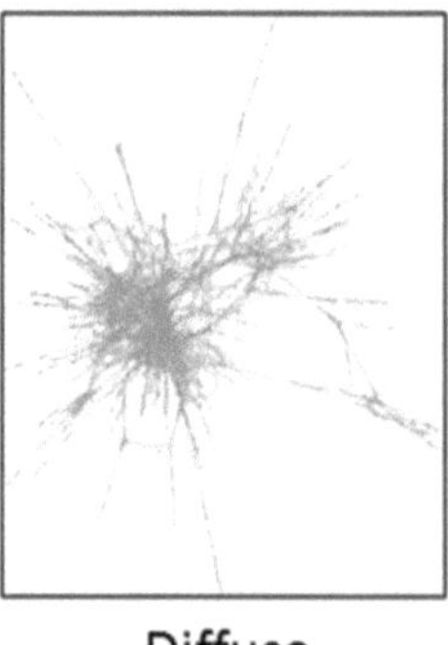

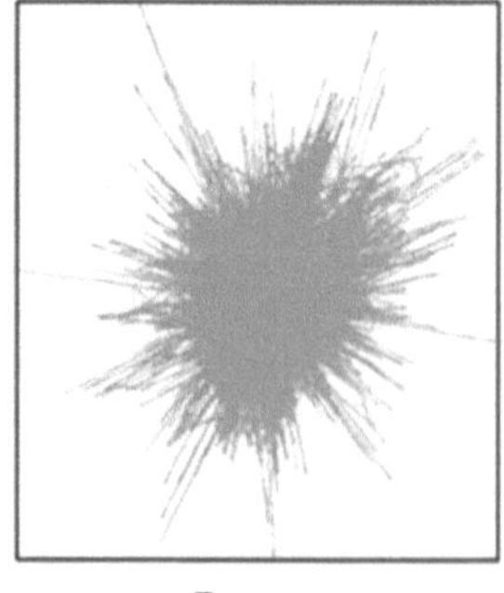

Fig. 5 Diffuse versus dense communities

Understanding the density of the community and the structures influences the inflection point and the types of network centrality measures one would like to use and weight for analysis [23].

References

1. Balas EA, Boren SA (2000) Managing clinical knowledge for health care improvement. Yearb Med Inform 65–70
2. Nair HS, Manchanda P, Bhatia T (2010) Asymmetric social interactions in physician prescription behavior: the role of opinion leaders. J Market Res 47(5):883–895
3. Valente TW, Pumpuang P (2006) Identifying opinion leaders to promote behavior change. Health Educ Behav 34(6):881–896
4. Bales M, Johnson SB, Weng C (2008) Social network analysis of interdisciplinarity in obesity research. AMIA Annu Symp Proc 870
5. Merrill J, Hripcsak G (2008) Using social network analysis within a department of biomedical informatics to induce a discussion of academic communities of practice. J Am Med Informat Assoc 15(6):780–782
6. Bordea G (2010) Concept extraction applied to the task of expert finding. In: Aroyo L, Antoniou G, Hyvönen E, Teije A, Stuckenschmidt H, Cabral L, Tudorache T (eds) The semantic web: research and applications, vol 6089. Springer, Berlin, pp 451–456
7. Buitelaar P, Eigner T (2009) Expertise mining from scientific literature In: Proceedings of the fifth international conference on knowledge capture. ACM, New York, NY, USA, pp 171–172
8. Christakis NA, Fowler JH (2007) The spread of obesity in a large social network over 32 years. [Research Support, N.I.H., Extramural]. N Engl J Med 357(4):370–379. doi:10.1056/NEJMsa066082
9. Borthakur D (2007) The hadoop distributed file system: architecture and design. http://hadoop.apache.org
10. Jonnalagadda S, Peeler R, Topham P (2012) Discovering opinion leaders for medical topics using news articles. J Biomed Semantics 3(1):2. doi:10.1186/2041-1480-3-2
11. Jonnalagadda SR, Topham P (2010) NEMO: extraction and normalization of organization names from PubMed affiliations. J Biomed Discov Collab 5:50–75
12. Lnx-Pharma (2007) Using large scale social network analysis—a comparative analysis for finding key opinion leaders. http://www.lnxpharma.com/Whitepapers/Lnx_Whitepaper_2.pdf
13. Bastian M, Heymann S, Jacomy M (2009) Gephi: an open source software for exploring and manipulating networks. Paper presented at the international AAAI conference on weblogs and social media
14. Shannon P, Markiel A, Ozier O, Baliga NS, Wang JT, Ramage D, Amin N, Schwikowski B, Ideker T (2003) Cytoscape: a software environment for integrated models of biomolecular interaction networks. Genome Res 13(11):2498–2504
15. Handcock MS, Hunter DR, Butts CT, Goodreau SM, Morris M (2008) Statnet: software tools for the representation, visualization, analysis and simulation of network data. J Stat Softw 24(1):1548
16. NLM (2013) PubMed interface. http://www.ncbi.nlm.nih.gov/pubmed
17. Li YN, Wang YP, Du JT (2013) E-FFC: an enhanced form-focused crawler for domain-specific deep web databases. J Intell Inf Syst 40(1):159–184. doi:10.1007/s10844-012-0221-8
18. Boldi P, Codenotti B, Santini M, Vigna S (2004) UbiCrawler: a scalable fully distributed Web crawler. Software Pract Ex 34(8):711–726. doi:10.1002/Spe.587
19. McCallum AK (2002) MALLET: a machine learning for language toolkit. http://mallet.cs.umass.edu
20. Torvik VI, Smalheiser NR (2009) Author name disambiguation in MEDLINE. ACM Trans Knowl Discov Data 3(3)
21. Fraley C, Raftery AE (2002) Model-based clustering, discriminant analysis, and density estimation. J Am Stat Assoc 97(458):611–631
22. Kriegel H-P, Kröger P, Zimek A (2009) Clustering high-dimensional data: a survey on subspace clustering, pattern-based clustering, and correlation clustering. ACM Trans Knowl Discov Data 3(1):1
23. Hanneman RA, Riddle M (eds) (2005) Introduction to social network methods. University of California, Riverside, Riverside, CA

Chapter 10

Predicting Future Discoveries from Current Scientific Literature

Ingrid Petrič and Bojan Cestnik

Abstract

Knowledge discovery in biomedicine is a time-consuming process starting from the basic research, through preclinical testing, towards possible clinical applications. Crossing of conceptual boundaries is often needed for groundbreaking biomedical research that generates highly inventive discoveries. We demonstrate the ability of a creative literature mining method to advance valuable new discoveries based on rare ideas from existing literature. When emerging ideas from scientific literature are put together as fragments of knowledge in a systematic way, they may lead to original, sometimes surprising, research findings. If enough scientific evidence is already published for the association of such findings, they can be considered as scientific hypotheses. In this chapter, we describe a method for the computer-aided generation of such hypotheses based on the existing scientific literature. Our literature-based discovery of NF-kappaB with its possible connections to autism was recently approved by scientific community, which confirms the ability of our literature mining methodology to accelerate future discoveries based on rare ideas from existing literature.

Key words Literature mining, Creative computing, Knowledge discovery, Rare terms, Outliers

1 Introduction

In research, as in many other fields, we are constantly confronted with the problem of an overwhelming amount of information to process and understand. Textual information contained in numerous professional articles, many of them available online through large multidisciplinary bibliographic databases, is of immense help, but has, on the other hand, made manual literature-based knowledge discovery a very time-consuming and arduous task. A famous example is Medline, the bibliographic database provided by the US National Library of Medicine which, at the beginning of 2013, contained over 20 million citations from 1949 to the present and which increases for more than 2,000 complete references to biomedical and life sciences journal articles daily [1]. In these circumstances, efficient software tools are highly desirable to support researchers in their literature-based discovery.

Vinod D. Kumar and Hannah Jane Tipney (eds.), *Biomedical Literature Mining*, Methods in Molecular Biology, vol. 1159, DOI 10.1007/978-1-4939-0709-0_10, © Springer Science+Business Media New York 2014

Excessive specialization of scientists and other professionals in their respective fields sometimes renders discovery processes even harder. One manifestation of this aspect can be traced in publications and cross-references that in many cases turn out to be closed within specific professional communities.

The process of scientific discovery usually involves different types of cognitive abilities, such as finding analogies or selecting and interrelating ideas that originate from different contexts of various disciplines. If such separated ideas or fragments of knowledge are combined together from different sources in a systematic way, they may lead to previously unknown and original, sometimes surprising, findings that provide the guidance for future research and exploration. When there is considerable evidence to reasonably support such findings, they can be regarded as candidate hypotheses. If these hypotheses can be proved as valid by the methods approved by scientific community, they represent new discoveries. Such context-crossing connections are called bisociations [2] and are often needed for creative cutting-edge scientific discoveries. Such discoveries are of particular importance in complex interdisciplinary research settings.

However, with the conventional associative approach it is difficult to identify and connect the information that is related across different contexts [3]. Specialized techniques supported by software tools are required to help researchers in crossing these boundaries, helping them in putting dispersed pieces of knowledge together into a meaningful, coherent whole. In 1986 it was demonstrated by Swanson that bibliographic databases such as Medline can serve to create new scientific discoveries [4]. Swanson suggested a simple but effective method for generating hypotheses which span over previously disjoint sets of literature. To examine whether two distinct areas of study *a* and *c* could be associated in spite of the fact that no direct evidence in favor of this suggestion has yet been presented in literature, he proposed bringing these two literatures together and searching for intermediate terms *b* connected with *a* in some articles and with *c* in some others. Bringing these terms together and exploring their meaning sometimes reveal interesting connections between *a* and *c* that are worthy of further investigation. For example, if a literature *A*, i.e., a set of records about *a* in the database serving as a source of data (*see* **Note 1**), associates a term *a* with a term *b*, and another literature *C* reports about a term *c* being in association with the term *b*, we may suggest a novel *A*–*C* relationship by connecting these sets of literature according to the simple *ABC* model (Fig. 1).

Recently, various literature-based discovery approaches have been successfully applied to the hypothesis generation task in the so-called open discovery process [5], e.g., in the detection of disease–drug interactions [6], as well as for novel associations between genes, pathways, and diseases [7]. Among them a literature-based discovery

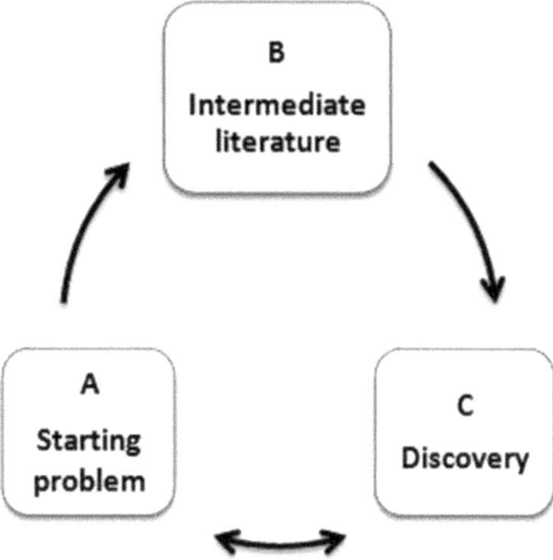

Fig. 1 Swanson's ABC model of knowledge discovery

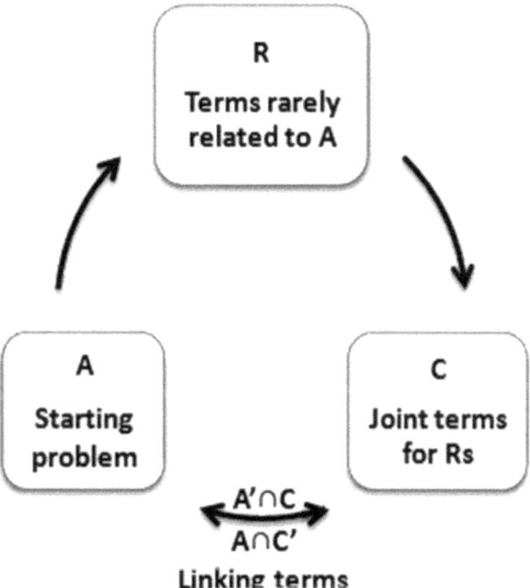

Fig. 2 RaJoLink model of knowledge discovery

method RaJoLink that was introduced already in 2007 [8] and described in more detail in Petrič et al. [9] was successfully applied to the autism domain [10], and its findings were subsequently supported by a clinical research [11]. The RaJoLink method, illustrated in Fig. 2, uses rarity of terms as a means of open knowledge discovery (*see* **Note 2**). It is named after the key elements of each step: rare terms, joint terms, and linking terms, respectively. Accordingly, we call the steps Ra, Jo, and Link.

This chapter describes how this particular method can be employed to investigate the abstracts (or titles) of scientific articles available in Medline in order to guide future discoveries.

In Subheading 2 we give an overview of materials used for literature-based knowledge discovery. In Subheading 3 we present the specifics of the RaJoLink approach to the literature-based knowledge discovery. In Subheading 4 of this chapter, we address some conceptual and methodological issues that need to be taken into account.

2 Materials

2.1 Scientific Literature Collection

1. A Medline literature search with specific search terms is performed at each of the RaJoLink steps (*see* **Note 3**).
2. Records in XML format are retrieved from the Medline database.
3. Abstracts of Medline citations are collected in a text file on the basis of search criteria (*see* **Note 4**).
4. Terms are extracted from the text collection and used as described in Subheading 3.

2.2 Medical Subject Headings

1. Medical Subject Headings (MeSH) descriptors [12] are used for classification of terms identified in text collections.
2. Text analysis results can be filtered according to the categories selected from the MeSH tree structure (*see* **Note 5**).

3 Methods

3.1 Open Knowledge Discovery

If we are investigating a phenomenon a, the open discovery starts with having only the term a and the corresponding set of articles in which the term a appears (i.e., a set of literature A), without knowing the target candidate c, which is discovered later as a result of such process. RaJoLink method involves two steps in the open discovery process for hypotheses generation: step Ra and step Jo, respectively.

3.1.1 Step Ra

1. Search Medline for citations in which the starting term a appears (*see* **Note 6**).
2. Retrieve records in XML format from the Medline database.
3. Collect abstracts of Medline citations in a single text file.
4. Extract terms from the text collection using lemmatization to eliminate various forms of a single term (*see* **Note 7**).
5. Exclude terms that are predictably of no interest by using a stoplist (*see* **Note 8**).
6. For each term r_i identified in the set of records (A), calculate $n(r_i)$ using frequency statistics (*see* **Note 9**).

7. Compare each term r_i identified in the set of records (A) with MeSH terms to classify them according to the MeSH hierarchy.
8. Sort terms by frequency in order to focus on rare terms, i.e., the terms rarely related to the starting problem a (*see* **Note 10**).
9. Filter terms by choosing MeSH categories of interest for identification of potentially interesting rare terms $r_1, r_2, \ldots, r_i$ (*see* **Note 11**).
10. Select at least two rare terms as intermediate results towards the discovery of the target candidate for c (*see* **Note 12**).

3.1.2 Step Jo

Search Medline for sets of citations in which each of the terms r_i selected in the Jo step appears (*see* **Note 13**).

1. Retrieve records in XML format from the Medline database.
2. Collect abstracts of Medline citations in a single text file for each of the r_i terms.
3. Extract terms from all text collections using lemmatization to eliminate various forms of a single term.
4. Exclude terms that are predictably of no interest by using a stoplist.
5. For each term c_j identified in a set of records (R_i), calculate $n(c_j)$ using frequency statistics.
6. A term c_j qualifies as a joint term if it appears in at least two sets of records (R_i) about the rare terms. At the same time, it has to be absent from the set of records about the term a, generated in step Ra.
7. Compare each term c_j identified in the intersections of the sets of records (R_i) with MeSH terms to classify them according to the MeSH hierarchy.
8. Sort terms by frequency in order to focus on the most frequent terms, i.e., the terms that are strongly related to the majority of the rare terms (*see* **Note 14**).
9. Filter terms by choosing MeSH categories of interest for identification of potentially interesting joint terms $c_1, c_2, \ldots, c_j$ (*see* **Note 15**).
10. Select one or more targets (joint terms $c_1, c_2, \ldots, c_j$) for formulation of scientific hypotheses.

3.2 Closed Knowledge Discovery

The search for the linking terms b in the closed discovery process is equivalent to Swanson's hypothesis testing in the closed discovery approach [13] that consists of looking for terms b that can be found in the literature A as well as in the literature C. However, the recent RaJoLink closed discovery approach [14] contains a unique aspect in comparison to the literature-based discovery investigated by others. It is the visual analysis of neighboring texts in the documents' similarity graphs which are defined over the combined datasets consisting of literatures *A* and *C*. Such analysis can focus on linking terms b_1, b_2, …, b_k in order to find pairs of articles, one from literature *A* and the other from literature *C*, both mentioning b_k. Such previously unseen relations between the domains *A* and *C* provide new knowledge. The RaJoLink's closed knowledge discovery approach [14] gives priority to terms *b* from two disparate literature sources *A* and *C* based on exploring outlier articles of the two domains as follows (*see* **Note 16**).

3.2.1 Step Link

1. Search Medline for citations in which the starting term *a* appears (*see* **Note 17**).
2. Retrieve records in XML format from the Medline database.
3. Collect titles or abstracts of Medline citations in a single text file.
4. Search Medline for citations in which the candidate term *c* appears (the term selected as a result in the step Jo) (*see* **Note 18**).
5. Retrieve records about the selected term *c* in XML format from the Medline database.
6. Collect titles or abstracts of Medline citations in a single text file.
7. Join the records collected in **steps 3** and **6** of Subheading 3.2.1 in a single text file, i.e., a joint document set *AC* (*see* **Note 19**).
8. Automatically generate two document clusters, A' and C', according to the documents' similarity measure where $A' \cup C' = AC$ (*see* **Note 20**).
9. Based on the original domains of interest *A* and *C*, each cluster (i.e., the cluster A' and the cluster C') is automatically further divided into two document sub-clusters (*see* **Note 21**). Cluster A' is divided into sub-clusters $A' \cap A$ and $A' \cap C$, while cluster C' is divided into $C' \cap A$ and $C' \cap C$.
10. Sub-clusters $A' \cap C$ (outliers of *C*, consisting of documents of domain *C* only) and $C' \cap A$ (outliers of *A*, consisting of documents of domain *A* only) are the two document sets where we find linking terms b_1, b_2, …, b_k between literature *A* and literature *C*, which can prove the hypothesis of connecting *a* with *c*.
11. The domain expert needs to be involved in exploring the potential *b* terms and confirming them as relevant only in the sub-cluster of outlier documents and he/she does not need to explore all the documents.

4 Notes

1. We would like to emphasize that the notations *A*, *B*, and *C* (uppercase symbols) are used to represent a set of terms (e.g., literature, a set of records, or a list of terms), while the notations *a*, *b*, and *c* (lowercase symbols) represent a single term.
2. Open discovery is the primary model of discovery. Open discovery process proceeds from the topic of a scientific problem *a* towards unknown concept *c* for generation of new hypothesis. Closed discovery process, on the other hand, involves finding novel intermediate concept *b* that could verify the initial hypothesis, where both the starting problem *a* and the target concept *c* are known at the onset of a research.
3. Swanson's model sees scientific articles as units of knowledge or sets of specialized literatures, where common matters are considered within each set [13].
4. The <AbstractText> field is used, which consists of English-language abstract that is taken rigorously from a published article. The present maximum length of abstracts is 10,000 characters for a record.
5. The MeSH terminology is organized into the hierarchical structure. The broader medical headings are situated at the most general levels of the hierarchy, while the more specific medical headings are found at more narrow levels. The top-level categories in the MeSH hierarchy include the following: Anatomy—A, Organisms—B, Diseases—C, Chemicals and Drugs—D, Analytical, Diagnostic and Therapeutic Techniques and Equipment—E, Psychiatry and Psychology—F, Biological Sciences—G, Natural Sciences—H, Anthropology, Education, Sociology, and Social Phenomena—I, Technology, Industry, Agriculture—J, Humanities—K, Information Science—L, Named Groups—M, Health Care—N, Publication Characteristics—V, and Geographicals—Z. We use the second-level categories from the 2008 MeSH tree structure (e.g., Behavior and Behavior Mechanisms—F01, Psychological Phenomena and Processes—F02, Mental Disorders—F03, Behavioral Disciplines and Activities—F04) to classify terms from the input text collection.
6. For the automatic access to the Medline data, which has to be performed outside of the regular web query interface, called Entrez, we use the ESearch tool of the Entrez Programming Utilities [15]. In particular, we operated with the following utility parameters:
   (a) Database name (DbName), where the PubMed database is the default value of the parameter DbName.
   (b) Date ranges to limit query results bounded by two specific dates, namely, the mindate and the maxdate parameters.

(c) Date type that limits dates to a specific date field in a database. Actually, we use the edat type of dates, which limits query results according to the date when a citation was added to PubMed.

The ESearch tool responds to a text query and returns data corresponding to the results of the query submitted to the Entrez system. As the results from ESearch are maintained in the user's environment, the maximum number of retrieved records (URL parameter retmax) is 10,000 records for a search. If there is a need to retrieve more than 10,000 records from Medline we suggest running a combination of queries with different date ranges.

7. We employ the lemmatizing software from the Lemma-Gen library developed by Juršič and colleagues [16].
8. We use a list of 571 English stop words (i.e., generic words such as *a*, *able*, *about*, *above*, and *according*).
9. We use frequency statistics based on bag of words (BoW) text representation [17], wherefore we employ the Txt2Bow utility from the TextGarden library [18].
10. By focusing on rare terms we want to identify candidates that are most likely to lead towards meaningful still unpublished relations with phenomenon a.
11. Note that i can be regarded as a parameter of the method. When interesting rare terms cannot be found by using this minimum frequency of term's occurrence, it is necessary to return back to the beginning of this step and repeat the process by taking into account larger number of input documents or by choosing higher value of the parameter $n(r_i)$.
12. The aim of the subsequent step Jo is to identify joint terms c_1, c_2, …, c_j at the intersections of the literatures on rare terms; therefore, at least two rare terms are needed as intermediate results towards the discovery of the target candidate for c.
13. For automatic access to the Medline data proceed as in step Ra but repeat the procedure for each of the selected r terms.
14. The higher count and sum of total frequencies for a joint term c_j (i.e., the higher number of occurrences of a joint term in the sets of records about the selected rare terms) means the better matching of a term as the striking joint term c_j. The obtained list of candidate joint terms c_j with their frequencies is arbitrary, since each term can be taken as relevant for the generation of new hypothesis only if it does not appear within the set of articles on the starting term (term a).
15. If no significant joint terms are obtained via the rare terms selected in the step Ra, it is necessary to return back to the results of the previous step and broaden or change the actual selection of rare terms and repeat the process.

16. The method assumes that by exploring outlier documents it is easier to discover linking *b* terms that establish previously unknown links between literature *A* and literature *C*.
17. This is the same search as described in the first three stages of the step Ra with the only difference that we may focus on titles instead of abstracts of citations in the third stage.
18. One or more selected joint terms c_j can be considered in the link step (i.e., the hypothesis testing) in subsequent iterations for detection of implicit links with the problem domain denoted by term *a*.
19. Each line of the joint document set *AC* is interpreted as a document with the first word in the line being its title.
20. OntoGen's [19] two-means clustering algorithm is used to automatically generate two document clusters, A′ and C′.
21. Again, the OntoGen tool [19] is used.

Acknowledgement

This work was performed within the Creative Core project (AHA-MOMENT), partially supported by the Ministry of Education, Science and Sport, Republic of Slovenia, and European Regional Development Fund.

References

1. US National Library of Medicine. Fact sheet. Medline, PubMed, and PMC (PubMed Central): how are they different? http://www.nlm.nih.gov/pubs/factsheets/dif_med_pub.html. Accessed 5 Aug 2013
2. Koestler A (1964) The act of creation. Macmillan, New York, p 751
3. Feldman R, Sanger J (2007) The text mining handbook: advanced approaches in analyzing un-structured data. Cambridge University Press, Cambridge, p 410
4. Swanson DR (1986) Fish oil, Raynaud's syndrome, and undiscovered public knowledge. Perspect Biol Med 30(1):7–18
5. Weeber M, Vos R, Klein H, de Jong-van den Berg LTW (2001) Using concepts in literature-based discovery: simulating Swanson's Raynaud–fish oil and migraine–magnesium discoveries. J Am Soc Inf Sci Tech 52(7): 548–557
6. Hristovski D, Rindflesch T, Peterlin B (2013) Using literature-based discovery to identify novel therapeutic approaches. Cardiovasc Hematol Agents Med Chem 11(1):14–24
7. Frijters R, van Vugt M, Smeets R, van Schaik R, de Vlieg J, Alkema W (2010) Literature mining for the discovery of hidden connections between drugs, genes and diseases. PLoS Comput Biol 6(9):e1000943
8. Petrič I, Urbančič T, Cestnik B (2007) Discovering hidden knowledge from biomedical literature. Informatica 31(1):15–20
9. Petrič I, Urbančič T, Cestnik B, Macedoni-Lukšič M (2009) Literature mining method RaJoLink for uncovering relations between biomedical concepts. J Biomed Inform 42(2): 219–227
10. Urbančič T, Petrič I, Cestnik B, Macedoni-Lukšič M (2007) Literature mining: towards better understanding of autism. In: Bellazzi R, Abu-Hanna A, Hunter J (eds) Proceedings of the 11th conference on artificial intelligence in medicine in Europe, AIME 2007, July 7–11, Amsterdam, The Netherlands, pp 217–226
11. Naik US, Gangadharan C, Abbagani K, Nagalla B, Dasari N, Manna SK (2011) A study of nuclear transcription factor-kappa B in childhood autism. PLoS One 6(5):e19488

12. Nelson SJ, Johnston D, Humphreys BL (2001) Relationships in medical subject headings. In: Bean CA, Green R (eds) Relationships in the organization of knowledge. Kluwer Academic Publishers, New York, pp 171–184
13. Swanson DR (1990) Medical literature as a potential source of new knowledge. Bull Med Libr Assoc 78(1):29–37
14. Petrič I, Cestnik B, Lavrač N, Urbančič T (2012) Outlier detection in cross-context link discovery for creative literature mining. Comput J 55(1):47–61
15. Sayers E, Wheeler D (2004) Building customized data pipelines using the entrez programming utilities (eUtils) In: U.S. National Library of Medicine. NCBI short courses
16. Juršič M, Mozetič I, Lavrač N (2007) Learning ripple down rules for efficient lemmatization. In: Bohanec M et al (eds) Proc. 10th intl. multiconference information society IS 2007, vol A. pp 206–209
17. Sebastiani F (2002) Machine learning in automated text categorization. ACM Comput Surv 34(1):1–47
18. Grobelnik M, Mladenić D (2004) Extracting human expertise from existing ontologies. In: EU-IST Project IST-2003-506826 SEKT
19. Fortuna B, Grobelnik M, Mladenić D (2006) Semi-automatic data-driven ontology construction system. In: Bohanec M et al (eds) Proc. 9th intl. multiconference information society IS 2006, vol A. pp 223–226

Part III

From Electronic Biology to Drug Discovery and Development

Chapter 11

Mining Emerging Biomedical Literature for Understanding Disease Associations in Drug Discovery

Deepak K. Rajpal, Xiaoyan A. Qu, Johannes M. Freudenberg, and Vinod D. Kumar

Abstract

Systematically evaluating the exponentially growing body of scientific literature has become a critical task that every drug discovery organization must engage in in order to understand emerging trends for scientific investment and strategy development. Developing trends analysis uses the number of publications within a 3-year window to determine concepts derived from well-established disease and gene ontologies to aid in recognizing and predicting emerging areas of scientific discoveries relevant to that space. In this chapter, we describe such a method and use obesity and psoriasis as use-case examples by analyzing the frequency of disease-related MeSH terms in PubMed abstracts over time. We share how our system can be used to predict emerging trends at a relatively early stage and we analyze the literature-identified genes for genetic associations, druggability, and biological pathways to explore any potential biological connections between the two diseases that could be utilized for drug discovery.

Key words Biomedical literature, PubMed, MEDLINE, MeSH, Genes, Diseases, Pathways, Drug discovery, Trends

1 Introduction

In a high-pace R & D environment, a comprehensive understanding of emerging scientific trends and capturing the knowledge from emerging human disease biology is critical to sense new scientific opportunities and potential novel implications to enable employment of available resources for effective translational research. Previously, it has been suggested that the knowledge acquired through studying these trends could potentially help develop long-term investment strategies as well as help in identifying near-term opportunities that could benefit the drug discovery process [1, 2]. Scientific literature is a great source of studying such scientific trends, and creating that knowledge space. Literature search and evaluation is thus one of the basic and necessary preoccupations for scientists in the research arena. However, the growth of scientific

Vinod D. Kumar and Hannah Jane Tipney (eds.), *Biomedical Literature Mining*, Methods in Molecular Biology, vol. 1159, DOI 10.1007/978-1-4939-0709-0_11, © Springer Science+Business Media New York 2014

literature in biomedical and life sciences necessitated the development of computational methodologies to evaluate the exponentially growing contributions from the scientific community. In response, numerous methodologies in text mining and knowledge based on automated information retrieval, extraction, and inference have been developed. In addition, various approaches to "bibliometrics" in surveying literatures are in general use across drug discovery research efforts [1, 2].

Systematically evaluating and employing biomedical literature for various tasks relevant to life sciences and drug discovery range from information gathering on molecular entities (e.g., genes, proteins), to discovery of gene–disease, gene–chemical, protein–protein, and disease–drug associations, and more extensively, to efforts to construct putative biological networks based on co-occurrence of entities within scientific articles [1–8]. New implications discovered or predictions made from using several literature mining methodologies and information retrieval have provided new hypotheses for assessment and some of these new concepts are subsequently validated experimentally [9, 10].

Methods of information retrieval from published literature rely on employing ontologies that represent essential and critical concepts organized in meaningful hierarchies to capture the core domain knowledge in scientific space. PubMed (http://www.ncbi.nlm.nih.gov/pubmed) contains more than 22 million citations from the MEDLINE database at the National Library of Medicine. The literature in the MEDLINE database is indexed by utilizing keywords from Medical Subject Headings (MeSH), a controlled vocabulary, to help index, catalogue, and search for biomedical and health-related information and literature [3, 11].

In this chapter we describe a method for evaluating emerging scientific trends in relation to obesity and psoriasis by analyzing the frequency of various MeSH terms from PubMed abstracts for a given time period. We show how our system can be used to predict emerging trends at a relatively early stage, and describe some of these trends. For the remainder of the chapter we will focus on how one can use a similar approach to predict emerging disease relevant genes by evaluating the frequency of genes/proteins in published scientific literature. In other words, given the pace with which the disease relevant knowledge is accumulating, we predict whether a gene becomes a disease relevant biomarker in the near future, given the observed frequency. The principle strength of our proposed method is the ability to anticipate emerging scientific trends that can help in drug discovery. Focusing on a drug discovery perspective, we then analyze the literature identified genes for genetic associations and their druggability. Additionally, we analyze disease associated genes, and the pathways enriched by these gene sets to explore any potential biological associations between the two diseases.

1.1 Obesity

Obesity is a medical condition characterized by the accumulation of excess body fat to a degree that is adversely affecting health and well-being and overall life expectancy. Obesity rates have increased substantially over the past few decades, and today obesity has reached epidemic proportions on a global scale. Obesity has been described as one of the potentially most important diseases of twenty-first century [12]. In 2008, the World Health Organization (WHO) estimated that more than 1.4 billion adults (aged 20 and above) worldwide were overweight [i.e., with a body mass index (BMI) $\geq$ 25 kg/m^2], which included over 200 million and 300 million men and women, respectively, being obese (BMI $\geq$ 30 kg/m^2). In 2010, an estimated more than 40 million children under the age of 5 years were overweight [13]. Approximately 10 % of global child population is now either overweight or obese [14]. Not only are these numbers expected to increase in the coming years, but also there are serious concerns regarding the epidemic of childhood obesity. Obesity is a risk factor for diabetes mellitus, hypertension, coronary artery disease, stroke, osteoarthritis, complications in pregnancy, and surgical outcomes for unrelated medical conditions. The prevalence of some of the common digestive disorders such as gastroesophageal reflux disease (GERD), esophagitis, gallstones, nonalcoholic fatty liver disease (NAFLD), sleep apnea, and cirrhosis is significantly higher in obese individuals compared to normal-weight individuals. Obesity has also been associated with various cancers, and among nonsmokers about 10 % of all cancer deaths are reportedly related to obesity [14, 15].

Rising rates of obesity and associated comorbidities place a severe burden on medical spending. A recent study estimates that medical costs of obesity have risen to $147 billion in 2008 compared to estimates of $78.5 billion for 1998 in the USA [16]. This represents as much as 10 % of all medical spending in the USA [14, 16]. Gastrointestinal weight-loss surgery has been proven to be superior in treating obesity and associated comorbidity conditions [12, 17]. This is despite the inherent risks associated with surgical procedures and the significant lifestyle changes one will have to make to achieve the optimal outcomes [18, 19]. There are very limited options for pharmacotherapy for obesity. Orlistat, a lipase inhibitor, which prevents breakdown of triglycerides by inhibiting pancreatic and intestinal lipases, Qsymia™, a combination of the drugs phentermine as well as topiramate, and Belviq™ (lorcaserin), which is an option for subjects who are either obese (BMI $\geq$ 30) or overweight (BMI $\geq$ 27) with various comorbidities such as type II diabetes mellitus, increased levels of cholesterol, and hypertension. For short-term treatment of obesity, the US FDA has previously approved the use of phentermine, which is an adrenergic reuptake inhibitor that increases adrenergic signaling [20–23]. With the available pharmacotherapy approaches, the weight reductions achieved are modest, especially when compared with gastrointestinal weight-loss surgery [18]. Therefore, there is a crucial need for drug discovery programs to find effective and less invasive therapeutic interventions for obesity, and potential new knowledge is contributing to studies for such development [24].

Recent clinical observations and cohort study have suggested obesity as one of the emerging comorbidities of psoriasis and obesity itself as a potential risk factor for the development of psoriasis [25, 26]. While analyzing data from more than 40,000 patients, it was observed that systemic disorders such as obesity, diabetes, and heart insufficiency were occurring significantly more often in patients with psoriasis, when compared to control subjects [27]. Also, psoriasis-affected individuals reportedly are at an increased risk for metabolic syndrome including insulin resistance, obesity, dyslipidemia, and hypertension [28, 29]. There have been case reports of remission of psoriasis after gastric bypass surgery [30, 31]. All these observations suggest a potential mechanistic link between the two conditions.

1.2 Psoriasis

Psoriasis is a chronic, immune-mediated disease that manifests itself in the skin of patients. Its most common form, plaque psoriasis, presents as inflamed red lesions or patches of skin that are covered with a silvery white buildup of dead skin cells, known as scale. Severity varies among individuals, and can range from only a few lesions to moderate or large areas of affected skin [32]. It is estimated that psoriasis affects approximately 2–3 % of the population in the USA and over 125 million subjects across the globe [33, 34].

The disease is characterized by scaly, erythematous, and inflammatory skin plaques, resulting from hyperproliferation of epidermis with incomplete differentiation of keratinocytes and abnormal formation of horn cells of the epidermis with persistence of nuclei. Lesions often display inflammatory cell infiltration and neovascularization. The pathophysiology of psoriasis remains largely unclear [35]. Population and genome-wide association studies suggest a genetic component that predisposes individuals to the disease [36–39]. A complex interaction between genetic, environmental and systemic factors, then leads to a wide spectrum of psoriasis disease in these genetically predisposed individuals [36].

None of the currently available therapies or combination of therapies offers a potential cure; thus, treatment is aimed at reducing the burden of disease and achieving an improvement in its signs and symptoms. The choice of employing topical therapies, phototherapy or systemic agents, is influenced by (1) the type of psoriasis, severity (mild to severe), (2) extent of involvement of various regions of the body, (3) symptoms, and (4) additional comorbidity conditions [40, 41]. Available therapeutic options for psoriasis continue to expand with the development of biologic agents aimed at modulating immune-mediated functions (TNF-α) and anti-cytokine approaches (for example IL-12, IL-23, IL-17, IL-22) [42]. Combination therapy is frequently employed and is often aimed at individualization of treatment to patient needs [43]. As mentioned earlier, despite the availability of a number of therapeutic options, there remains no cure for psoriasis. Furthermore, several therapeutic agents have been associated with risk of severe adverse reactions and, occasionally, significant morbidities such as skin cancers, serious infections, lymphomas, and hepatic conditions [44–46]. Given the risk of adverse treatment outcomes and the considerable variability in response to current treatments, there is an urgent need for new therapies and an effective disease management of psoriasis.

1.3 Obesity and Psoriasis: Are There Any Common Linkages?

Observational and cohort studies suggest patients with psoriasis have higher risk of metabolic syndrome, type II diabetes, and cardiovascular mortality [47–50]. The prevalence of obesity has been reported as almost doubled in psoriasis patients as in general population. Multiple prior studies also suggest that weight-loss or bariatric surgery in obese patients with psoriasis improves the severity of psoriasis or response to treatment [51, 52]. While the underlying linkage between the psoriasis and obesity remains unclear, these accumulating clinical findings strongly implicate mechanistic associations between the two conditions.

Intriguingly, it has been proposed that the metabolic agent metformin alone or as an add-on therapy to methotrexate could be useful for the treatment of psoriasis [53]. Metformin is efficacious in the treatment of prediabetes and contributes to a sustained weight loss in overweight individuals. Methotrexate, initially introduced as an antineoplastic agent, later found entry into dermatology and has been used for treatment of psoriasis and other skin conditions [54]. Recently, multiple observational studies indicate that psoriasis patients treated by methotrexate may be associated with lower cardiovascular risk, thus uncovering additional role of methotrexate in anti-inflammation and vasocardio-protection [55]. Metformin and methotrexate may derive their pharmacological effects through modulating the same common cellular target, AMP-activated protein kinase (AMPK). The AMPK enzyme is a critical master regulator of metabolism and regulates a number of downstream targets that are, for example, important for cellular growth or function in many tissues including T-lymphocytes. The dual- or multiple-benefits of these agents in the management of psoriasis and obesity further support the hypothesis that both diseases perhaps share common mechanistic biology. For example, inflammatory signaling mechanisms play an important role in psoriasis and may further contribute to the development comorbidities such as obesity, diabetes, or higher cardiovascular risk [56].

In addition, clinical transcriptomics analyses comparing psoriasis lesional skin with healthy controls also suggest some potential the mechanistic links to pathways involved in atherosclerosis, fatty acid metabolism, and diabetes [57, 58].

2 Materials

MEDLINE 2013, the US National Library of Medicine (NLM) biomedical abstract repository contains approximately 22 million reference articles from around 5,640 journals published worldwide. MEDLINE remains a central point of access to biomedical research. Each year under a licensing agreement with the US National Library of Medicine, the entire set of PubMed baseline databases are downloaded onto one of our a local servers.

An example of MEDLINE record, describing a full-text article, is shown in Fig. 1. Each record contains the article title, abstract, author's names, MeSH headings, affiliations, publication date, journal name, and other information.

2.1 MEDLINE and MeSH

A unique feature of MEDLINE is that the records are indexed with NLM's controlled vocabulary, i.e., the Medical Subject Headings (MeSH) thesaurus for retrieval and classification. MeSH headings consist of sets of descriptors in a hierarchical structure. Subheadings, or qualifiers, provide additional specificity for each descriptor. The major MeSH headings (denoted MH) indicate the main contents of the article, and the minor MeSH headings are

PMID - 24048425
DP - 2013 Sep
TI - The role of Fcgamma receptor polymorphisms in the response to anti- tumor necrosis factor therapy in psoriasis A pharmacogenetic study.
AB - Variability in genes encoding proteins involved in the immunological pathways of biological therapy may account for the differences
AU - Julia M
AU - Guilabert A
PT - Journal Article
TA - JAMA Dermatol
JID – 101589530
RN - 185243-69-0 (TNFR-Fc fusion protein)
RN - FYS6T7F842 (adalimumab)
SB - AIM
MH - Antibodies, Monoclonal/pharmacology/therapeutic use
MH - Dermatologic Agents/pharmacology/*therapeutic use
MH - Psoriasis/*drug therapy/genetics/pathology
MH - Receptors, IgG/*genetics
MH - Tumor Necrosis Factor-alpha/*antagonists & inhibitors
RN - 185243-69-0 (TNFR-Fc fusion protein)
RN - FYS6T7F842 (adalimumab)

Fig. 1 A MEDLINE record example—PMID: PubMed ID, *TI* title, *AB* abstract, *AU* author, *MH* MeSH term

used to describe secondary topics. The MeSH terms are organized into concept hierarchies that represent "is-a" and "part-whole" relationships. For example in the MeSH disease hierarchy, Multiple Sclerosis is an example of Autoimmune Demyelinating Disease, which in turn is an example of Nervous System Disease.

A MEDLINE MeSH field is a combination of a MeSH descriptor with zero or more MeSH qualifiers. The MeSH Vocabulary is available for download at http://www.nlm.nih.gov/mesh/filelist.html. The MeSH 2013 vocabulary file includes 26,853 descriptors, 83 qualifiers, and 215,463 supplementary concepts. Descriptors are the main elements of the vocabulary. Qualifiers are assigned to descriptors inside the MeSH fields to express a special aspect of the concept. Both descriptors and qualifiers are organized in several hierarchies. On an average there are roughly 10 MeSH indexing terms applied for each MEDLINE citation by professional indexers, who choose these descriptors after reading the full length text article. From each PubMed article, the following features are extracted: PubMed identifier, year of publication, title, author list, affiliation, MeSH terms (with flag indicating if major) and substance names as described by Agarwal and Searls [1, 2]. This yields ~260 million article-to-MeSH term or substance-name mappings.

2.2 MEDLINE and Genes

The Gene name list is built by integrating names and descriptions from multiple fields within EntrezGene, HUGO, and UniProt. All mouse and rat gene synonyms are mapped onto the orthologous human EntrezGene using Homologene [59], a system developed for the detection of homologs among annotated genes of several completely sequenced eukaryotic genomes. These mappings yielded a total of 557K synonyms, though only ~131K of the synonyms was found in PubMed abstracts. Gene synonyms that refer to multiple human EntrezGenes are flagged, while gene names that correspond to common English or those that are likely to be other medical terms or abbreviations are discarded. Two additional data files from EntrezGene are used to augment Gene to PubMed mappings: Gene2pubmed and GeneRIF. GeneRIF (Gene Reference into Function) provides a quality functional annotation that may extend beyond the genes mentioned in the abstract [60]. GeneRIF provides functional annotation of the gene mentioned in the abstract, and is usually produced by manual curation. Each MEDLINE article's title and abstract is scanned for all high-quality human, mouse, and rat gene names along with their synonyms as described by Agarwal and Searls [1, 2]. In total, the system contains ~7.4 million PubMed-to-human-gene mappings covering ~7.5M articles and ~19.9K human genes.

3 Methods

3.1 Obesity and Psoriasis: Trends

We counted the number of publications since 2001 for both "Obesity [majr]" and "Psoriasis [majr]" and then also searched for publications that represent clinical trials with Obesity or Psoriais as an indexed subject term (using the PubMed query: "Clinical Trial [pt] Obesity or Psoriasis [mh]"). These results are shown in Fig. 2.

3.2 Obesity and Psoriasis: "Hot" and "Cold" Concepts and Genes

To evaluate trends of scientific activity in relation to obesity and psoriasis, we analyzed the number of publications on various MeSH terms. We compared the number of publications in the time frame of 2010–2012 to 2007–2009 to get 3-year trends. All these comparisons are done within the context of obesity and psoriasis publications, i.e., publications between 2010 and 2012 that are indexed with obesity and psoriasis as a major MeSH term. This trend analysis corrects for increase in the overall number of publications in obesity from 25,967 between 2007 and 2009 to 37,241 for the years between 2010 and 2012. Similar analysis for Psoriasis corrects for increase in the overall number of publications in Psoriasis from 4,773 between 2010 and 2012 and 3,373 for 2007–2009. Thus, a "hot" trend is a concept that is growing statistically faster than the background rise in obesity or Psoriasis publications (Fig. 3). We use a hypergeometric distribution or a two-sided Fisher's exact test to determine if the MeSH term under

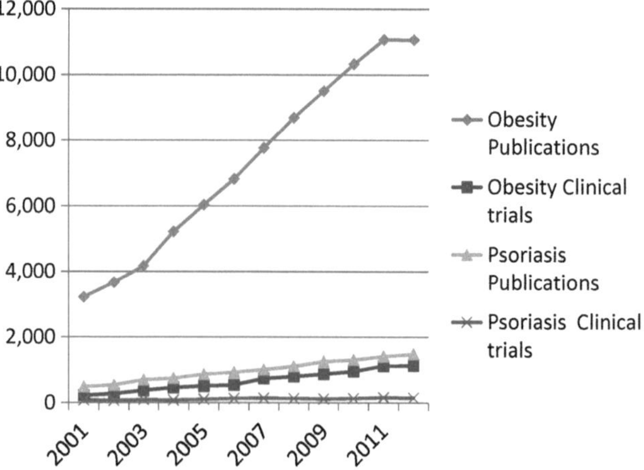

Fig. 2 Number of scientific publications associated with obesity or psoriasis over time showing the years 2001–2012. For comparison, the graph also shows the subset of publications that discuss clinical trials related to either disease. The citation numbers were derived using the described literature mining methodology. The annual number of citations for each of the diseases increased between twofold and fivefold during that period with obesity clinical trials having the largest relative increase. Specifically, the number of publications on obesity increased 3.4-fold, on obesity clinical trials 5.1-fold, on psoriasis 3.1-fold, and on psoriasis clinical trials 2.0-fold, respectively

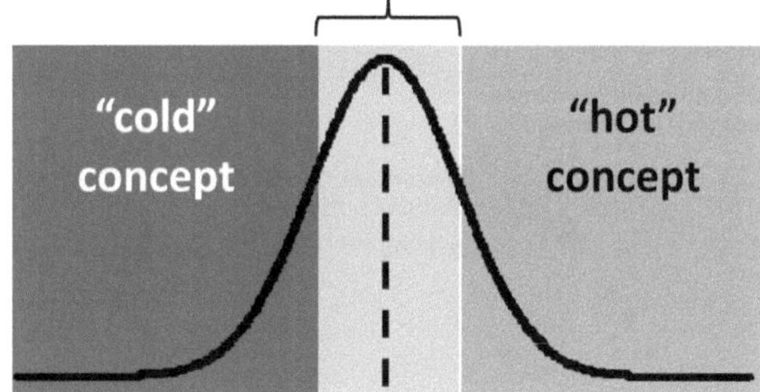

Fig. 3 Method to determine whether a concept or MeSH term was "hot" or "cold" in the last 3 years relative to the preceding 3-year period. Significantly higher or lower than expected citation counts are interpreted as "hot" and "cold," respectively. The normalized citation count associated with the concept is expected remain constant over each 3-year period. Statistical significance is determined using a two-tailed Fisher's exact test

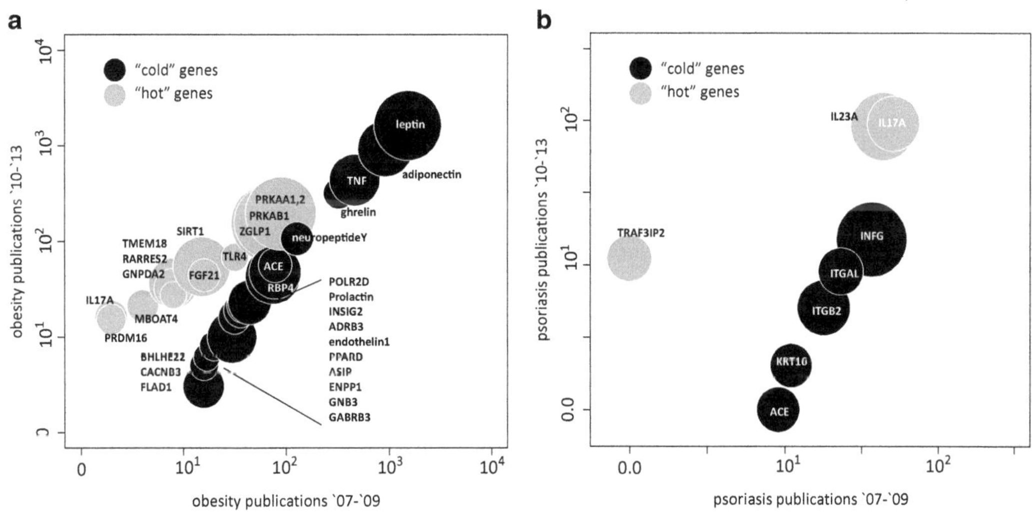

Fig. 4 *Bubble plots* highlighting genes that show significant trends ("hot" *grey bubbles* or "cold" *black bubbles*) in their publications within the obesity (**a**) or psoriasis (**b**) literature in the last 3 years (2010–2012) compared to 2007–2009. *Larger circles* indicate more significant Fisher *p*-values

consideration had more publications than would be expected by chance [2]. We restricted the analysis to articles with the search term as a major MeSH annotation, and excluded search terms that were descendants of the MeSH tree. The reason for the exclusion was to deemphasize obvious relationships between parent terms and their children (such as Diabetes Mellitus and Type II Diabetes Mellitus). The results of this analysis with regards to Genes are presented in Fig. 4, and with MeSH terms are presented in Fig. 5.

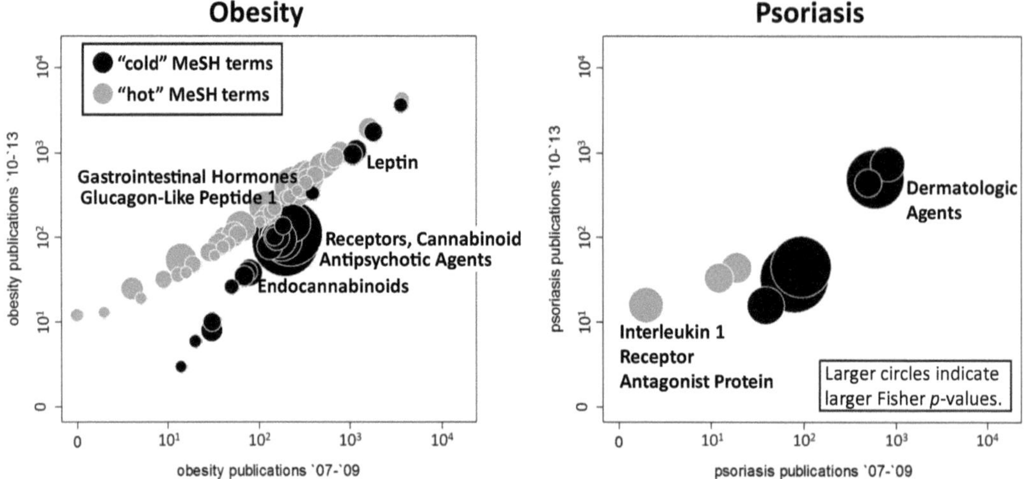

Fig. 5 *Bubble plots* highlighting MeSH terms that show significant trends ("hot" *grey bubbles* or "cold" *black bubbles*) in their publications within the obesity (*left panel*) or psoriasis literature (*right panel*) in the last 3 years compared to 2007–2009. *Larger circles* indicate more significant Fisher *p*-values

3.3 Obesity and Psoriasis: Genes

One of the first analyses for biological understanding of a disease is investigating the genes that are implicated in the disease biology. Biomedical literature is still a major source of information for identifying such gene–disease associations. In a drug discovery context, the value of categorizing literature by gene or disease is manifold. Based on the method described by Agarwal and Searls [1], we can immediately get a ranked list of genes associated with a disease, for results based on the PubMed query "Obesity [majr]" and "Psoriasis[majr]" respectively. The resulting gene list for each query includes *p*-values which are computed independently for each gene name using Fisher's exact test and can help with prioritizing the gene–disease level associations. Based on this approach, we identified genes in literature associated with obesity and psoriasis, respectively. Selection of the significant genes was based on geneRIF records or statistical significance (p-value ≤ 0.05). Functional Enrichment Analysis was performed as previously described [61] using a false discovery cutoff of 0.1 and employing the functional gene annotation packages from Bioconductor version 2.22 [62].

3.4 Obesity and Psoriasis: Canonical Pathways

The large number of genes preferentially expressed in each disease suggests that they differ in many fundamental signaling and metabolic pathways. To more systematically assess canonical signaling and metabolic pathways enriched in each disease as well as those common to both, we used the IPA tool from Ingenuity Systems (http://www.ingenuity.com). Given a list of genes, IPA performs a statistical test for enrichment of these genes in its manually curated canonical pathway database. Each individual IPA signaling pathway includes extracellular signaling components, cell membrane receptors, downstream effectors, and transcription factors that have been described to interact in the published scientific literature, with each pathways often including from 30 to 100 or more individual genes. The IPA metabolic pathways are derived from the KEGG metabolic pathways. These tools allow us to determine the top canonical signaling and metabolic pathways represented among the genes that are significantly enriched for any given disease or diseases. Figure 7 summarizes the top signaling pathways and metabolic pathways enriched at $p < 0.05$ for obesity and psoriasis, respectively. Comparative pathway analysis is conducted to identify the common significantly enriched pathways for both diseases. These enriched signaling pathways are likely influenced by cell–cell interactions, perturbed by disease processes, and affected by drugs to yield physiological effects.

4 Results and Discussion

4.1 Obesity and Psoriasis Trends: Growing Health Concerns

It is reasonable to assume that publication frequency represents one measure of scientific activity in a specific area of scientific research. The number of publications associated with obesity grew from 3,216 to 11,065, and those associated with Psoriasis grew from 484 to 1,481, during the years 2001 to 2012. The number of publications on clinical trials associated with obesity grew from 223 to 1,131, and those associated with Psoriasis grew from 75 to 151, during the same period of 2001–2012 (Fig. 2). Analysis of scientific activity trends through bibliometrics or related text-mining methodologies offer opportunities to evaluate growth in certain areas, which may suggest growing areas of knowledge as well as areas of important health concerns that may create new avenues for therapeutic intervention. The growing numbers of publications associated with obesity over the last 12 years are consistent with increased concerns about the health impacts of obesity reaching epidemic proportions globally [14]. Similarly, the increase in the number of psoriasis publications potentially reflects a similar health concern due to this disease. In terms of drug discovery, clinical trials represent a major area of direct investigation of drug effects in human subjects. When we searched for publications that represent clinical trials with obesity or psoriasis as an indexed subject term (using the PubMed query: "Clinical Trial [pt] Obesity or psoriasis [mh]"), we see an increase in the number of publications in this category suggesting an increase in investigation activity of therapeutic interventions in human subjects for obesity (Fig. 2). These counts are not necessarily the number of obesity clinical trials, but represent publications that are of type "clinical trial" and have obesity as an indexed MeSH term.

Collectively, the publication trends point to obesity being a growing health concern, and psoriasis being a major health concern that are being evaluated by clinical studies to potentially find effective therapeutic intervention strategies.

4.2 Obesity and Psoriasis: "Hot" and "Cold" MeSH Concepts and Genes

4.2.1 Obesity: "Hot" and "Cold" Genes

As mentioned in Subheading 3, with regards to measuring scientific activity as trends, and evaluating them as "hot" or "cold" concepts, for obesity and psoriasis, we analyzed the number of publications on various MeSH terms as mentioned above, in the time frame 2010–2012, when compared to 2007–2009. Thus, a "hot" trend is a concept that is growing statistically faster than the background rise in total publications on obesity or psoriasis. For example (Fig. 4a), presents "Leptin" as one of the "cold" concept genes, while "SIRT1 (Sirtuin-1)" and "PRKAA1 (Protein kinase, AMP-activated, alpha 1 catalytic subunit) and PRKAA2 (Protein kinase, AMP-activated, alpha 2 catalytic subunit)" as some of the "hot" concept genes. The leptin gene (LEP) encodes a 16-kDa circulating peptide that binds to the leptin receptor encoded by the leptin receptor gene (LEPR). Leptin is primarily secreted by white adipocytes and plays a role in signaling pathways involved in appetite regulation as well as other physiological functions. The discovery of leptin was one of the exciting aspects of obesity research, which led to intense focus in the late 1990s to the early 2000s and included many physiological studies that suggested a critical axis between body fat and the hypothalamus [63, 64]. This may have resulted in an elevated baseline for this gene, and lack of direct pharmacological modulation for antiobesity, may potentially have played a role in exploration into this space along with unexpected outcomes with cannabinoid systems modulation, as described later in this chapter.

Sirtuin 1 is a protein deacetylase that targets endothelial nitric oxide synthase resulting in an enhanced state of bioavailability for nitric oxide. SIRT1, the most studied member among the family of sirtuins, responds to overfeeding, starvation, energy homeostasis, and exercise similarly as that of AMP-activated protein kinase, although on a different scale of time. SIRT1 and AMP-activated protein Kinase appear to have a closely regulated role in metabolic homeostasis, and it is one of the intense areas of scientific investigation for therapeutic intervention [65]. This hot concept is a potential reflection of that. Another set of "hot" genes are "PRKAA1 (Protein kinase, AMP-activated, alpha 1 catalytic subunit) and PRKAA2 (Protein kinase, AMP-activated, alpha 2 catalytic subunit)." The proteins encoded by these two genes belong to the ser/thr protein kinase family. AMP-activated protein kinase is a heterotrimer consisting of an alpha catalytic subunit, and non-catalytic beta and gamma subunits. The proteins encoded by PRKAA1 and PRKAA2 form the catalytic subunits of the 5′-prime-AMP-activated protein kinase (AMPK). AMPK's critical role is to function as a cellular energy sensor and protect cells from stresses that cause ATP depletion is to switch off ATP-consuming biosynthetic pathways by regulating the activity of a number of key metabolic enzymes through phosphorylation. Increase in the cellular AMP–ATP ratio serves as a signal for activation of the kinase activity of AMPK. Activated AMPK phosphorylates and inactivates acetyl-CoA carboxylase (ACC) and beta-hydroxy beta-methylglutaryl-CoA reductase (HMGCR), key enzymes involved in regulating de novo biosynthesis of fatty acid and cholesterol. The role of AMPK in energy homeostasis and in relevant pathologies is an active area of scientific research for therapeutic modulation [66].

4.2.2 Obesity: "Hot" and "Cold" MeSH Concepts

Figure 5a presents "Cannabinoid Receptors" as a "cold" concept and "Gastrointestinal Hormones" as a "hot" concept. The endocannabinoid system is a critical regulator of feeding behavior and energy homeostasis. The discovery of the endocannabinoid system, with CNR1 (Cannabinoid receptor 1) and CNR2 (cannabinoid receptor 2) as well as their ligands and enzymatic systems, has been a critical research area in a growing number of physiological and pathological functions. This research has been instrumental in exploring therapeutic options of modulating cannabinoid receptors for a beneficial aspect in broad range of diseases such as obesity as well as various inflammatory, neurodegenerative and cardiovascular conditions. However, clinical studies with CNR1 antagonists for obesity and metabolic conditions as well as other modulations of CNR1 and CNR2 with agonism resulted in unexpected complexities, potentially suggesting that an increased understanding of endocannabinoid system is required before realizing therapeutically beneficial aspects of its modulation. The cold concept is a possible reflection of that realization [67]. On the other hand, there is increasing evidence suggesting that peptide hormones from the L cells of the small intestine that form the core of enteroendocrine system, have critical roles to play in metabolic status of individuals. Specifically, they serve as incretins or through their function in central nervous system act as mediators of satiety and appetite. Additionally, increasing evidence suggests that various gastrointestinal hormones such as glucagon-like peptide-1 (GLP-1), peptide YY (PYY), and others are elevated after bariatric surgery and may play a role in modulating bile acids by potentially regulating key sensors in the gastrointestinal tract [68]. GLP-1 is a clinically relevant pharmacological analogue, and along with other peptide hormones is an intense area of investigation for therapeutic intervention strategies for metabolic diseases. Perhaps, this "hot" concept is a reflection of scientific interest in elucidating novel functions of these hormones for pharmacologic development of peptide analogues for obesity and other relevant metabolic diseases [69].

4.2.3 Psoriasis: "Hot" and "Cold" Genes

Figure 4b highlights some of the "hot" and "cold" genes in relation to the searches for Psoriasis associated genes. For example, this analysis suggests, "IL17A (Interleukin 17A)," "IL23A (Interleukin 23, alpha subunit p19)," and "TRAF3IP2 (TRAF3 interacting protein 2)" as some of the "hot" genes, while "IFNG" (Interferon, gamma) as a "cold" gene, among some other "cold" genes. Interestingly, TRAF3IP2 encodes a signaling adaptor involved in the regulation of humoral and cellular immunity. This protein, interestingly, also represents a major link between IL-17 mediated adaptive immune responses and NF-κB as the master regulator of innate immunity. In terms of genetic basis, GWAS studies (genome-wide association studies) have identified genetic association of TRAF3IP2 with susceptibility to psoriasis and psoriatic arthritis [70, 71]. Reduced binding of TRAF3IP2 variant to TRAF6 (TNF receptor-associated factor 6, E3 ubiquitin protein ligase), suggesting alternated modulation of immunoregulatory signal through altered TRAF interactions [70].

Psoriasis lesions develop through sustained immune dysregulation of pathways involving T helper cell responses mainly belonging to the Thl, Th17 and Th22 lineages. Various studies suggested strong association and causality between the T-cell helper 1 inductor cytokine IL-12 and a range of immune-mediated disorders including psoriasis. It was later found that IL-23, another cytokine important to the development and maintenance of Th17 cells, is in fact sharing a subunit of p40 with IL-12. Subsequent studies further established that IL12/IL23 p40 is playing an integral role in the pathologies of psoriasis and other autoimmune diseases [72]. While the role of these individual T-cell subsets as well as cytokine interactions in human disease still remain to be further understood, IL-23 is known to promote Th17 response and regulate Th17 T cells, which secrete several proinflammatory cytokines includingIL-17A. IL-17A, which is expressed by Th17 cells, has a direct effect on the regulation of genes expressed by keratinocytes that play roles in innate immunity. Subsequent gene expression studies demonstrated that the levels of IL12, IL17, and their target genes are elevated in psoriasis skin lesions [73, 74]. Notably, the scientific understanding from various studies has contributed to the development of therapeutic agents around IL23/IL12 and IL17. For example, Ustekinumab™, one of the first human monoclonal antibody that binds to the p40 subunit of IL12 and IL23, was developed to treat moderate to severe plague psoriasis and later on approved for the treatment of psoriatic arthritis (FDA approval on Sept. 2013, Ref. Stelera Label). Another promising agent, Ixekizumab™, a humanized anit-IL17 monoclonal antibody, is under active clinical evaluation for psoriasis treatment and other relevant autoimmune disorders. Given the overlap between human Thl and Th17 pathways and the plasticity between human Th lineages, it is likely that more research investment will continue in this direction for enhanced understanding of disease biology, and therapeutic intervention.

4.2.4 Psoriasis: "Hot" and "Cold" MeSH Concepts

The notion of psoriasis being a local skin disease has been challenged by several large epidemiological studies and it is now recognized as a systemic disease. Figure 5b shows some of the "Hot" and "Cold" MeSH concepts in relation to the searches for psoriasis. For example, one of the "cold" concepts is "Dermatologic Agents." It has been estimated that between 7% and 40 % of patients with psoriasis eventually go on to develop psoriatic arthritis [75]. In addition, several important diseases including metabolic syndrome, Crohn's disease are comorbid or associated with psoriasis. The increasing knowledge led to a potential shift of therapeutic intervention strategy from conventional topical "Dermatologic Agents" to "Systemic Agents" for immune modulators. In particular, since the first drug that interferes with the functions of Th17 cells was approved in 2009 [76], more therapeutic efforts and investments have been put into the development of biologics. Further research and development, led to several novel biologic candidates, such as monoclonal antibodies targeting IL12/IL23 p40 subunit and IL-17 systems that show promise not only for the management for psoriasis but also for various other autoimmune diseases, Along with this, IL-1 system emerged to be another intriguing avenue for potential therapeutic modulation. IL-1 is a cytokine abundantly present in human epidermis and influences cutaneous inflammation and keratinocyte proliferation. The pathophysiological relevance of IL-1 in psoriasis has been evidenced by in vivo models and clinical studies [77, 78]. The IL-1 epidermal system is an important biological system for modulation with both IL-1 agonism as well as IL-1 receptor antagonism to influence keratinocyte differentiation and cutaneous inflammation, which are two hallmarks of psoriasis.

4.3 Obesity and Psoriasis: Genes

Next, we evaluated genes that are implicated in the disease biology. Our literature mining methodology, identified 2,744 and 684 genes in literature associated with obesity, and psoriasis, respectively (Fig. 6). The selected genes were based on geneRIF records or having a *p*-value less than the cutoff of 0.05. 169 genes were common between the two searches (Fig. 6), suggesting potentially common biological pathways and molecular mechanisms associated with these diseases. Interestingly, some of the common genes include well-known obesity-related genes including leptin, insulin, several interleukins (IL1A, IL4, IL8, IL10, IL13, IL20, etc.), glucagon, angiotensin converting enzyme, IGF1, and sirtuin 1 (Table 1). Similarly, some of the well-known psoriasis relevant genes such as IL17A, various interleukins, IFNG, and various others also make to this list of intersected genes between obesity and psoriasis, suggesting some potential linkages at gene level, and may be at pathway level, for potential therapeutic interventions (Table 1). Some of the genes from these analyses have also been proposed as potential targets for further evaluation by drug discovery programs through integrative analysis approaches [79].

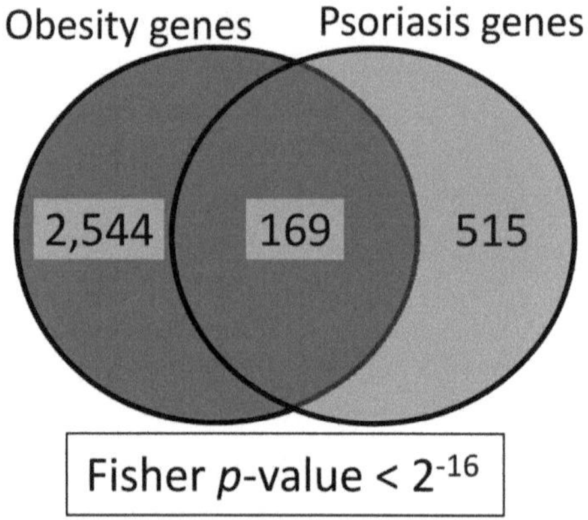

Fig. 6 Venn diagram showing the number genes derived using the described literature mining methodology in order to identify genes associated with obesity and/or psoriasis in the scientific literature. Statistical significance is computed using Fisher's exact test

4.4 Obesity and Psoriasis: Pathways

Pathway enrichment analysis was conducted using genes associated with obesity and psoriasis identified from literature mining. Top enriched pathways are illustrated in Fig. 7. A strong functional theme from metabolic signaling pathways known to be associated with metabolic syndromes have been identified for obesity gene list, including *Type II Diabetes Mellitus signaling*, *AMPK signaling*, *insulin signaling*, *PPARα/RXRα activation*, *triacylglycerol degradation*, *and leptin signaling* (Fig. 7—left column). In parallel, a strong functional theme of immune and inflammatory-related pathways is identified for psoriasis gene list, e.g., *cytokines in mediating communication between immune cells*, *IL17A in psoriasis*, *altered T cell and B cell signaling*, *cross talk between dendritic cell s and natural killer cells.* These findings are consistent with our current understandings of disease biology of both diseases, which in turn, supports the validity of the genes identified from literature mining using our methodology.

Notably, when we compared pathways enriched for each disease associated gene sets, we observed some pathways of interest that both diseases shared (Fig. 7—middle column). This suggested that lipid metabolism (e.g., *triacylglycerol biosynthesis*) and related signaling pathways (*FXR/RXR/LXR activation and PPAR signaling*) may have a role in obesity as well as psoriasis pathogenesis. This may potentially point to further understandings of the clinical comorbidity of the two conditions at a mechanistic level. Consistently, Functional Enrichment Analysis [61] using the list of common genes between obesity and psoriasis also supports the observation that these genes are related to response to lipid, inflammatory response, response to insulin stimulus among a number of other significantly over-represented gene ontology categories and molecular pathways (Table 2).

Table 1
List of 169 genes derived using the described literature mining methodology in order to identify genes associated with both obesity and psoriasis in the scientific literature

Entrez gene ID	HUGO symbol	Description	Druggable genome (score 1-5) [82]	GWAS (obesity) [83, 84]	GWAS (psoriasis) [83, 84]	Mouse phenotypes [85]
133	ADM	Adrenomedullin	2	Diabetes; metabolism		
177	AGER	Advanced glycosylation end product-specific receptor	1	Diabetes; insulin resistance	Psoriasis	Glucose; weight gain/loss
196	AHR	Aryl hydrocarbon receptor	2	Metabolism		Weight gain/loss
199	AIF1	Allograft inflammatory factor 1	2	Obesity; diabetes; weight gain/loss		
207	AKT1	v-akt murine thymoma viral oncogene homolog 1	5	Obesity; diabetes; metabolism; insulin resistance		Weight gain/loss
213	ALB	Albumin	3	Diabetes		
239	ALOX12	Arachidonate 12-lipoxygenase	3	Diabetes; metabolism		
467	ATF3	Activating transcription factor 3	1	Diabetes		
596	BCL2	B-cell CLL/lymphoma 2	5	Diabetes		Weight gain/loss
629	CFB	Complement factor 3	3	Diabetes; metabolism		
685	BTC	Betacellulin	2	Diabetes		
718	C3	Complement component 3	2	Diabetes; metabolism		Adipose; energy homeostasis; glucose; insulin; weight gain/loss

(continued)

Table 1
(continued)

Entrez gene ID	HUGO symbol	Description	Druggable genome (score 1-5) [82]	GWAS (obesity) [83, 84]	GWAS (psoriasis) [83, 84]	Mouse phenotypes [85]
760	CA2	Carbonic anhydrase II	4			
834	CASP1	Caspase 1, apoptosis-related cysteine peptidase	5	Diabetes		
847	CAT	Catalase	3	Diabetes; insulin resistance		Energy homeostasis
857	CAV1	Caveolin 1, caveolae protein, 22 kDa	2	Diabetes; metabolism; insulin resistance		Obesity; adipose; insulin; weight gain/loss
894	CCND2	Cyclin D2	1	Diabetes		
896	CCND3	Cyclin D3	2			
912	CD1D	CD1d molecule	3			
920	CD4	CD4 molecule	1	Diabetes		Diabetes
958	CD40	CD40 molecule, TNF receptor superfamily member 5	2	Diabetes		
983	CDK1	Cyclin-dependent kinase 1	3			
1113	CHGA	Chromogranin A (parathyroid secretory protein 1)	1			
1116	CHI3L1	Chitinase 3-like 1 (cartilage glycoprotein-39)		Diabetes; insulin resistance		
1139	CHRNA7	Cholinergic receptor, nicotinic, alpha 7 (neuronal)	4	Obesity; weight gain/loss		
1277	COL1A1	Collagen, type I, alpha 1	4	Diabetes; metabolism; weight gain/loss; insulin resistance		Adipose; weight gain/loss

1373	CPS1	Carbamoyl-phosphate synthase 1, mitochondrial	1		Diabetes	Psoriasis	
1392	CRH	Corticotropin releasing hormone	3		Obesity		Glucose; weight gain/loss
1509	CTSD	Cathepsin D	4		Diabetes; metabolism; insulin resistance		Weight gain/loss
1543	CYP1A1	Cytochrome P450, family 1, subfamily A, polypeptide 1	4		Obesity; diabetes; weight gain/loss	Psoriasis	
1634	DCN	Decorin	3		Diabetes		
1636	ACE	Angiotensin I converting enzyme	4		Obesity; diabetes; metabolism; weight gain/loss; insulin resistance	Psoriasis	Insulin; weight gain/loss
1728	NQO1	NAD(P)H dehydrogenase, quinone 1	4		Diabetes	Psoriasis	Adipose; glucose; insulin; weight gain/loss
1956	EGFR	Epidermal growth factor receptor	5		Diabetes		Weight gain/loss
1958	EGR1	Early growth response 1	2				Weight gain/loss
2064	ERBB2	v-erb-b2 avian erythroblastic leukemia viral oncogene homolog 2	5		Obesity		
2068	ERCC2	Excision repair cross-complementing rodent repair deficiency, complementation group 2	1		Obesity; diabetes		
2147	F2	Coagulation factor II (thrombin)	4		Obesity; diabetes; metabolism; weight gain/loss; insulin resistance		
2162	F13A1	Coagulation factor XIII, A1 polypeptide	4		Obesity; diabetes; metabolism; insulin resistance		

(continued)

Table 1
(continued)

Entrez gene ID	HUGO symbol	Description	Druggable genome (score 1-5) [82]	GWAS (obesity) [83, 84]	GWAS (psoriasis) [83, 84]	Mouse phenotypes [85]
2167	FABP4	Fatty acid binding protein 4, adipocyte	3	Obesity; diabetes; metabolism; insulin resistance		Glucose; insulin
2169	FABP2	Fatty acid binding protein 2, intestinal	1	Obesity; diabetes; metabolism; weight gain/ loss; insulin resistance		Obesity; insulin
2170	FABP3	Fatty acid binding protein 3, muscle and heart (mammary-derived growth inhibitor)	2	Obesity; diabetes		Glucose; insulin
2171	FABP5	Fatty acid binding protein 5 (psoriasis-associated)	2			Obesity; adipose; glucose; insulin
2243	FGA	Fibrinogen alpha chain	4	Obesity; diabetes; metabolism		
2244	FGB	Fibrinogen beta chain	5	Obesity; diabetes; metabolism	Psoriasis	
2246	FGF1	Fibroblast growth factor 1 (acidic)	4	Metabolism	Psoriasis	
2247	FGF2	Fibroblast growth factor 2 (basic)	3	Diabetes; metabolism	Psoriasis	
2252	FGF7	Fibroblast growth factor 7	2			
2255	FGF10	Fibroblast growth factor 10	1			Adipose
2308	FOXO1	Forkhead box O1		Obesity; diabetes; insulin resistance		
2641	GCG	Glucagon	4	Diabetes		
2740	GLP1R	Glucagon-like peptide 1 receptor	4	Diabetes		Glucose; insulin

2833	CXCR3	Chemokine (C-X-C motif) receptor 3	4			Glucose
2952	GSTT1	Glutathione S-transferase theta 1	1	Obesity; diabetes	Psoriasis	
3075	CFH	Complement factor H	2	Obesity; diabetes		
3077	HFE	Hemochromatosis	1	Obesity; diabetes; metabolism; insulin resistance	Psoriasis	
3117	HLA-DQA1	Major histocompatibility complex, class II, DQ alpha 1	1	Diabetes	Psoriasis	
3162	HMOX1	Heme oxygenase (decycling) 1	3	Diabetes; metabolism		
3163	HMOX2	Heme oxygenase (decycling) 2	1			
3315	HSPB1	Heat shock 27 kDa protein 1	1			
3458	IFNG	Interferon, gamma	4	Diabetes	Psoriasis	Diabetes
3479	IGF1	Insulin-like growth factor 1 (somatomedin C)	3	Obesity; diabetes; metabolism; weight gain/loss; insulin resistance		Glucose; insulin; weight gain/loss
3481	IGF2	Insulin-like growth factor 2 (somatomedin A)	1	Obesity; diabetes; metabolism; weight gain/loss; insulin resistance		Weight gain/loss
3552	IL1A	Interleukin 1, alpha	1	Obesity; diabetes; metabolism	Psoriasis	
3557	IL1RN	Interleukin 1 receptor antagonist	2	Obesity; diabetes; metabolism; weight gain/loss	Psoriasis	Weight gain/loss
3565	IL4	Interleukin 4	2	Obesity; diabetes; metabolism	Psoriasis	Glucose

(continued)

Table 1
(continued)

Entrez gene ID	HUGO symbol	Description	Druggable genome (score 1-5) [82]	GWAS (obesity) [83, 84]	GWAS (psoriasis) [83, 84]	Mouse phenotypes [85]
3570	IL6R	Interleukin 6 receptor	1	Obesity; diabetes; metabolism; insulin resistance	Psoriasis	
3576	IL8	Interleukin 8	4	Obesity; diabetes; metabolism	Psoriasis	
3586	IL10	Interleukin 10	2	Obesity; diabetes; metabolism; insulin resistance	Psoriasis	
3596	IL13	Interleukin 13	2	Obesity; diabetes	Psoriasis	
3600	IL15	Interleukin 15	2	Obesity; diabetes; metabolism	Psoriasis	Weight gain/loss
3605	IL17A	Interleukin 17A	2			
3627	CXCL10	Chemokine (C-X-C motif) ligand 10	3	Diabetes		
3630	INS	Insulin	2	Obesity; diabetes; metabolism; weight gain/ loss; insulin resistance		Adipose; diabetes; glucose; insulin; weight gain/loss
3685	ITGAV	Integrin, alpha V	4	Diabetes		
3725	JUN	Jun proto-oncogene	4	Diabetes		Weight gain/loss
3738	KCNA3	Potassium voltage-gated channel, shaker-related subfamily, member 3	4	Insulin resistance		Obesity; glucose; insulin; weight gain/loss

3802	KIR2DL1	Killer cell immunoglobulin-like receptor, two domains, long cytoplasmic tail, 1		Obesity; diabetes	Psoriasis	
3804	KIR2DL3	Killer cell immunoglobulin-like receptor, two domains, long cytoplasmic tail, 3		Obesity; diabetes	Psoriasis	
3809	KIR2DS4	Killer cell immunoglobulin-like receptor, two domains, short cytoplasmic tail, 4		Obesity; diabetes	Psoriasis	
3811	KIR3DL1	Killer cell immunoglobulin-like receptor, three domains, long cytoplasmic tail, 1	1	Diabetes	Psoriasis	
3815	KIT	v-kit Hardy-Zuckerman 4 feline sarcoma viral oncogene homolog	5			Weight gain/loss
3832	KIF11	Kinesin family member 11	3	Diabetes		
3929	LBP	Lipopolysaccharide binding protein	3	Diabetes		
3934	LCN2	Lipocalin 2	2			
3949	LDLR	Low density lipoprotein receptor	4	Obesity; diabetes; metabolism; weight gain/loss; insulin resistance		Glucose; insulin; weight gain/loss
3952	LEP	Leptin	2	Obesity; diabetes; metabolism; weight gain/loss; insulin resistance	Psoriasis	Obesity; adipose; glucose; insulin; weight gain/loss
4018	LPA	Lipoprotein, Lp(a)	5	Obesity; diabetes		
4067	LYN	v-yes-1 Yamaguchi sarcoma viral related oncogene homolog	3	Diabetes		

(continued)

Table 1
(continued)

Entrez gene ID	HUGO symbol	Description	Druggable genome (score 1-5) [82]	GWAS (obesity) [83, 84]	GWAS (psoriasis) [83, 84]	Mouse phenotypes [85]
4091	SMAD6	SMAD family member 6	2	Obesity; diabetes		
4282	MIF	Macrophage migration inhibitory factor (glycosylation-inhibiting factor)	2	Obesity; diabetes	Psoriasis	
4286	MITF	Microphthalmia-associated transcription factor	2			
4311	MME	Membrane metallo-endopeptidase	5	Metabolism		
4327	MMP19	Matrix metallopeptidase 19	1	Diabetes; metabolism		Obesity; weight gain/loss
4524	MTHFR	Methylenetetrahydrofolate reductase (NAD(P)H)	4	Obesity; diabetes; metabolism; weight gain/loss; insulin resistance	Psoriasis	Weight gain/loss
4615	MYD88	Myeloid differentiation primary response 88	2	Diabetes		
4804	NGFR	Nerve growth factor receptor	2	Weight gain/loss		
4846	NOS3	Nitric oxide synthase 3 (endothelial cell)	4	Obesity; diabetes; metabolism; weight gain/loss; insulin resistance		Glucose; insulin; weight gain/loss
4853	NOTCH2	Notch 2	1	Obesity; Diabetes; Weight gain/loss; Insulin resistance		
4929	NR4A2	Nuclear receptor subfamily 4, group A, member 2	1	Obesity; diabetes		
5444	PON1	Paraoxonase 1	4	Obesity; diabetes; metabolism; weight gain/loss; insulin resistance		

5465	PPARA	Peroxisome proliferator-activated receptor alpha	5	Obesity; diabetes; metabolism; weight gain/loss; insulin resistance	Psoriasis	Glucose; insulin; weight gain/loss
5467	PPARD	Peroxisome proliferator-activated receptor delta	4	Obesity; diabetes; metabolism; weight gain/loss; insulin resistance		Adipose; energy homeostasis
5532	PPP3CB	Protein phosphatase 3, catalytic subunit, beta isozyme	3	Diabetes		Weight gain/loss
5551	PRF1	Perforin 1 (pore forming protein)	2	Obesity; diabetes		Diabetes; glucose
5595	MAPK3	Mitogen-activated protein kinase 3	4			
5617	PRL	Prolactin	2	Obesity		
5660	PSAP	Prosaposin	3	Diabetes		Weight gain/loss
5724	PTAFR	Platelet-activating factor receptor	1	Diabetes		
5727	PTCH1	Patched 1	1	Diabetes		
5896	RAG1	Recombination activating gene 1	1	Diabetes		
5919	RARRES2	Retinoic acid receptor responder (tazarotene induced) 2	1	Diabetes		
5981	RFC1	Replication factor C (activator 1) 1, 145 kDa	1	Diabetes	Psoriasis	
6275	S100A4	S100 calcium binding protein A4	1			
6280	S100A9	S100 calcium binding protein A9	1			
6319	SCD	Stearoyl-CoA desaturase (delta-9-desaturase)		Diabetes; metabolism		Obesity; adipose; glucose; insulin
6351	CCL4	Chemokine (C-C motif) ligand 4	2			

(continued)

Table 1
(continued)

Entrez gene ID	HUGO symbol	Description	Druggable genome (score 1-5) [82]	GWAS (obesity) [83, 84]	GWAS (psoriasis) [83, 84]	Mouse phenotypes [85]
6364	CCL20	Chemokine (C-C motif) ligand 20	3	Diabetes		
6376	CX3CL1	Chemokine (C-X3-C motif) ligand 1	2	Diabetes; metabolism	Psoriasis	
6404	SELPLG	Selectin P ligand	2	Diabetes		
6462	SHBG	Sex hormone-binding globulin	4	Obesity; diabetes; insulin resistance		
6648	SOD2	Superoxide dismutase 2, mitochondrial	3	Obesity; diabetes; weight gain/loss; insulin resistance	Psoriasis	Energy homeostasis; glucose; weight gain/loss
6696	SPP1	Secreted phosphoprotein 1	2	Diabetes; metabolism		
6775	STAT4	Signal transducer and activator of transcription 4	3	Diabetes	Psoriasis	Diabetes; glucose; insulin
6868	ADAM17	ADAM metallopeptidase domain 17	3	Diabetes		
7021	TFAP2B	Transcription factor AP-2 beta (activating enhancer binding protein 2 beta)	1	Obesity; diabetes; metabolism; insulin resistance		
7046	TGFBR1	Transforming growth factor, beta receptor 1	4	Obesity; diabetes		
7078	TIMP3	TIMP metallopeptidase inhibitor 3	2	Diabetes		Weight gain/loss
7082	TJP1	Tight junction protein 1	3			

7126	TNFAIP1	Tumor necrosis factor, alpha-induced protein 1 (endothelial)	2	Obesity; metabolism		
7177	TPSAB1	Tryptase alpha/beta 1	3			
7248	TSC1	Tuberous sclerosis 1	1		Psoriasis	
7351	UCP2	Uncoupling protein 2 (mitochondrial, proton carrier)	3	Obesity; diabetes; metabolism; weight gain/loss; insulin resistance	Psoriasis	Glucose
7376	NR1H2	Nuclear receptor subfamily 1, group H, member 2	3	Obesity; diabetes; metabolism; insulin resistance		
7421	VDR	Vitamin D (1,25-dihydroxyvitamin D3) receptor	5	Obesity; diabetes; metabolism; weight gain/loss; insulin resistance	Psoriasis	Diabetes; glucose; weight gain/loss
7422	VEGFA	Vascular endothelial growth factor A	3	Obesity; diabetes; metabolism; insulin resistance	Psoriasis	Weight gain/loss
8455	ATRN	Attractin	1			
8600	TNFSF11	Tumor necrosis factor (ligand) superfamily, member 11	2	Obesity; diabetes; metabolism		
8608	RDH16	Retinol dehydrogenase 16 (all-trans)	3			
8742	TNFSF12	Tumor necrosis factor (ligand) superfamily, member 12	2			
8792	TNFRSF11A	Tumor necrosis factor receptor superfamily, member 11a, NFKB activator	2	Obesity; diabetes; metabolism		

(continued)

Table 1
(continued)

Entrez gene ID	HUGO symbol	Description	Druggable genome (score 1-5) [82]	GWAS (obesity) [83, 84]	GWAS (psoriasis) [83, 84]	Mouse phenotypes [85]
8809	IL18R1	Interleukin 18 receptor 1	1	Obesity; diabetes; metabolism		
9021	SOCS3	Suppressor of cytokine signaling 3	2	Obesity; diabetes; weight gain/loss; insulin resistance		Obesity; glucose; insulin; weight gain/loss
9088	PKMYT1	Protein kinase, membrane associated tyrosine/threonine 1	2			
9173	IL1RL1	Interleukin 1 receptor-like 1	2	Obesity; diabetes		
9252	RPS6KA5	Ribosomal protein S6 kinase, 90 kDa, polypeptide 5	3			
9314	KLF4	Kruppel-like factor 4 (gut)	1	Diabetes		
9517	SPTLC2	Serine palmitoyltransferase, long chain base subunit 2	2			
9663	LPIN2	Lipin 2		Diabetes		
10170	DHRS9	Dehydrogenase/reductase (SDR family) member 9	4			
10216	PRG4	Proteoglycan 4	2			
10558	SPTLC1	Serine palmitoyltransferase, long chain base subunit 1	3			
10890	RAB10	RAB10, member RAS oncogene family	1			
22954	TRIM32	Tripartite motif containing 32		Diabetes		

23411	SIRT1	Sirtuin 1	2	Obesity; diabetes; metabolism; weight gain/loss; insulin resistance		Glucose; insulin; weight gain/loss
23581	CASP14	Caspase 14, apoptosis-related cysteine peptidase	1			
25819	CCRN4L	CCR4 carbon catabolite repression 4-like (*S. cerevisiae*)	1	Diabetes		Obesity; glucose; insulin
26191	PTPN22	Protein tyrosine phosphatase, non-receptor type 22 (lymphoid)	4	Diabetes	Psoriasis	
50604	IL20	Interleukin 20	2		Psoriasis	
54901	CDKAL1	CDK5 regulatory subunit associated protein 1-like 1		Obesity; diabetes; metabolism; weight gain/loss; insulin resistance	Psoriasis	
55806	HR	Hair growth associated	1			
56729	RETN	Resistin	1	Obesity; diabetes; metabolism; weight gain/loss; insulin resistance		Glucose
57521	RPTOR	Regulatory associated protein of MTOR, complex 1				
79071	ELOVL6	ELOVL fatty acid elongase 6	1			
84649	DGAT2	Diacylglycerol O-acyltransferase 2	1	Obesity; diabetes; insulin resistance		Adipose; weight gain/loss
90865	IL33	Interleukin 33		Obesity		
117159	DCD	Dermcidin	2	Diabetes		
148022	TICAM1	Toll-like receptor adaptor molecule 1	1			
406991	MIR21	microRNA 21				

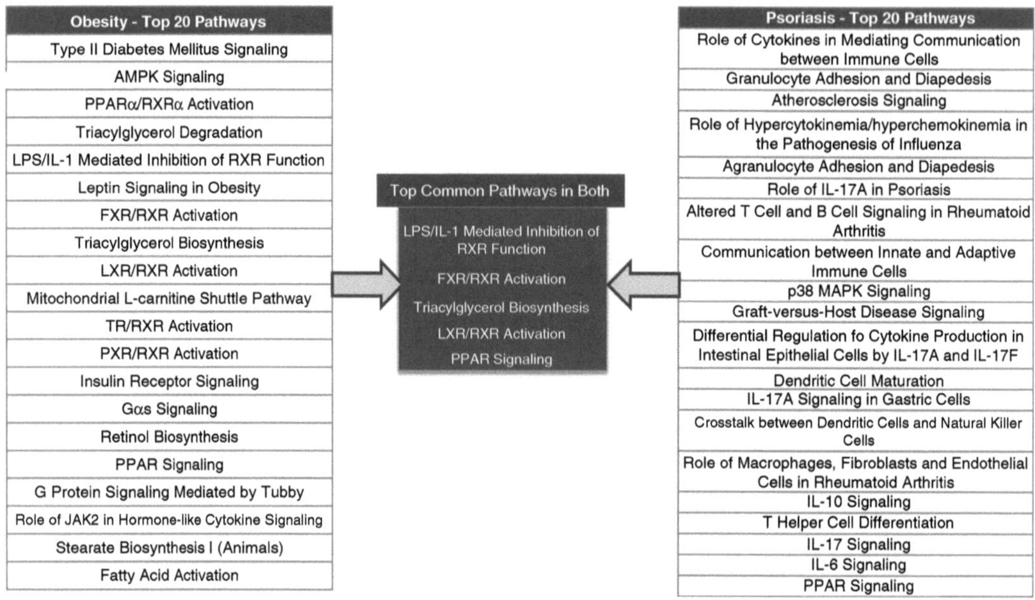

Fig. 7 Pathway evaluation and comparison. Top 20 enriched pathways ($p < 0.05$) for obesity (*left column*) and psoriasis (*right column*) are identified based on genes extracted from literature mining. *Middle table* highlights the top common enrichment pathways between the two diseases

Table 2

Subset of gene ontology categories and molecular pathways significantly enriched in the list of 169 literature derived genes common to both obesity and psoriasis. Enrichment analysis was performed as previously described [61] using gene ontology biological processes (GO:BP), cellular compartments (GO:CC), and molecular pathways (KEGG, REACTOME)

ID	Type	Description	Fisher FDR	Log odds ratio
GO:0005615	GO:CC	Extracellular space	5.6E-23	2.06
GO:0033993	GO:BP	Response to lipid	9.4E-23	2.28
GO:0006954	GO:BP	Inflammatory response	2.3E-22	2.30
GO:0001934	GO:BP	Positive regulation of protein phosphorylation	1.5E-20	2.16
GO:0007243	GO:BP	Intracellular protein kinase cascade	3.7E-18	1.85
GO:0045321	GO:BP	Leukocyte activation	4.8E-16	1.97
GO:0032868	GO:BP	Response to insulin stimulus	6.9E-14	2.35
GO:0007259	GO:BP	JAK-STAT cascade	7.9E-13	2.87
hsa04060	KEGG	Cytokine-cytokine receptor interaction	8.8E-10	1.88
168249	REACTOME	Innate immune system	4.6E-05	1.33
2219530	REACTOME	Constitutive PI3K/AKT signaling in cancer	3.6E-04	2.11

Additionally, employing integrated methodologies with emerging knowledge space could be extremely valuable in identifying new areas for therapeutic intervention [80, 81].

4.5 Obesity and Psoriasis: Drug Discovery

Similar to the approach outlined in [79], we further analyzed the list of 169 genes common to both obesity and psoriasis literature searches (Table 1). The additional data sources to annotate these genes include the Druggable Genome [82], human genetic disease associations from two public databases [83, 84], and mouse phenotypes [85]. 157 out of the 169 genes were represented in the Druggable Genome database and 40 of those had the highest confidence scores of 4 or 5. 129 genes had obesity related genetic associations and 39 out of the 169 genes are reported to have genetic associations with psoriasis. As an example, vitamin D receptor (VDR) is among the high-confidence druggable genes, has genetic association with psoriasis and obesity in both mouse and human. VDR and relevant pathways are of interest in not only metabolic homeostasis but also for other conditions [86]. Thorough analysis of the information presented in Table 1 along with additional information from other sources, could potentially be valuable in identifying some drug discovery targets for further prosecution, and for exploring common mechanistic understandings between psoriasis and obesity. Along with the enhanced knowledge and results from observational and intervention studies, there may potentially be opportunities for repurposing some of the existing molecules to help manage these two diseases [80].

At least 12 clinical trials are found (based on http://www.clinicaltrial.gov November, 2013) to assess interventions or behavior impacts for human subjects with dual conditions of Obesity and Psoriasis, including 6 actively ongoing studies (Table 3). Whereas further insights will be gained once these studies start generating data and results, promising findings have started to surface. For example, it has been reported from several trials that obese patients with moderate to severe psoriasis can show improvements in the severity of psoriasis and/or increase their response to treatment if controlled diet is included in the treatment regimen (clinical trials NCT00512187, NCT00537212, NCT01876875, and NCT01137188).

5 Concluding Remarks

Obesity and psoriasis are major health concerns. In this chapter, we present a literature-mining methodology that evaluates current trends in these fields, and points to gene–disease associations that can be employed in helping make scientific and strategic decisions in drug discovery efforts. Psoriasis has evolved into a disease model

Table 3
Clinical trials with both "Obesity" and "Psoriasis" as disease conditions

NCT number	Title	Status
NCT00512187	Moderate weight loss makes obese patients with severe chronic plaque psoriasis responsive to suboptimal dose of cyclosporine: an investigator blinded, controlled, randomized clinical trial	Completed
NCT01876875	n-3 Polyunsaturated fatty acid-rich diet in psoriasis	Completed
NCT01137448	The effect of weight loss on psoriasis area severity index in adult psoriasis patients	Recruiting
NCT01856647	A pilot study to characterize adipose tissue leukocytes by flow cytometry and microscopy in lean, obese and psoriatic subjects (lean/obese)	Recruiting
NCT00477191	Effects of TNF-alpha antagonism (Etanercept) in patients with the metabolic syndrome and psoriasis	Recruiting
NCT01122095	Cross-sectional evaluation of biological markers of cardiovascular disease in children and adolescents with psoriasis	Recruiting
NCT01439425	Weight reduction alone may not be sufficient to maintain disease remission in obese patients with psoriasis	Not yet recruiting
NCT00800982	Open label study etanercept's maintenance dose in obese patients with moderate to severe plaque type psoriasis	Completed
NCT00537212	Study of the effect of diet in overweight or obese patients with psoriasis on light therapy	Completed
NCT01137188	Effect of weight loss on psoriasis	Completed
NCT01181570	Efficacy and safety of adalimumab in patients with psoriasis and obstructive sleep apnea	Completed
NCT00879944	Impact of the severity of pediatric psoriasis on childhood body mass index	Recruiting

for what are classified as immune mediated inflammatory disorders and its association with metabolic syndrome implicates an underlying common link between these conditions in the form of protein-coding genes and molecular mechanisms. Uncovering these genes and biological pathways offers a chance to further investigate the two diseases in order to gain a new understanding of the disease pathologies and to realize potential opportunities for therapeutic intervention by drug discovery programs. Moreover, understanding comorbidity conditions may lead to new disease treatment strategies and applications of existing or new medicines for effective management of both diseases.

References

1. Agarwal P, Searls DB (2008) Literature mining in support of drug discovery. Brief Bioinform 9:479–492
2. Agarwal P, Searls DB (2009) Can literature analysis identify innovation drivers in drug discovery? Nat Rev Drug Discov 8:865–878
3. Andrade MA, Valencia A (1997) Automatic annotation for biological sequences by extraction of keywords from MEDLINE abstracts. Development of a prototype system. Proc Int Conf Intell Syst Mol Biol 5:25–32
4. Davis AP, Wiegers TC, Johnson RJ, Lay JM, Lennon-Hopkins K, Saraceni-Richards C, Sciaky D, Murphy CG, Mattingly CJ (2013) Text mining effectively scores and ranks the literature for improving chemical-gene-disease curation at the comparative toxicogenomics database. PLoS One 8:e58201
5. Hanisch D, Fluck J, Mevissen HT, Zimmer R (2003) Playing biology's name game: identifying protein names in scientific text. Pac Symp Biocomput 403–414
6. Jenssen TK, Laegreid A, Komorowski J, Hovig E (2001) A literature network of human genes for high-throughput analysis of gene expression. Nat Genet 28:21–28
7. Shatkay H, Edwards S, Wilbur WJ, Boguski M (2000) Genes, themes and microarrays: using information retrieval for large-scale gene analysis. Proc Int Conf Intell Syst Mol Biol 8:317–328
8. Yandell MD, Majoros WH (2002) Genomics and natural language processing. Nat Rev Genet 3:601–610
9. van Haagen HH, 't Hoen PA, de Morrée A, van Roon-Mom WM, Peters DJ, Roos M, Mons B, van Ommen GJ, Schuemie MJ (2011) In silico discovery and experimental validation of new protein-protein interactions. Proteomics 11:843–853
10. van Haagen HH, 't Hoen PA, Mons B, Schultes EA (2013) Generic information can retrieve known biological associations: implications for biomedical knowledge discovery. PLoS One 8:e78665
11. Srinivasan P, Hristovski D (2004) Distilling conceptual connections from MeSH co-occurrences. Stud Health Technol Inform 107:808–812
12. O'Brien PE (2010) Bariatric surgery: mechanisms, indications and outcomes. J Gastroenterol Hepatol 25:1358–1365
13. World Health Organization (2012) Obesity and overweight—fact sheet no. 311. WHO Media Centre Geneva, Switzerland. http://www.who.int/mediacentre/factsheets/fs311/en/
14. Haslam DW, James WP (2005) Obesity. Lancet 366:1197–1209
15. Foxx-Orenstein AE (2010) Gastrointestinal symptoms and diseases related to obesity: an overview. Gastroenterol Clin North Am 39: 23–37
16. Finkelstein EA, Trogdon JG, Cohen JW, Dietz W (2009) Annual medical spending attributable to obesity: payer-and service-specific estimates. Health Aff (Millwood) 28:W822–W831
17. Courcoulas AP, Christian NJ, Belle SH, Berk PD, Flum DR, Garcia L, Horlick M, Kalarchian MA, King WC, Mitchell JE, Patterson EJ, Pender JR, Pomp A, Pories WJ, Thirlby RC, Yanovski SZ, Wolfe BM (2013) Weight change and health outcomes at 3 years after bariatric surgery among individuals with severe obesity. JAMA 310(22):2416–2425
18. Fontana MA, Wohlgemuth SD (2010) The surgical treatment of metabolic disease and morbid obesity. Gastroenterol Clin North Am 39:125–133
19. Rubino F, Schauer PR, Kaplan LM, Cummings DE (2010) Metabolic surgery to treat type 2 diabetes: clinical outcomes and mechanisms of action. Annu Rev Med 61:393–411
20. Derosa G, Maffioli P (2012) Anti-obesity drugs: a review about their effects and their safety. Expert Opin Drug Saf 11:459–471
21. Kaplan LM (2010) Pharmacologic therapies for obesity. Gastroenterol Clin North Am 39:69–79
22. O'Neil PM, Smith SR, Weissman NJ, Fidler MC, Sanchez M, Zhang J, Raether B, Anderson CM, Shanahan WR (2012) Randomized placebo-controlled clinical trial of lorcaserin for weight loss in type 2 diabetes mellitus: the BLOOM-DM study. Obesity (Silver Spring) 20:1426–1436
23. Wong D, Sullivan K, Heap G (2012) The pharmaceutical market for obesity therapies. Nat Rev Drug Discov 11:669–670
24. Kim GW, Lin JE, Blomain ES, Waldman SA (2013) New advances in models and strategies for developing anti-obesity drugs. Expert Opin Drug Discov 8:655–671
25. Herron MD, Hinckley M, Hoffman MS, Papenfuss J, Hansen CB, Callis KP, Krueger GG (2005) Impact of obesity and smoking on psoriasis presentation and management. Arch Dermatol 141:1527–1534
26. Naldi L, Addis A, Chimenti S, Giannetti A, Picardo M, Tomino C, Maccarone M, Chatenoud L, Bertuccio P, Caggese E, Cuscito R (2008) Impact of body mass index and

obesity on clinical response to systemic treatment for psoriasis. Evidence from the Psocare project. Dermatology 217:365–373

27. Henseler T, Christophers E (1995) Disease concomitance in psoriasis. J Am Acad Dermatol 32:982–986
28. Gottlieb AB, Chao C, Dann F (2008) Psoriasis comorbidities. J Dermatolog Treat 19:5–21
29. Gottlieb AB, Dann F, Menter A (2008) Psoriasis and the metabolic syndrome. J Drugs Dermatol 7:563–572
30. Hossler EW, Maroon MS, Mowad CM (2011) Gastric bypass surgery improves psoriasis. J Am Acad Dermatol 65:198–200
31. Farias MM, Achurra P, Boza C, Vega A, de la Cruz C (2012) Psoriasis following bariatric surgery: clinical evolution and impact on quality of life on 10 patients. Obes Surg 22:877–880
32. Krueger G, Ellis CN (2005) Psoriasis—recent advances in understanding its pathogenesis and treatment. J Am Acad Dermatol 53:S94–S100
33. Kurd SK, Gelfand JM (2009) The prevalence of previously diagnosed and undiagnosed psoriasis in US adults: results from NHANES 2003–2004. J Am Acad Dermatol 60:218–224
34. Stern RS, Nijsten T, Feldman SR, Margolis DJ, Rolstad T (2004) Psoriasis is common, carries a substantial burden even when not extensive, and is associated with widespread treatment dissatisfaction. J Investig Dermatol Symp Proc 9:136–139
35. Gudjonsson JE, Elder JT (2007) Psoriasis: epidemiology. Clin Dermatol 25:535–546
36. Elder JT (2009) Genome-wide association scan yields new insights into the immunopathogenesis of psoriasis. Genes Immun 10:201–209
37. Gudjonsson JE (2007) Analysis of global gene expression and genetic variation in psoriasis. J Am Acad Dermatol 57:365
38. Shaiq PA, Stuart PE, Latif A, Schmotzer C, Kazmi AH, Khan MS, Azam M, Tejasvi T, Voorhees JJ, Raja GK, Elder JT, Qamar R, Nair RP (2013) Genetic associations of psoriasis in a Pakistani population. Br J Dermatol 169:406–411
39. Stuart PE, Nair RP, Hiremagalore R, Kullavanijaya P, Kullavanijaya P, Tejasvi T, Lim HW, Voorhees JJ, Elder JT (2010) Comparison of MHC class I risk haplotypes in Thai and Caucasian psoriatics shows locus heterogeneity at PSORS1. Tissue Antigens 76:387–397
40. Shah KN (2013) Diagnosis and treatment of pediatric psoriasis: current and future. Am J Clin Dermatol 14(3):195–213
41. Gan EY, Chong WS, Tey HL (2013) Therapeutic strategies in psoriasis patients with psoriatic arthritis: focus on new agents. BioDrugs 27(4):359–373
42. Kupetsky EA, Mathers AR, Ferris LK (2013) Anti-cytokine therapy in the treatment of psoriasis. Cytokine 61:704–712
43. Davis SA, Feldman SR (2013) Combination therapy for psoriasis in the United States. J Drugs Dermatol 12:546–550
44. Mudigonda P, Mudigonda T, Feneran AN, Alamdari HS, Sandoval L, Feldman SR (2012) Interleukin-23 and interleukin-17: importance in pathogenesis and therapy of psoriasis. Dermatol Online J 18:1
45. Dommasch E, Gelfand JM (2009) Is there truly a risk of lymphoma from biologic therapies? Dermatol Ther 22:418–430
46. Kamangar F, Neuhaus IM, Koo JY (2012) An evidence-based review of skin cancer rates on biologic therapies. J Dermatolog Treat 23: 305–315
47. Gisondi P, Tessari G, Conti A, Piaserico S, Schianchi S, Peserico A, Giannetti A, Girolomoni G (2007) Prevalence of metabolic syndrome in patients with psoriasis: a hospital-based case-control study. Br J Dermatol 157:68–73
48. Langan SM, Seminara NM, Shin DB, Troxel AB, Kimmel SE, Mehta NN, Margolis DJ, Gelfand JM (2012) Prevalence of metabolic syndrome in patients with psoriasis: a population-based study in the United Kingdom. J Invest Dermatol 132:556–562
49. Davidovici BB, Sattar N, Prinz J, Puig L, Emery P, Barker JN, van de Kerkhof P, Stahle M, Nestle FO, Girolomoni G, Krueger JG (2010) Psoriasis and systemic inflammatory diseases: potential mechanistic links between skin disease and co-morbid conditions. J Invest Dermatol 130:1785–1796
50. Azfar RS, Seminara NM, Shin DB, Troxel AB, Margolis DJ, Gelfand JM (2012) Increased risk of diabetes mellitus and likelihood of receiving diabetes mellitus treatment in patients with psoriasis. Arch Dermatol 148:995–1000
51. Higa-Sansone G, Szomstein S, Soto F, Brasecsco O, Cohen C, Rosenthal RJ (2004) Psoriasis remission after laparoscopic Roux-en-Y gastric bypass for morbid obesity. Obes Surg 14:1132–1134
52. Gisondi P, Del GM, Di Francesco V, Zamboni M, Girolomoni G (2008) Weight loss improves the response of obese patients with moderate-to-severe chronic plaque psoriasis to low-dose cyclosporine therapy: a randomized, controlled, investigator-blinded clinical trial. Am J Clin Nutr 88:1242–1247
53. Glossmann H, Reider N (2013) A marriage of two "Methusalem" drugs for the treatment of psoriasis?: arguments for a pilot trial with metformin as add-on for methotrexate. Dermatoendocrinol 5:252–263

54. Fiehn C (2010) Methotrexate transport mechanisms: the basis for targeted drug delivery and ss-folate-receptor-specific treatment. Clin Exp Rheumatol 28:S40–S45
55. Ahlehoff O, Skov L, Gislason G, Lindhardsen J, Kristensen SL, Iversen L, Lasthein S, Gniadecki R, Dam TN, Torp-Pedersen C, Hansen PR (2013) Cardiovascular disease event rates in patients with severe psoriasis treated with systemic anti-inflammatory drugs: a Danish real-world cohort study. J Intern Med 273:197–204
56. Perera GK, Di MP, Nestle FO (2012) Psoriasis. Annu Rev Pathol 7:385–422
57. Tian S, Krueger JG, Li K, Jabbari A, Brodmerkel C, Lowes MA, Suarez-Farinas M (2012) Meta-analysis derived (MAD) transcriptome of psoriasis defines the "core" pathogenesis of disease. PLoS One 7:e44274
58. Suarez-Farinas M, Li K, Fuentes-Duculan J, Hayden K, Brodmerkel C, Krueger JG (2012) Expanding the psoriasis disease profile: interrogation of the skin and serum of patients with moderate-to-severe psoriasis. J Invest Dermatol 132:2552–2564
59. Wheeler DL, Barrett T, Benson DA, Bryant SH, Canese K, Chetvernin V, Church DM, Dicuccio M, Edgar R, Federhen S, Feolo M, Geer LY, Helmberg W, Kapustin Y, Khovayko O, Landsman D, Lipman DJ, Madden TL, Maglott DR, Miller V, Ostell J, Pruitt KD, Schuler GD, Shumway M, Sequeira E, Sherry ST, Sirotkin K, Souvorov A, Starchenko G, Tatusov RL, Tatusova TA, Wagner L, Yaschenko E (2008) Database resources of the National Center for Biotechnology Information. Nucleic Acids Res 36:D13–D21
60. Mitchell JA, Aronson AR, Mork JG, Folk LC, Humphrey SM, Ward JM (2003) Gene indexing: characterization and analysis of NLM's GeneRIFs. AMIA Annu Symp Proc 460–464
61. Freudenberg JM, Joshi VK, Hu Z, Medvedovic M (2009) CLEAN: CLustering Enrichment ANalysis. BMC Bioinformatics 10:234
62. Gentleman RC, Carey VJ, Bates DM, Bolstad B, Dettling M, Dudoit S, Ellis B, Gautier L, Ge Y, Gentry J, Hornik K, Hothorn T, Huber W, Iacus S, Irizarry R, Leisch F, Li C, Maechler M, Rossini AJ, Sawitzki G, Smith C, Smyth G, Tierney L, Yang JY, Zhang J (2004) Bioconductor: open software development for computational biology and bioinformatics. Genome Biol 5:R80
63. Leibel RL (2008) Molecular physiology of weight regulation in mice and humans. Int J Obes (Lond) 32 Suppl 7: S98–S108
64. Valentino MA, Lin JE, Waldman SA (2010) Central and peripheral molecular targets for antiobesity pharmacotherapy. Clin Pharmacol Ther 87:652–662
65. Ruderman NB, Carling D, Prentki M, Cacicedo JM (2013) AMPK, insulin resistance, and the metabolic syndrome. J Clin Invest 123: 2764–2772
66. Lindholm CR, Ertel RL, Bauwens JD, Schmuck EG, Mulligan JD, Saupe KW (2013) A high-fat diet decreases AMPK activity in multiple tissues in the absence of hyperglycemia or systemic inflammation in rats. J Physiol Biochem 69:165–175
67. Pacher P, Kunos G (2013) Modulating the endocannabinoid system in human health and disease—successes and failures. FEBS J 280: 1918–1943
68. Ionut V, Burch M, Youdim A, Bergman RN (2013) Gastrointestinal hormones and bariatric surgery-induced weight loss. Obesity (Silver Spring) 21:1093–1103
69. Mells JE, Anania FA (2013) The role of gastrointestinal hormones in hepatic lipid metabolism. Semin Liver Dis 33:343–357
70. Huffmeier U, Uebe S, Ekici AB, Bowes J, Giardina E, Korendowych E, Juneblad K, Apel M, McManus R, Ho P, Bruce IN, Ryan AW, Behrens F, Lascorz J, Bohm B, Traupe H, Lohmann J, Gieger C, Wichmann HE, Herold C, Steffens M, Klareskog L, Wienker TF, Fitzgerald O, Alenius GM, McHugh NJ, Novelli G, Burkhardt H, Barton A, Reis A (2010) Common variants at TRAF3IP2 are associated with susceptibility to psoriatic arthritis and psoriasis. Nat Genet 42:996–999
71. Bohm B, Burkhardt H, Uebe S, Apel M, Behrens F, Reis A, Huffmeier U (2012) Identification of low-frequency TRAF3IP2 coding variants in psoriatic arthritis patients and functional characterization. Arthritis Res Ther 14:R84
72. Benson JM, Sachs CW, Treacy G, Zhou H, Pendley CE, Brodmerkel CM, Shankar G, Mascelli MA (2011) Therapeutic targeting of the IL-12/23 pathways: generation and characterization of ustekinumab. Nat Biotechnol 29:615–624
73. Yawalkar N, Karlen S, Hunger R, Brand CU, Braathen LR (1998) Expression of interleukin-12 is increased in psoriatic skin. J Invest Dermatol 111:1053–1057
74. Martin DA, Towne JE, Kricorian G, Klekotka P, Gudjonsson JE, Krueger JG, Russell CB (2013) The emerging role of IL-17 in the pathogenesis of psoriasis: preclinical and clinical findings. J Invest Dermatol 133:17–26
75. Gisondi P, Girolomoni G, Sampogna F, Tabolli S, Abeni D (2005) Prevalence of psoriatic arthritis and joint complaints in a large population

of Italian patients hospitalised for psoriasis. Eur J Dermatol 15:279–283

76. Nestle FO, Kaplan DH, Barker J (2009) Psoriasis. N Engl J Med 361:496–509
77. Debets R, Hegmans JP, Croughs P, Troost RJ, Prins JB, Benner R, Prens EP (1997) The IL-1 system in psoriatic skin: IL-1 antagonist sphere of influence in lesional psoriatic epidermis. J Immunol 158:2955–2963
78. Groves RW, Kapahi P, Barker JN, Haskard DO, MacDonald DM (1995) Detection of circulating adhesion molecules in erythrodermic skin disease. J Am Acad Dermatol 32:32–36
79. Freudenberg JM, Rajpal N, Way JM, Magid-Slav M, Rajpal DK (2013) Gastrointestinal weight-loss surgery: glimpses at the molecular level. Drug Discov Today 18(13–14):625–636
80. Hurle MR, Yang L, Xie Q, Rajpal DK, Sanseau P, Agarwal P (2013) Computational drug repositioning: from data to therapeutics. Clin Pharmacol Ther 93:335–341
81. Buchan NS, Rajpal DK, Webster Y, Alatorre C, Gudivada RC, Zheng C, Sanseau P, Koehler J (2011) The role of translational bioinformatics in drug discovery. Drug Discov Today 16:426–434
82. Hopkins AL, Groom CR (2002) The druggable genome. Nat Rev Drug Discov 1:727–730
83. Hindorff LA, Sethupathy P, Junkins HA, Ramos EM, Mehta JP, Collins FS, Manolio TA (2009) Potential etiologic and functional implications of genome-wide association loci for human diseases and traits. Proc Natl Acad Sci U S A 106:9362–9367
84. Zhang Y, De S, Garner JR, Smith K, Wang SA, Becker KG (2010) Systematic analysis, comparison, and integration of disease based human genetic association data and mouse genetic phenotypic information. BMC Med Genomics 3:1
85. Eppig JT, Blake JA, Bult CJ, Kadin JA, Richardson JE (2012) The Mouse Genome Database (MGD): comprehensive resource for genetics and genomics of the laboratory mouse. Nucleic Acids Res 40:D881–D886
86. Carlberg C, Molnar F (2012) Current status of vitamin D signaling and its therapeutic applications. Curr Top Med Chem 12:528–547

Chapter 12

Integrative Literature and Data Mining to Rank Disease Candidate Genes

Chao Wu, Cheng Zhu, and Anil G. Jegga

Abstract

While the genomics-derived discoveries promise benefits to basic research and health care, the speed and affordability of sequencing following recent technological advances has further aggravated the data deluge. Seamless integration of the ever-increasing clinical, genomic, and experimental data and efficient mining for knowledge extraction, delivering actionable insight and generating testable hypotheses are therefore critical for the needs of biomedical research. For instance, high-throughput techniques are frequently applied to detect disease candidate genes. Experimental validation of these candidates however is both time-consuming and expensive. Hence, several computational approaches based on literature and data mining have been developed to identify the most promising candidates for follow-up studies. Based on "guilt by association" principle, most of these methods use prior knowledge about a disease of interest to discover and rank novel candidate genes. In this chapter, we provide a brief overview of recent advances made in literature- and data-mining-based approaches for candidate gene prioritization. As a case study, we focus on a Web-based computational approach that uses integrated heterogeneous data sources including gene–literature associations for ranking disease candidate genes and explain how to run typical queries using this system.

Key words Literature mining, Data mining, Candidate gene ranking, Disease gene ranking, Integrative genomics, Systems biology

1 Introduction

Recent advances in high-throughput techniques in biomedical domains are generating humongous amounts of heterogeneous data at a pace that undermines the strategies to deal with it effectively. The rapid accumulation of literature data which is beyond the reach of manual curation precipitates the information deluge problem further. According to an estimate, the number of published articles continues to increase further at a rate exceeding 600,000 articles per year [1]. As a result, often, there is a significant delay between the publication of research finding and the extraction of this information and using it in the discovery pipelines. To improve the efficiency and the consistency of the overall

Vinod D. Kumar and Hannah Jane Tipney (eds.), *Biomedical Literature Mining*, Methods in Molecular Biology, vol. 1159, DOI 10.1007/978-1-4939-0709-0_12,

knowledge extraction process and to minimize the workload of expert or manual curation, text- and data-mining tools are used as potential alternatives with varying results. Comprehensive literature-based analyses are now an integral part of complex biomedical research and knowledge extraction pipelines. Further, the scope of literature analysis now encompasses not just the abstracts of published articles but the full texts and other sources of biomedical information including gene expressions, biochemistry, and genome biology [2]. Literature-mining and integrated data-mining approaches have been particularly effective discovering and prioritizing disease-causal genes and targets of bioactive compounds.

Integrated literature and data-mining-based analyses can broadly be classified into following independent but related procedures that include *information retrieval* (IR) and *information extraction* (IE), building entities associations, and knowledge inference. An IR engine responds to a user-generated query comprising of specified terms of interest (e.g., keywords), and retrieves a collection of documents across multiple scientific disciplines that are related to the query. The retrieval of text is usually performed on the basis of keyword matching. Therefore, a challenging task is to identify and extract such entities (present often as synonyms), and this process is referred to as *information extraction* (IE). In the biomedical domains, IE typically comprises the identification of bio-entities (e.g., genes, diseases or bioactive compounds) and the relationships between these entities. An IE solution typically begins with entity recognition. Given a stream of text, the *named entity recognition* (NER) determines which items in the text map to proper names that are stored in a controlled vocabulary. Different IE solutions may apply various dictionaries or controlled vocabularies, such as Unified Medical Language System (UMLS) [3], Mammalian Phenotype Ontology [4], MeSH [5], OMIM [6], Gene Ontology [7], etc. To index a scientific publication, the text will usually undergo a series of procedures starting with the segmentation and labeling of the text into different sections such as abstract, introduction, results, etc. The partitioned text is further segmented into sentences which will be further decomposed into a bag of tokens where performing specific steps (e.g., stemming) may be advantageous [8]. Once the preprocessing of each document of the corpus is completed, an IR engine, such as PubMed/MEDLINE, will be able to map the keywords of the query to the indexed documents, retrieving relevant documents. Table 1 lists some of the resources and tools related to information retrieval and information extraction from biomedical literature (see BioNLP.org for more details).

1.1 Literature-Mining-Based Biomedical Discoveries

Starting with the Swanson's ABC model [25] to the more recent semantic-Web-based approaches (e.g., PosMed [26]), several biomedical literature mining systems have been developed for knowledge extraction and hypotheses generation directly from the literature. Several applications have also been developed where literature mining

Table 1
Information retrieval and information extraction tools (*Note*: *this list is not exhaustive and is primarily meant to provide a list of examples. We apologize for any oversights*)

Resource/tool	URL	Summary
CTD [9]	http://ctd.mdibl.org/	Provides insights into chemical actions, disease susceptibility, toxicity, and therapeutic drug interactions by curating and integrating data describing relationships between chemicals, genes/proteins, and human diseases
PharmGKB [10]	http://www.pharmgkb.org/	A comprehensive resource that curates knowledge about the impact of genetic variation on drug response for clinicians and researchers
STITCH [11]	http://stitch.embl.de/	Contains interactions between 1.5 million genes and over 68,000 bioactive compounds
iHOP [12]	http://www.ihop-net.org/UniPub/iHOP/	Processes Medline abstracts and generate hyperlinked sets of protein interactions
Predictive networks [13]	http://predictivenetworks.org/	Integrates gene interactions and networks information from PubMed literature and other online biological databases and presents it in an accessible and efficient user interface
GeneWays [14]	http://anya.igsb.anl.gov/Geneways/GeneWays.html	A system for automatically extracting, analyzing, visualizing, and integrating molecular pathway data from the research literature
CoPub [15]	http://services.nbic.nl/copub/portal	Retrieves co-occurring biomedical concepts that are divided into curated classes from MEDLINE abstracts
Textpresso [16]	http://www.textpresso.org	Provides extracted statements containing entities of interest on a subset of full text articles. The entities are organized in hierarchical structures
Reflect [17]	http://reflect.embl.de	Reflects tags gene, protein, and small molecule names in any Web page, typically within a few seconds, and without affecting document layout. Clicking on the tagged concepts then presents detailed knowledge associated with the concepts
Bio2RDF [18]	http://bio2rdf.org	Annotates documents from public bioinformatics databases into Resource Description Framework (RDF) format with normalized URIs and ontologies

(continued)

Table 1 (continued)

Resource/tool	URL	Summary
MetaMap [19]	http://metamap.nlm.nih.gov/	Maps biomedical text to the UMLS Metathesaurus or, equivalently, to discover Metathesaurus concepts referred to in text
BITOLA [20]	http://ibmi3.mf.uni-lj.si/bitola/	An interactive literature-based biomedical discovery support system. The purpose of the system is to help the biomedical researchers make new discoveries by discovering potentially new relations between biomedical concepts in both closed and open discovery systems
GLAD4U [21]	http://bioinfo.vanderbilt.edu/glad4u/	The gene retrieval and prioritization tool takes advantage of existing resources of NCBI to identify genes associated with the query and rank candidate genes based upon statistical significance between genes and the query entity
BioGraph [22]	http://biograph.be/	Offers prioritizations of putative disease genes, supported by functional hypotheses
PosMed [23]	http://biosparql.org/PosMed/	Ranks biomedical resources such as genes, metabolites, diseases, and drugs, based on the statistical significance of associations between a user-specified phenotypic keyword and resources connected directly or inferentially through a Semantic Web of biological databases
Arrowsmith [24]	http://arrowsmith.psych.uic.edu	Identifies and prioritizes concepts from two literature searches that may be relevant to both entities of the searches

is used in an integrated manner complementing other bioinformatics methods (e.g., genomic and gene annotation data mining) for identifying and ranking candidate genes for diseases and drugs (e.g., [26–28]). The classical triangulation principle wherein two concepts without explicit communications may be associated through a third concept being their common intermediate serves as a base to exploit existing concept co-occurring relationships in literature to build novel and potentially meaningful associations. For example, Arrowsmith [29], which was built on the closed discovery mode of the original ABC model, relates two articles to each other even if they did not share any authors and represent disparate topics. Likewise, CoPub Discovery [30], directly mines the concept co-occurrence from biomedical literature and prioritizes potential concepts related to diseases or other genes. Another class of literature-mining-based methods uses lexical statistics (e.g., relative token counts, inverse document frequency, etc.) of the key terms (e.g., genes, diseases, etc.) to discover hidden connections in the biomedical literature [31]. These statistics are analyzed together with co-occurrence statistics and intermediate terms that are similar to the query terms are identified and prioritized [32]. Box 1 presents some examples from published literature where literature-mining-based methods have been used successfully to make biomedical discoveries.

Box 1
Literature-Mining-Based Biomedical Discoveries: Examples and Tools

Molecular triangulation: Bridging linkage and molecular-network information for identifying candidate genes in Alzheimer's disease (AD): Mining in 25 scientific journals using GeneWays [33] a literature-derived interaction network was generated. A list of 60 known AD candidate genes was used as a set of seeds to search subnetworks that might harbor entities related to AD [34].

Literature mined gene-interaction network analyses: Starting with a list of 15 prostate cancer-related genes (seed genes), a disease-specific gene-interaction network was built and mined using dependency parsing and support vector machine-based approach. The extended list of genes in the gene-interaction network was then ranked and prioritized according to the closeness centrality in the literature- mined network [35].

Semantic networks for pharmacogenomics: Ontology of pharmacogenomics (GPx), built from a lexicon of key pharmacogenomics entities and a syntactic parsing of abstracts. The hierarchical structure was then used to systematically extract commonly occurring relationships and map them to common schemas. The result was a network of 40,000 relationships between more than 200 entity types with clear semantics [36].

Drug-drug interactions discovery: This study was based on the hypothesis that drug-drug interactions (DDIs) can occur when two drugs interact with the same gene product. First, a collection of normalized gene–drug relationships was derived through literature mining. Using a training set of established DDIs, a random forest classifier was trained to score potential DDIs based on the features of the normalized assertions extracted from the literature that relate two drugs to a gene product [37].

PhenomeNET: Phenotypes are investigated in model organisms to understand and reveal the molecular mechanisms underlying disease. This study applied literature-mining techniques to transform phenotype ontologies into a formal representation under the framework of Web Ontology Language (OWL). The system compared phenotypes in animal models with the phenotypes of human diseases and had revealed causal mutations with high accuracy. The system also exhibited potential to rank genes for diseases with unknown molecular basis [38].

Discovering hidden connections between drugs, genes and diseases: CoPub Discovery is based on the observation of the ABC-principle, where A and C have no direct relationship but are connected via B as an intermediate. This method extracts concept entities such as genes, diseases, and drugs from published literature. An *R*-scaled score is then calculated based on the mutual information measure ranging from 1 to 100 [30].

Traditional prioritization approaches for disease gene identification based on literature review however can rapidly become overwhelming. Alternatively, computational integration of different criteria can be performed to create a ranking function to identify and rank potential disease candidate genes. As outlined in Table 2, several gene prioritization methods have been developed to overcome the limitations of high-throughput, genome-wide studies like linkage analysis and gene expression profiling, both of which typically result in the identification of hundreds of potential candidate genes [27, 28, 39–47]. While all of these tools are based on the assumption that similar phenotypes are caused by genes with similar or related functions [27, 40, 48–50], they differ by the strategy adopted in calculating similarity and by the data sources utilized [51]. Using an example of recently discovered disease gene, we demonstrate here the applicability of integrative literature and heterogeneous data-mining-based approaches to identify and prioritize disease candidate genes incorporating known disease-gene knowledge. Specifically, we describe ToppGene (http://toppgene.cchmc.org) [27] which facilitates integration of several heterogeneous data sources including published literature to perform candidate gene prioritization.

2 Materials

Since the application we present here is Web-based, a computer with Internet connection and a compatible Web browser is needed. The system we are going to present here (Toppgene: http://toppgene.cchmc.org) is tested regularly on a number of browsers and operating systems and does not have any compatibility issues to the best of our knowledge. For any "guilt by association" based disease gene prioritization approach, a training set (also referred to as "seed" set) representing known knowledge in the form of genes associated/related to disease of interest is critical.

3 Methods

The methods described here, and the screenshots used to illustrate them, are correct for the servers/databases as it were at the time of writing (June 2013). From time to time, interfaces and query/search options may be developed in response to users' feedback and details may change. For compiling training set genes for disease, we will use comparative toxicogenomics database (CTD) [52] and for ranking the disease genes we will use the Toppgene application from the ToppGene Suite [27].

Table 2
List of some of the bioinformatics approaches and tools to rank human disease candidate genes. The *first column* has the source or the name of the tool (including reference, if available). The *second column* has the URL of the corresponding Web application, when available. If there is no Web application, information regarding either the project home page or links to the corresponding supplementary material are provided. The *third column* is the genomic annotation types/features used by each of the methods. The *last column* has details of the training or the input data, if used (*Note*: *this list is neither extensive nor exhaustive and we apologize for any oversights. The primary purpose of this list is to provide a list of tools as examples*). *See* also Gene Prioritization Portal (http://www.esat.kuleuven.be/gpp) which provides detailed information for more than 40 gene prioritization tools and helps users in selecting a gene prioritization strategy that suits best their needs [63]

Approach	Online availability	Data types used	Training set (input)
Approaches using links between genes and phenotypes			
Genes2Diseases [64, 65]	http://www.ogic.ca/projects/g2d_2/	Sequence, GO, literature mining	Phenotype GO terms Known genes
BITOLA [20]	http://www.mf.uni-lj.si/bitola/	Literature mining	Concept
Tiffin et al. [41]	Article supplementary data available at http://www.sanbi.ac.za/tiffin_et_al/	Expression, literature mining	Disease
GeneSeeker [66, 67]	http://www.cmbi.ru.nl/GeneSeeker/	Expression, phenotype, literature mining	N/A
GFINDer [68, 69]	http://www.bioinformatics.polimi.it/GFINDer/	Expression, phenotype	N/A
TOM [70]	http://www-micrel.deis.unibo.it/~tom/	Expression, GO	Known genes and/or disease loci
Approaches using functional relatedness between candidate genes			
OMIM phenome map [71]	http://www.cmbi.ru.nl/MimMiner/	Phenotype, sequence, GO, protein interactions	N/A
SUSPECTS [46]	http://www.genetics.med.ed.ac.uk/suspects/	Sequence, expression, GO	Known genes
Prioritizer [72]	http://www.prioritizer.nl/	Expression, GO, protein interactions	Disease loci
Endeavour [28]	http://www.esat.kuleuven.be/endeavour/	Sequence, expression, GO, pathways, literature mining	Known genes
ToppGene [27]	http://toppgene.cchmc.org	Mouse phenotype, expression, GO, pathways, literature mining	Known genes

3.1 ToppGene: Functional Annotations-Based Candidate Gene Ranking

The backend knowledgebase of ToppGene consists of 17 gene feature types compiled from different publicly available the public domain. These include disease-dependent and disease-independent information such as known disease-genes, previous linkage regions, association studies, human and mouse phenotypes, known drug-targets, and microarray expression results, gene regulatory regions (transcription factor target genes and microRNA targets), protein domains, protein interactions, pathways, biological processes, literature co-citations, etc. Each of these sources is used in an integrated manner to prioritize disease candidate genes.

As part of the workflow, a representative profile of the training genes (functional enrichment) using 17 different features (as listed above) is generated first. From the functional enrichment profile of the training genes, over-representative terms are identified. The test set genes are then compared to these overrepresented terms for all categorical annotations and the average vector for the expression values. For each of the test set genes, a similarity score to the training profile for each of the 17 features is then derived and summarized (17 similarity scores). If a test gene lacks one or more annotations, the score is set to −1; otherwise, it is a real value in [0, 1]. For computing similarity measures of categorical (e.g., GO annotations) annotations, ToppGene uses a fuzzy-based similarity measure (*see* Popescu et al. [53] for additional details). In case of numeric annotations (e.g., microarray expression values), the similarity score is calculated as the Pearson correlation of the two expression vectors of the two genes. The 17 similarity scores are combined into an overall score using statistical meta-analysis and a *p-value* of each annotation of a test gene G is derived by random sampling of the whole genome. The *p-value* of the similarity score S_i is defined as:

$$p(S_i) = \frac{\text{Count of genes having score higher than G in the random sample}}{\text{Count of genes in the random sample containing annotation}}.$$

Fisher's inverse chi-square method, which states that $-2\sum_{i=1}^{n} \log p_i \sum \chi^2_{(2n)}$ (assuming the p_i values come from independent tests) is then applied to combine the *p-values* from multiple annotations into an overall *p-value.* The final similarity score of the test gene is then obtained by 1 minus the combined *p-value.* For additional details regarding the development of the ranking approach, validation and comparison with other related applications, readers are referred to previously published studies [27, 47].

3.2 Identifying and Ranking Disease Genes for Malformations of Cortical Development (MCD)

To illustrate the use of ToppGene, we will use the disease MCD (malformations of cortical development) and four recently reported pathogenic mutations in *TUBG1*, *DYNC1H1*, *KIF2A*, and *KIF5C* [54]. The goal is to demonstrate the effectiveness of integrated literature- and data-mining-based approaches in ranking novel disease candidate genes. The following sections describe the workflow and the results.

3.2.1 Compiling Disease Training Set Genes for MCD

Some of the resources which are commonly used to compile known disease-associated genes are OMIM [55], the Genetic Association Database [56], GWAS [57], and the Comparative Toxicogenomics Database (CTD) [52]. The latter, i.e., CTD also integrates diseases biomarkers derived from literature and specialized database mining. For the current example, we will use CTD to compile the training list for MCD.

1. On the home page of CTD, select "Diseases" from the drop-down list under "Keyword Search" and enter "Malformations of Cortical Development" and hit "Search" (Fig. 1); this will bring you to the corresponding disease entry (Fig. 2).
2. On the "Malformations of Cortical Development" click either on the "Genes" tab (from the navigation bar) or the small helix icon following the "Malformations of Cortical Development" (Fig. 2).

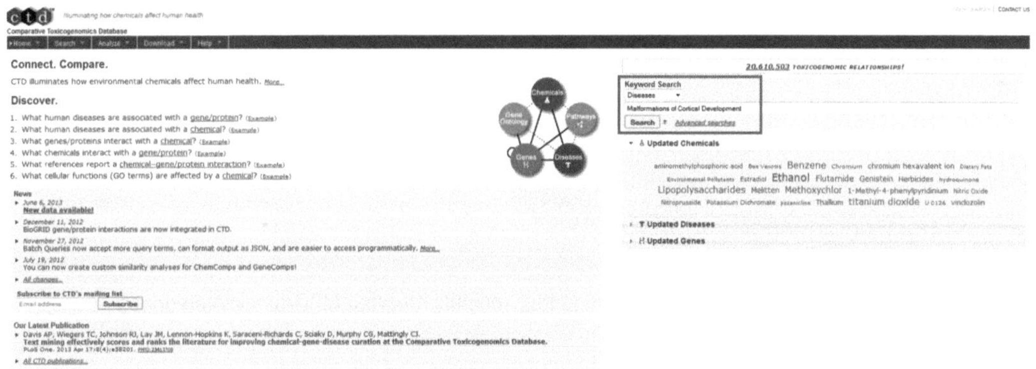

Fig. 1 The CTD home page (the *red box* indicates **step 1** of Subheading 3.2.1) (Color figure online)

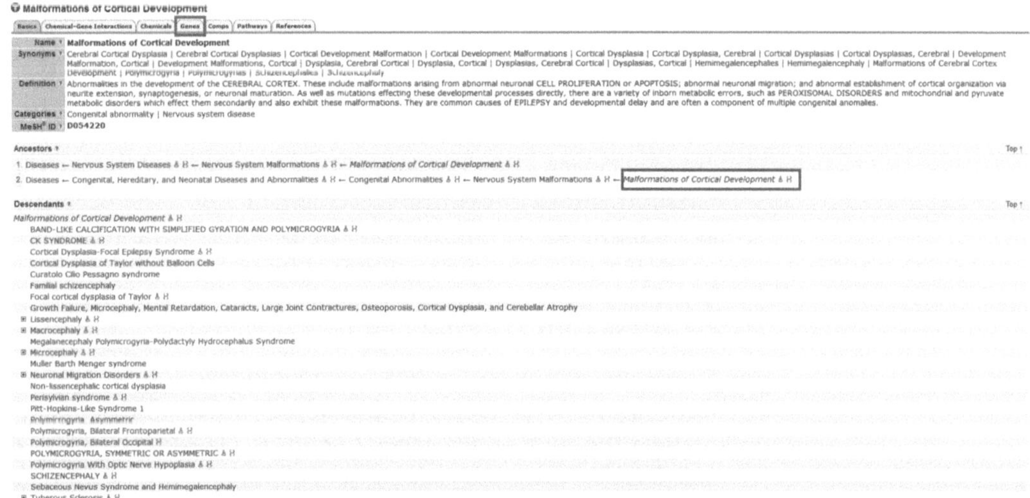

Fig. 2 The CTD results page showing the details of the queried disease

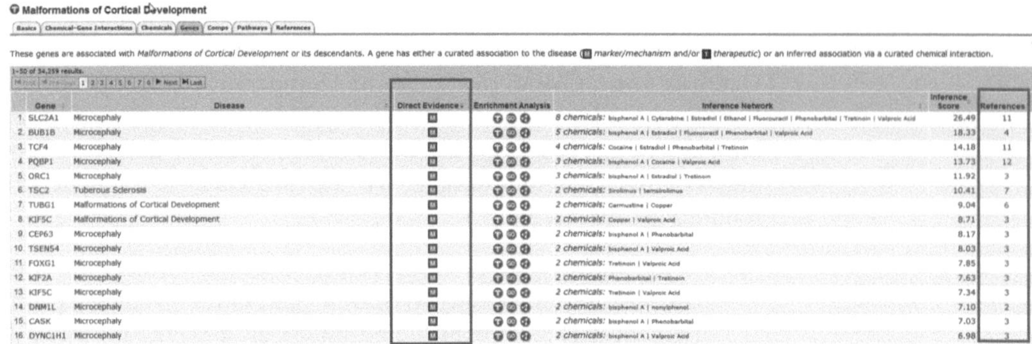

Malformations of Cortical Development

Basics | Chemical-Gene Interactions | Chemicals | Genes | Comps | Pathways | References

These genes are associated with *Malformations of Cortical Development* or its descendants. A gene has either a curated association to the disease (M *marker/mechanism* and/or T *therapeutic*) or an inferred association via a curated chemical interaction.

1-50 of 34,259 results.

First | Previous | 1 | 2 | 3 | 4 | 5 | 6 | 7 | 8 | Next | Last

	Gene	Disease	Direct Evidence	Enrichment Analysis	Inference Network	Inference Score	References
1.	SLC2A1	Microcephaly	M		*8 chemicals:* bisphenol A | Cytarabine | Estradiol | Ethanol | Fluorouracil | Phenobarbital | Tretinoin | Valproic Acid	26.49	11
2.	BUB1B	Microcephaly	M		*5 chemicals:* bisphenol A | Estradiol | Fluorouracil | Phenobarbital | Valproic Acid	18.33	4
3.	TCF4	Microcephaly	M		*4 chemicals:* Cocaine | Estradiol | Phenobarbital | Tretinoin	14.18	11
4.	PQBP1	Microcephaly	M		*3 chemicals:* bisphenol A | Cocaine | Valproic Acid	13.73	12
5.	ORC1	Microcephaly	M		*3 chemicals:* bisphenol A | Estradiol | Tretinoin	11.92	3
6.	TSC2	Tuberous Sclerosis	M		*2 chemicals:* Sirolimus | temsirolimus	10.41	7
7.	TUBG1	Malformations of Cortical Development	M		*2 chemicals:* Carmustine | Copper	9.04	6
8.	KIF5C	Malformations of Cortical Development	M		*2 chemicals:* Copper | Valproic Acid	8.71	3
9.	CEP63	Microcephaly	M		*2 chemicals:* bisphenol A | Phenobarbital	8.17	3
10.	TSEN54	Microcephaly	M		*2 chemicals:* bisphenol A | Valproic Acid	8.03	3
11.	FOXG1	Microcephaly	M		*2 chemicals:* Tretinoin | Valproic Acid	7.85	3
12.	KIF2A	Microcephaly	M		*2 chemicals:* Phenobarbital | Tretinoin	7.63	3
13.	KIF5C	Microcephaly	M		*2 chemicals:* Tretinoin | Valproic Acid	7.34	3
14.	DNM1L	Microcephaly	M		*2 chemicals:* bisphenol A | nonylphenol	7.10	2
15.	CASK	Microcephaly	M		*2 chemicals:* bisphenol A | Phenobarbital	7.03	3
16.	DYNC1H1	Microcephaly	M		*2 chemicals:* bisphenol A | Valproic Acid	6.98	3

Fig. 3 The CTD results page showing the details of the queried disease and its gene associations

Nat Genet. 2013 Apr 21;45(6):639-47. doi: 10.1038/ng.2613. Epub 2013 Apr 21.

Mutations in TUBG1, DYNC1H1, KIF5C and KIF2A cause malformations of cortical development and microcephaly.

Poirier K, Lebrun N, Broix L, Tian G, Saillour Y, Boscheron C, Parrini E, Valence S, Pierre BS, Oger M, Lacombe D, Geneviève D, Fontana E, Darra F, Cances C, Barth M, Bonneau D, Bernadina BD, N'guyen S, Gitiaux C, Parent P, des Portes V, Pedespan JM, Legrez V, Castelnau-Ptakine L, Nitschke P, Hieu T, Masson C, Zelenika D, Andrieux A, Francis F, Guerrini R, Cowan NJ, Bahi-Buisson N, Chelly J.

1] Institut Cochin, Université Paris-Descartes, Centre National de la Recherche Scientifique (CNRS), UMR 8104, Paris, France. [2] Institut National de la Santé et de la Recherche Médicale (INSERM) U1016, Paris, France.

Abstract

The genetic causes of malformations of cortical development (MCD) remain largely unknown. Here we report the discovery of multiple pathogenic missense mutations in TUBG1, DYNC1H1 and KIF2A, as well as a single germline mosaic mutation in KIF5C, in subjects with MCD. We found a frequent recurrence of mutations in DYNC1H1, implying that this gene is a major locus for unexplained MCD. We further show that the mutations in KIF5C, KIF2A and DYNC1H1 affect ATP hydrolysis, productive protein folding and microtubule binding, respectively. In addition, we show that suppression of mouse Tubg1 expression in vivo interferes with proper neuronal migration, whereas expression of altered γ-tubulin proteins in Saccharomyces cerevisiae disrupts normal microtubule behavior. Our data reinforce the importance of centrosomal and microtubule-related proteins in cortical development and strongly suggest that microtubule-dependent mitotic and postmitotic processes are major contributors to the pathogenesis of MCD.

PMID: 23603762 [PubMed - in process]

Fig. 4 Screenshot of PubMed abstract highlighting four recently reported MCD-associated genes

3. This will lead you to a gene–disease relationship table page with additional details supporting the relationship along with the related references (Fig. 3). The column marked "Direct Evidence" (green icon "M") as the name suggests indicates that a gene has a curated association to the disease. Download this results table by using any of the "Download" options present at the end on the page.
4. Use all the genes that have "Direct Evidence" (indicated by "marker/mechanism") and "Disease Name" as either "Malformations of Cortical Development" or "Microcephaly" as a training set for MCD. You can use either gene symbols or gene IDs (columns one and two) as input for further analyses. There are 34 unique genes related to MCD or microcephaly (as on June 2013 in CTD).
5. Since our goal is to see how computational approaches perform in ranking recently discovered novel disease genes, remove the four target genes (*TUBG1*, *DYNC1H1*, *KIF2A*, and *KIF5C*) from the training set resulting in 30 MCD genes (*see* MCD Training Set; Supplementary File 1). These four genes are from a recently publication (Fig. 4).

3.2.2 Compilation of Test Set Genes for MCD

To the four target genes, add 396 random human genes such that the test set is made of 400 genes (*see* MCD Test Set; Supplementary File 1). Alternatively, you can analyze one target gene at a time (e.g., one target gene plus 99 random genes). You can also add the neighboring genes of the target genes (i.e., genes occurring in the same loci as the target genes) instead of random genes. Irrespective of how you compile the test set, the goal is to prioritize the test set and check whether the target genes are ranked higher.

3.2.3 Prioritization of MCD Disease Candidate Genes

Now that you have the training and test set genes for MCD ready, proceed with the ranking of test set genes (target genes plus random genes) using ToppGene [27].

1. From the ToppGene Suite homepage (http://toppgene.cchmc.org) click on the second link ("ToppGene: Candidate gene prioritization") (Fig. 5).
2. On the following page ("ToppGene: Candidate gene prioritization"), enter either gene symbols or Entrez gene IDs of the training and test set genes (30 and 400 genes) respectively in this case; (*see* MCD Training Set, MCD Test Set; Supplementary File 1), and click "Submit query" (Figs. 6 and 7).
3. Select the appropriate statistical parameters. For the "Training Parameters" and "Test Parameter" you can use the default parameters (i.e., Bonferroni correction and p-value cutoff set to 0.05; *see* "Help" section of ToppGene Suite Web site for more details). Under the "Test parameter", the "random sampling size" option is for selecting the background gene set from the genome for computing the p-value while the "Min. feature count" represents the number of features that need to be considered for prioritization. The default options are 6 % of the genome (or 1,500 genes from a total of 25,000 genes) for random sampling size and feature count is 2 (Fig. 8).

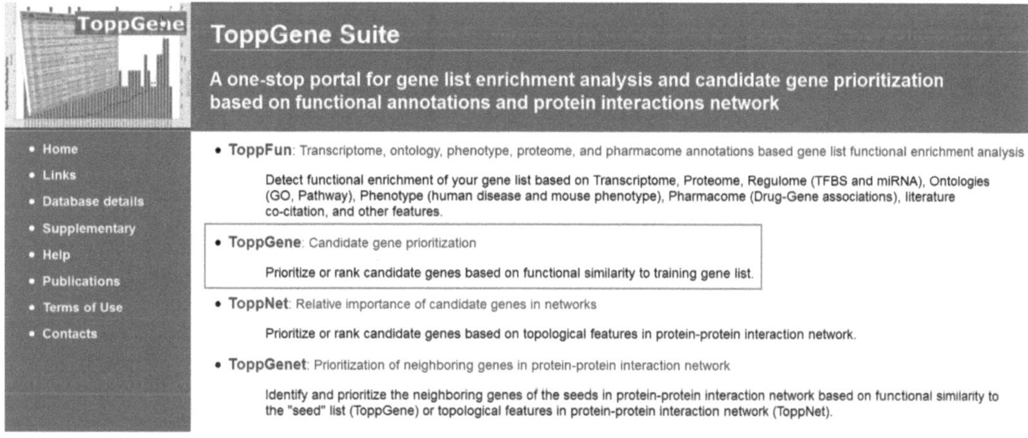

Fig. 5 The ToppGene Suite home page

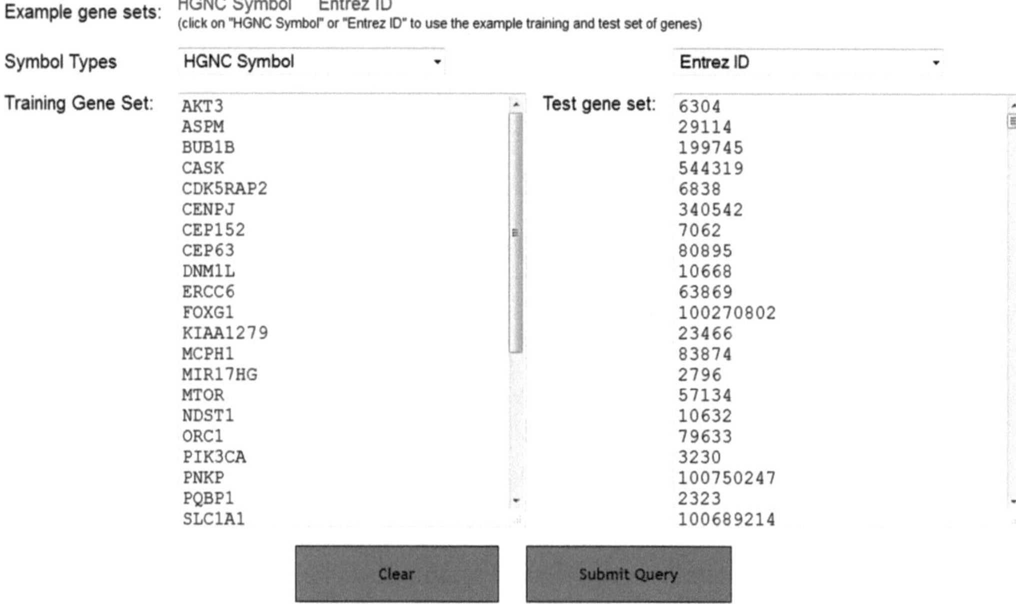

Fig. 6 The ToppGene entry page for launching gene prioritization

Training set (30 / 30)

Entered	Human Symbol	Gene ID
AKT3	AKT3	10000
ASPM	ASPM	259266
BUB1B	BUB1B	701
CASK	CASK	8573
CDK5RAP2	CDK5RAP2	55755
CENPJ	CENPJ	55835
CEP152	CEP152	22995
CEP63	CEP63	80254
DNM1L	DNM1L	10059
ERCC6	ERCC6	2074
FOXG1	FOXG1	2290
KIAA1279	KIAA1279	26128
MCPH1	MCPH1	79648
MIR17HG	MIR17HG	407975
MTOR	MTOR	2475
NDST1	NDST1	3340
ORC1	ORC1	4998
PIK3CA	PIK3CA	5290
PNKP	PNKP	11284
PQBP1	PQBP1	10084
SLC1A1	SLC1A1	6505

Test set (400 / 400)

Entered	Human Symbol	Gene ID
6304	SATB1	6304
29114	TAGLN3	29114
199745	THAP8	199745
544319	BP7	544319
6838	SURF6	6838
340542	BEX5	340542
7062	TCHH	7062
80895	ILKAP	80895
10668	CGRRF1	10668
63869	PSORS6	63869
100270802	STQTL16	100270802
23466	CBX6	23466
83874	TBC1D10A	83874
2796	GNRH1	2796
57134	MAN1C1	57134
10632	ATP5L	10632
79633	FAT4	79633
3230	HOXD@	3230
100750247	HIF1A-AS2	100750247
2323	FLT3LG	2323
100689214	HYSP4	100689214

Fig. 7 An example of training and test set genes

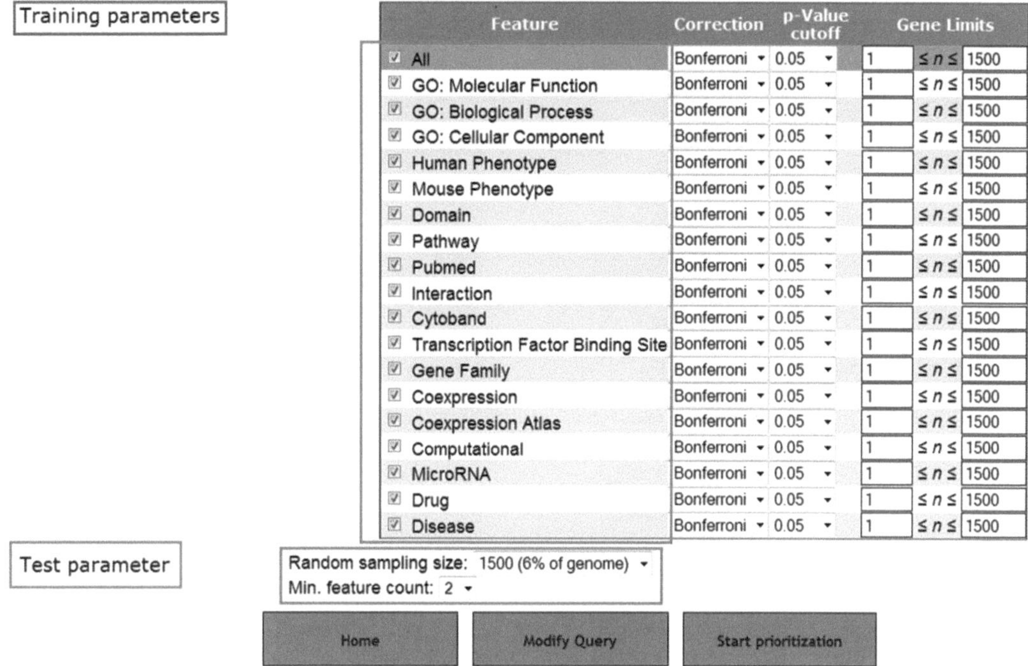

Fig. 8 The ToppGene entry page for selecting prioritization parameters

4. If your gene lists contain alternate symbols or duplicates or obsolete symbols, they are ignored or presented with the option to resolve them and add them back to your input list. Additionally, if there are common genes between training and test sets (i.e., test set genes which are also found in training set), these will be removed from the test set and no ranks will be assigned to them. After selecting the appropriate statistical parameters (training and test) click on the "Start prioritization" button.
5. Once the analysis is complete, the first half of the results page shows the enrichment results for the training set (Fig. 9).
6. The prioritized list of test set genes sorted according to their ranks based on the *p*-values are displayed in the lower half (Fig. 10). Each column indicates the features used to compute similarity between training and test set.
7. Additional details about the ranked genes can be obtained in both graphical and tabular format. For this, select the gene(s) you are interested in from the "Rank" column (Fig. 11), and click on the "Show Network" link. This will lead to the "Common Terms for selected genes and Training Set" page where you can see the details as to how a test set gene is connected to training set. The network view can be downloaded as

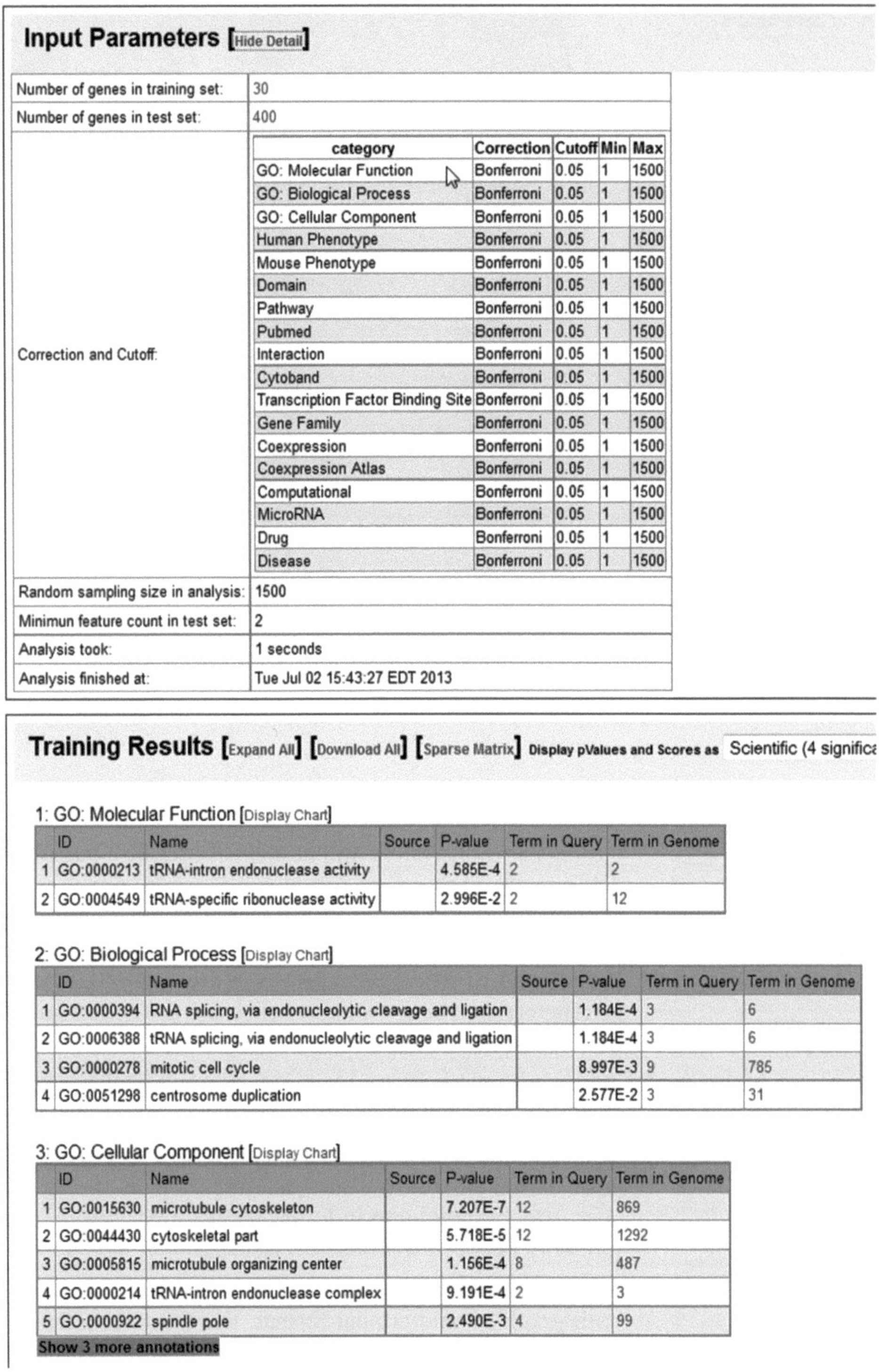

Input Parameters [Hide Detail]

Number of genes in training set:	30
Number of genes in test set:	400
Correction and Cutoff:	(see table below)
Random sampling size in analysis:	1500
Minimun feature count in test set:	2
Analysis took:	1 seconds
Analysis finished at:	Tue Jul 02 15:43:27 EDT 2013

category	Correction	Cutoff	Min	Max
GO: Molecular Function	Bonferroni	0.05	1	1500
GO: Biological Process	Bonferroni	0.05	1	1500
GO: Cellular Component	Bonferroni	0.05	1	1500
Human Phenotype	Bonferroni	0.05	1	1500
Mouse Phenotype	Bonferroni	0.05	1	1500
Domain	Bonferroni	0.05	1	1500
Pathway	Bonferroni	0.05	1	1500
Pubmed	Bonferroni	0.05	1	1500
Interaction	Bonferroni	0.05	1	1500
Cytoband	Bonferroni	0.05	1	1500
Transcription Factor Binding Site	Bonferroni	0.05	1	1500
Gene Family	Bonferroni	0.05	1	1500
Coexpression	Bonferroni	0.05	1	1500
Coexpression Atlas	Bonferroni	0.05	1	1500
Computational	Bonferroni	0.05	1	1500
MicroRNA	Bonferroni	0.05	1	1500
Drug	Bonferroni	0.05	1	1500
Disease	Bonferroni	0.05	1	1500

Training Results [Expand All] [Download All] [Sparse Matrix] Display pValues and Scores as Scientific (4 significa

1: GO: Molecular Function [Display Chart]

	ID	Name	Source	P-value	Term in Query	Term in Genome
1	GO:0000213	tRNA-intron endonuclease activity		4.585E-4	2	2
2	GO:0004549	tRNA-specific ribonuclease activity		2.996E-2	2	12

2: GO: Biological Process [Display Chart]

	ID	Name	Source	P-value	Term in Query	Term in Genome
1	GO:0000394	RNA splicing, via endonucleolytic cleavage and ligation		1.184E-4	3	6
2	GO:0006388	tRNA splicing, via endonucleolytic cleavage and ligation		1.184E-4	3	6
3	GO:0000278	mitotic cell cycle		8.997E-3	9	785
4	GO:0051298	centrosome duplication		2.577E-2	3	31

3: GO: Cellular Component [Display Chart]

	ID	Name	Source	P-value	Term in Query	Term in Genome
1	GO:0015630	microtubule cytoskeleton		7.207E-7	12	869
2	GO:0044430	cytoskeletal part		5.718E-5	12	1292
3	GO:0005815	microtubule organizing center		1.156E-4	8	487
4	GO:0000214	tRNA-intron endonuclease complex		9.191E-4	2	3
5	GO:0000922	spindle pole		2.490E-3	4	99

Show 3 more annotations

Fig. 9 Result of prioritization showing the input parameters and partial view of the training set enrichment results

Test Results [Hide Detail] [Download] [Show Network]

Rank (net)	Gene Symbol	Gene ID	GO: Molecular Function		GO: Biological Process		GO: Cellular Component		Human Phenotype		Mouse Phenotype		Domain		Pathway	
			Score	pValue	Score	pValue	Score	pValue	Score	pValue	Score	pValue	Score	pValue	Score	pValue
1	TUBG1	7283	0.000E0	5.003E-1	6.191E-1	1.308E-3	9.792E-1	1.000E-6			1.047E-1	6.606E-2	0.000E0	4.997E-1	6.929E-1	1.308E-3
2	KIF2A	3796	0.000E0	5.003E-1	6.191E-1	1.308E-3	9.721E-1	1.000E-6			9.218E-1	5.886E-3	0.000E0	4.997E-1	6.244E-2	4.578E-3
3	DYNC1H1	1778	0.000E0	5.003E-1	6.191E-1	1.308E-3	8.935E-1	3.924E-3	8.255E-1	3.205E-2	3.975E-1	4.382E-2	0.000E0	4.997E-1	7.107E-1	1.308E-3
4	CDKN1B	1027	0.000E0	5.003E-1	6.191E-1	1.308E-3	2.540E-1	1.014E-1	1.547E-1	5.690E-2	9.932E-1	6.540E-4	0.000E0	4.997E-1	4.085E-1	2.616E-3
5	NUP133	55746	0.000E0	5.003E-1	6.191E-1	1.308E-3	4.752E-1	3.401E-2			6.125E-1	2.812E-2	0.000E0	4.997E-1	6.244E-2	4.578E-3
6	TGM1	7051	0.000E0	5.003E-1	0.000E0	5.147E-1	3.493E-1	9.549E-2	8.820E-1	2.485E-2	0.000E0	5.409E-1	0.000E0	4.997E-1		
7	TPR	7175	0.000E0	5.003E-1	6.191E-1	1.308E-3	4.752E-1	3.401E-2					0.000E0	4.997E-1	0.000E0	5.036E-1
8	MAP2K1	5604	0.000E0	5.003E-1	6.191E-1	1.308E-3	9.540E-1	1.308E-3	9.997E-1	1.962E-3	6.021E-1	3.009E-2	0.000E0	4.997E-1	0.000E0	5.036E-1
9	CEP128	145508					9.721E-1	1.000E-6								
10	PAN2	9924	5.562E-1	1.000E-6	0.000E0	5.147E-1	0.000E0	5.945E-1					0.000E0	4.997E-1	0.000E0	5.036E-1

Fig. 10 Result of prioritization showing the ranked list of test set genes

Test Results [Hide Detail] [D

Rank (net)	Gene Symbol	Gene ID
1 ☑	TUBG1	7283
2 ☑	DYNC1H1	1778
3 ☑	KIF2A	3796
4 ☐	TGM1	7051
5 ☐	CDKN1B	1027
6 ☐	NUP133	55746
7 ☐	SETBP1	26040
8 ☐	PAN2	9924
9 ☐	TPR	7175
10 ☐	MAP2K1	5604

Fig. 11 List of top ranked genes for MCD showing three of the four target genes ranked among the top

an XGMML file and can be imported into Cytoscape [58] for visualization (Fig. 12).

8. The prioritized list can be downloaded as a table. Of the four MCD target genes, 3 of them are ranked at the top while one of them (KIF5A) is ranked lower (*see* ToppGeneRanking Set, Supplementary file 1)

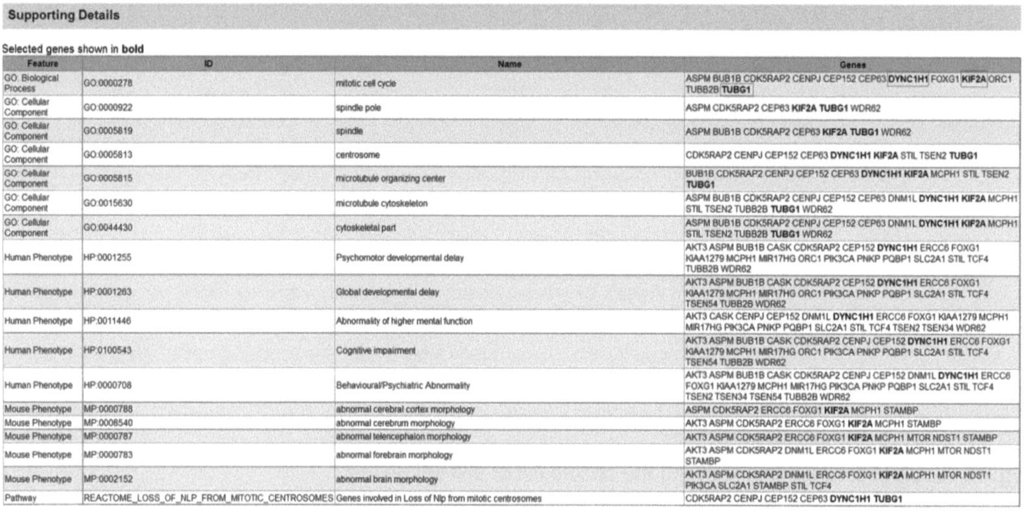

Supporting Details

Selected genes shown in **bold**

Feature	ID	Name	Genes
GO: Biological Process	GO:0000278	mitotic cell cycle	ASPM BUB1B CDK5RAP2 CENPJ CEP152 CEP63 **DYNC1H1** FOXG1 **KIF2A** ORC1 TUBB2B **TUBG1**
GO: Cellular Component	GO:0000922	spindle pole	ASPM CDK5RAP2 CEP63 **KIF2A** **TUBG1** WDR62
GO: Cellular Component	GO:0005819	spindle	ASPM BUB1B CDK5RAP2 CEP63 **KIF2A** **TUBG1** WDR62
GO: Cellular Component	GO:0005813	centrosome	CDK5RAP2 CENPJ CEP152 CEP63 **DYNC1H1** **KIF2A** STIL TSEN2 **TUBG1**
GO: Cellular Component	GO:0005815	microtubule organizing center	BUB1B CDK5RAP2 CENPJ CEP152 CEP63 **DYNC1H1** **KIF2A** MCPH1 STIL TSEN2 **TUBG1**
GO: Cellular Component	GO:0015630	microtubule cytoskeleton	ASPM BUB1B CDK5RAP2 CENPJ CEP152 CEP63 DNM1L **DYNC1H1** **KIF2A** MCPH1 STIL TSEN2 TUBB2B **TUBG1** WDR62
GO: Cellular Component	GO:0044430	cytoskeletal part	ASPM BUB1B CDK5RAP2 CENPJ CEP152 CEP63 DNM1L **DYNC1H1** **KIF2A** MCPH1 STIL TSEN2 TUBB2B **TUBG1** WDR62
Human Phenotype	HP:0001255	Psychomotor developmental delay	AKT3 ASPM BUB1B CASK CDK5RAP2 CEP152 **DYNC1H1** ERCC6 FOXG1 KIAA1279 MCPH1 MIR17HG ORC1 PIK3CA PNKP PQBP1 SLC2A1 STIL TCF4 TUBB2B WDR62
Human Phenotype	HP:0001263	Global developmental delay	AKT3 ASPM BUB1B CASK CDK5RAP2 CEP152 **DYNC1H1** ERCC6 FOXG1 KIAA1279 MCPH1 MIR17HG ORC1 PIK3CA PNKP PQBP1 SLC2A1 STIL TCF4 TSEN54 TUBB2B WDR62
Human Phenotype	HP:0011446	Abnormality of higher mental function	AKT3 CASK CENPJ CEP152 DNM1L **DYNC1H1** ERCC6 FOXG1 KIAA1279 MCPH1 MIR17HG PIK3CA PNKP PQBP1 SLC2A1 STIL TCF4 TSEN2 TSEN34 WDR62
Human Phenotype	HP:0100543	Cognitive impairment	AKT3 ASPM BUB1B CASK CDK5RAP2 CENPJ CEP152 **DYNC1H1** ERCC6 FOXG1 KIAA1279 MCPH1 MIR17HG ORC1 PIK3CA PNKP PQBP1 SLC2A1 STIL TCF4 TSEN54 TUBB2B WDR62
Human Phenotype	HP:0000708	Behavioural/Psychiatric Abnormality	AKT3 ASPM BUB1B CASK CDK5RAP2 CENPJ CEP152 DNM1L **DYNC1H1** ERCC6 FOXG1 KIAA1279 MCPH1 MIR17HG PIK3CA PNKP PQBP1 SLC2A1 STIL TCF4 TSEN2 TSEN34 TSEN54 TUBB2B WDR62
Mouse Phenotype	MP:0000788	abnormal cerebral cortex morphology	ASPM CDK5RAP2 ERCC6 FOXG1 **KIF2A** MCPH1 STAMBP
Mouse Phenotype	MP:0008540	abnormal cerebrum morphology	AKT3 ASPM CDK5RAP2 ERCC6 FOXG1 **KIF2A** MCPH1 STAMBP
Mouse Phenotype	MP:0000787	abnormal telencephalon morphology	AKT3 ASPM CDK5RAP2 ERCC6 FOXG1 **KIF2A** MCPH1 MTOR NDST1 STAMBP
Mouse Phenotype	MP:0000783	abnormal forebrain morphology	AKT3 ASPM CDK5RAP2 DNM1L ERCC6 FOXG1 **KIF2A** MCPH1 MTOR NDST1 STAMBP
Mouse Phenotype	MP:0002152	abnormal brain morphology	AKT3 ASPM CDK5RAP2 DNM1L ERCC6 FOXG1 **KIF2A** MCPH1 MTOR NDST1 PIK3CA SLC2A1 STAMBP STIL TCF4
Pathway	REACTOME_LOSS_OF_NLP_FROM_MITOTIC_CENTROSOMES	Genes involved in Loss of Nlp from mitotic centrosomes	CDK5RAP2 CENPJ CEP152 CEP63 **DYNC1H1** **TUBG1**

Fig. 12 Browsing details of the top ranked genes. *Red boxed genes* are target genes and the remaining are the training set genes (Color figure online)

4 Notes

1. Most of the literature-mining systems are based on processing PubMed records and therefore are often restricted to abstracts. Although ready access to all of the full-text articles is a bottleneck, those available in NCBI's PubMed Central can be used. Thus, when selecting a literature-mining approach it is important to consider as to what literature collection is being used and whether it is current. Additionally, as with any computational approach, it is always a good strategy to try out more than one literature-mining system for a same set of tasks.
2. The BioNLP.org resource provides useful links to resources that can help researchers in their work on natural language processing for articles in the biomedical literature.
3. Candidate gene ranking approaches based on integrated literature- and data-mining-based functional similarity have the following limitations:
   (a) Since the ranking of test set genes is dependent on a training set, these approaches are based on the assumption that novel disease genes yet to be discovered will be consistent with what is already known about a disease and/or its genetic basis which may not always be true.
   (b) Since most of the literature- and data-mining-based systems rely on publicly available data resources, it is important to note that the accuracy of the prioritization depends on the accuracy of the original literature-mining or database-derived annotations.

(c) Although it has been speculated that complex traits result more often from noncoding regulatory variants than from coding sequence variants [59–61], current disease gene identification and prioritization approaches predominantly are gene-centric. Interpreting the consequences of noncoding sequence variants however is relatively complex because the relationships among promoter, intergenic, or noncoding sequence variation, gene expression level, and trait phenotype are less well understood than the relationship between coding DNA sequence and protein function.

(d) Literature- or data-mining-based candidate gene prioritization approaches tend to be biased towards well studied or better annotated genes. For instance, a novel, "true" disease gene can be missed if it lacks sufficient annotations. On the whole, in spite of these methods being relatively unbiased (except the bias towards training or seed set—*see* **Note 3a** above) compared to the "cherry-picked" genes of interest by researchers, the final selection of candidate genes for further experimental analyses could be partly subjective and researcher-specific. This is can be mitigated to some extent if the disease under investigation is defined both molecularly and physiologically.

(e) Finally, different gene prioritization methods use different algorithms to integrate, mine, and rank, and objectively, there is no one methodology that is better than the others for all data inputs [62]. To increase the robustness of prioritization analysis, we recommend using at least three algorithmically different ranking approaches.

References

1. Cheung WA, Ouellette BF, Wasserman WW (2012) Inferring novel gene-disease associations using Medical Subject Heading Over-representation Profiles. Genome Med 4(9):75
2. Rebholz-Schuhmann D, Oellrich A, Hoehndorf R (2012) Text-mining solutions for biomedical research: enabling integrative biology. Nat Rev Genet 13(12):829–839
3. Bodenreider O (2004) The unified medical language system (UMLS): integrating biomedical terminology. Nucleic Acids Res 32(Suppl 1):D267–D270
4. Smith CL, Goldsmith C-A, Eppig JT (2005) The Mammalian Phenotype Ontology as a tool for annotating, analyzing and comparing phenotypic information. Genome Biol 6(1):R7
5. Gault LV, Shultz M, Davies KJ (2002) Variations in Medical Subject Headings (MeSH) mapping: from the natural language of patron terms to the controlled vocabulary of mapped lists. J Med Libr Assoc 90(2):173
6. McKusick VA (1998) Mendelian inheritance in man: a catalog of human genes and genetic disorders. Johns Hopkins University Press, Maryland, USA
7. Ashburner M, Ball CA, Blake JA, Botstein D, Butler H, Cherry JM, Davis AP, Dolinski K, Dwight SS, Eppig JT (2000) Gene Ontology: tool for the unification of biology. Nat Genet 25(1):25–29
8. Cohen KB, Hunter LE (2013) Text mining for translational bioinformatics. PLoS Comput Biol 9(4):e1003044
9. Mattingly CJ, Colby GT, Forrest JN, Boyer JL (2003) The Comparative Toxicogenomics Database (CTD). Environ Health Perspect 111(6):793

10. Klein T, Chang J, Cho M, Easton K, Fergerson R, Hewett M, Lin Z, Liu Y, Liu S, Oliver D (2001) Integrating genotype and phenotype information: an overview of the PharmGKB project. Pharmacogenomics J 1(3):167–170
11. Kuhn M, von Mering C, Campillos M, Jensen LJ, Bork P (2008) STITCH: interaction networks of chemicals and proteins. Nucleic Acids Res 36(Suppl 1):D684–D688
12. Hoffmann R, Valencia A (2004) A gene network for navigating the literature. Nat Genet 36(7):664
13. Olsen C, Djebbari A, Bontempi G, Correll M, Bouton C, Haibe-Kains B, Quackenbush J (2012) Predictive networks: a flexible, open source, web application for integration and analysis of human gene networks. Nucleic Acids Res 40(D1):D866–D875
14. Rzhetsky A, Koike T, Kalachikov S, Gomez SM, Krauthammer M, Kaplan SH, Kra P, Russo JJ, Friedman C (2000) A knowledge model for analysis and simulation of regulatory networks. Bioinformatics 16(12):1120–1128
15. Frijters R, Heupers B, van Beek P, Bouwhuis M, van Schaik R, de Vlieg J, Polman J, Alkema W (2008) CoPub: a literature-based keyword enrichment tool for microarray data analysis. Nucleic Acids Res 36(Suppl 2):W406–W410
16. Müller H-M, Kenny EE, Sternberg PW (2004) Textpresso: an ontology-based information retrieval and extraction system for biological literature. PLoS Biol 2(11):e309
17. Pafilis E, O'Donoghue SI, Jensen LJ, Horn H, Kuhn M, Brown NP, Schneider R (2009) Reflect: augmented browsing for the life scientist. Nat Biotechnol 27(6):508–510
18. Fo B, Nolin M-A, Tourigny N, Rigault P, Morissette J (2008) Bio2RDF: towards a mashup to build bioinformatics knowledge systems. J Biomed Inform 41(5):706–716
19. Aronson AR (2001) Effective mapping of biomedical text to the UMLS Metathesaurus: the MetaMap program. In: Proceedings of the AMIA symposium, 2001. American Medical Informatics Association, p 17
20. Hristovski D, Peterlin B, Mitchell JA, Humphrey SM (2005) Using literature-based discovery to identify disease candidate genes. Int J Med Inform 74(2–4):289–298
21. Jourquin J, Duncan D, Shi Z, Zhang B (2012) GLAD4U: deriving and prioritizing gene lists from PubMed literature. BMC Genomics 13(Suppl 8):S20
22. Liekens AM, De Knijf J, Daelemans W, Goethals B, De Rijk P, Del-Favero J (2011) BioGraph: unsupervised biomedical knowledge discovery via automated hypothesis generation. Genome Biol 12(6):R57
23. Yoshida Y, Makita Y, Heida N, Asano S, Matsushima A, Ishii M, Mochizuki Y, Masuya H, Wakana S, Kobayashi N (2009) PosMed (Positional Medline): prioritizing genes with an artificial neural network comprising medical documents to accelerate positional cloning. Nucleic Acids Res 37(Suppl 2):W147–W152
24. Swanson DR, Smalheiser NR (1997) An interactive system for finding complementary literatures: a stimulus to scientific discovery. Artif Intell 91(2):183–203
25. Swanson DR (1990) Medical literature as a potential source of new knowledge. Bull Med Libr Assoc 78(1):29
26. Makita Y, Kobayashi N, Yoshida Y, Doi K, Mochizuki Y, Nishikata K, Matsushima A, Takahashi S, Ishii M, Takatsuki T, Bhatia R, Khadbaatar Z, Watabe H, Masuya H, Toyoda T (2013) PosMed: ranking genes and bioresources based on Semantic Web Association Study. Nucleic Acids Res 41(Web Server issue): W109–W114
27. Chen J, Bardes EE, Aronow BJ, Jegga AG (2009) ToppGene suite for gene list enrichment analysis and candidate gene prioritization. Nucleic Acids Res 37(Web Server issue): W305–W311. doi:10.1093/nar/gkp427
28. Aerts S, Lambrechts D, Maity S, Van Loo P, Coessens B, De Smet F, Tranchevent LC, De Moor B, Marynen P, Hassan B, Carmeliet P, Moreau Y (2006) Gene prioritization through genomic data fusion. Nat Biotechnol 24(5): 537–544
29. Smalheiser NR, Torvik VI, Zhou W (2009) Arrowsmith two-node search interface: a tutorial on finding meaningful links between two disparate sets of articles in MEDLINE. Comput Methods Programs Biomed 94(2):190
30. Frijters R, van Vugt M, Smeets R, van Schaik R, de Vlieg J, Alkema W (2010) Literature mining for the discovery of hidden connections between drugs, genes and diseases. PLoS Comput Biol 6(9):e1000943
31. Lindsay RK, Gordon MD (1999) Literature-based discovery by lexical statistics. J Am Soc Inform Sci 50(7):574–587
32. Kemper B, Matsuzaki T, Matsuoka Y, Tsuruoka Y, Kitano H, Ananiadou S, Ji T (2010) PathText: a text mining integrator for biological pathway visualizations. Bioinformatics 26(12):i374–i381
33. Rzhetsky A, Iossifov I, Koike T, Krauthammer M, Kra P, Morris M, Yu H, Duboue PA, Weng W, Wilbur WJ, Hatzivassiloglou V, Friedman C (2004) GeneWays: a system for extracting, analyzing, visualizing, and integrating molecular pathway data. J Biomed Inform 37(1):43–53. doi:10.1016/j.jbi.2003.10.001

34. Krauthammer M, Kaufmann CA, Gilliam TC, Rzhetsky A (2004) Molecular triangulation: bridging linkage and molecular-network information for identifying candidate genes in Alzheimer's disease. Proc Natl Acad Sci U S A 101(42):15148–15153. doi:10.1073/pnas.0404315101
35. Özgür A, Vu T, Erkan G, Radev DR (2008) Identifying gene-disease associations using centrality on a literature mined gene-interaction network. Bioinformatics 24(13):i277–i285. doi:10.1093/bioinformatics/btn182
36. Coulet A, Shah NH, Garten Y, Musen M, Altman RB (2010) Using text to build semantic networks for pharmacogenomics. J Biomed Inform 43(6):1009–1019. doi:10.1016/j.jbi.2010.08.005
37. Percha B, Garten Y, Altman RB (2012) Discovery and explanation of drug-drug interactions via text mining. In: Pacific symposium on biocomputing. Pacific symposium on biocomputing, 2012. World Scientific, p 410
38. Hoehndorf R, Schofield PN, Gkoutos GV (2011) PhenomeNET: a whole-phenome approach to disease gene discovery. Nucleic Acids Res 39(18):e119. doi:10.1093/nar/gkr538
39. Freudenberg J, Propping P (2002) A similarity-based method for genome-wide prediction of disease-relevant human genes. Bioinformatics 18 Suppl 2:S110–S115
40. Turner FS, Clutterbuck DR, Semple CA (2003) POCUS: mining genomic sequence annotation to predict disease genes. Genome Biol 4(11):R75
41. Tiffin N, Kelso JF, Powell AR, Pan H, Bajic VB, Hide WA (2005) Integration of text- and data-mining using ontologies successfully selects disease gene candidates. Nucleic Acids Res 33(5):1544–1552
42. Adie EA, Adams RR, Evans KL, Porteous DJ, Pickard BS (2005) Speeding disease gene discovery by sequence based candidate prioritization. BMC Bioinformatics 6:55
43. Thornblad TA, Elliott KS, Jowett J, Visscher PM (2007) Prioritization of positional candidate genes using multiple web-based software tools. Twin Res Hum Genet 10(6):861–870
44. Zhu M, Zhao S (2007) Candidate gene identification approach: progress and challenges. Int J Biol Sci 3(7):420–427
45. Tiffin N, Adie E, Turner F, Brunner HG, van Driel MA, Oti M, Lopez-Bigas N, Ouzounis C, Perez-Iratxeta C, Andrade-Navarro MA, Adeyemo A, Patti ME, Semple CA, Hide W (2006) Computational disease gene identification: a concert of methods prioritizes type 2 diabetes and obesity candidate genes. Nucleic Acids Res 34(10):3067–3081
46. Adie EA, Adams RR, Evans KL, Porteous DJ, Pickard BS (2006) SUSPECTS: enabling fast and effective prioritization of positional candidates. Bioinformatics 22(6):773–774
47. Chen J, Xu H, Aronow BJ, Jegga AG (2007) Improved human disease candidate gene prioritization using mouse phenotype. BMC Bioinformatics 8:392
48. Goh KI, Cusick ME, Valle D, Childs B, Vidal M, Barabasi AL (2007) The human disease network. Proc Natl Acad Sci U S A 104(21):8685–8690. doi:10.1073/pnas.0701361104
49. Jimenez-Sanchez G, Childs B, Valle D (2001) Human disease genes. Nature 409(6822):853–855
50. Smith NG, Eyre-Walker A (2003) Human disease genes: patterns and predictions. Gene 318:169–175
51. Tranchevent LC, Barriot R, Yu S, Van Vooren S, Van Loo P, Coessens B, De Moor B, Aerts S, Moreau Y (2008) ENDEAVOUR update: a web resource for gene prioritization in multiple species. Nucleic Acids Res 36(Web Server issue):W377–W384
52. Davis AP, Murphy CG, Saraceni-Richards CA, Rosenstein MC, Wiegers TC, Mattingly CJ (2009) Comparative Toxicogenomics Database: a knowledgebase and discovery tool for chemical-gene-disease networks. Nucleic Acids Res 37(Database issue):D786–D792. doi:10.1093/nar/gkn580
53. Popescu M, Keller JM, Mitchell JA (2006) Fuzzy measures on the gene ontology for gene product similarity. IEEE/ACM Trans Comput Biol Bioinform 3(3):263–274
54. Poirier K, Lebrun N, Broix L, Tian G, Saillour Y, Boscheron C, Parrini E, Valence S, Pierre BS, Oger M, Lacombe D, Genevieve D, Fontana E, Darra F, Cances C, Barth M, Bonneau D, Bernadina BD, N'Guyen S, Gitiaux C, Parent P, des Portes V, Pedespan JM, Legrez V, Castelnau-Ptakine L, Nitschke P, Hieu T, Masson C, Zelenika D, Andrieux A, Francis F, Guerrini R, Cowan NJ, Bahi-Buisson N, Chelly J (2013) Mutations in TUBG1, DYNC1H1, KIF5C and KIF2A cause malformations of cortical development and microcephaly. Nat Genet 45(6):639–647. doi:10.1038/ng.2613
55. Hamosh A, Scott A, Amberger J, Bocchini C, McKusick V (2005) Online Mendelian Inheritance in Man (OMIM), a knowledgebase of human genes and genetic disorders. Nucleic Acids Res 33:D514–D517
56. Becker KG, Barnes KC, Bright TJ, Wang SA (2004) The genetic association database. Nat Genet 36(5):431–432. doi:10.1038/ng0504-431

57. Hindorff LA, Sethupathy P, Junkins HA, Ramos EM, Mehta JP, Collins FS, Manolio TA (2009) Potential etiologic and functional implications of genome-wide association loci for human diseases and traits. Proc Natl Acad Sci U S A 106(23):9362–9367. doi:10.1073/pnas.0903103106
58. Shannon P, Markiel A, Ozier O, Baliga NS, Wang JT, Ramage D, Amin N, Schwikowski B, Ideker T (2003) Cytoscape: a software environment for integrated models of biomolecular interaction networks. Genome Res 13(11):2498–2504. doi:10.1101/gr.1239303
59. King MC, Wilson AC (1975) Evolution at two levels in humans and chimpanzees. Science 188(4184):107–116
60. Korstanje R, Paigen B (2002) From QTL to gene: the harvest begins. Nat Genet 31(3): 235–236
61. Mackay TF (2001) Quantitative trait loci in Drosophila. Nat Rev Genet 2(1):11–20
62. Bromberg Y (2013) Chapter 15: disease gene prioritization. PLoS Comput Biol 9(4):e1002902. doi:10.1371/journal.pcbi.1002902
63. Tranchevent LC, Capdevila FB, Nitsch D, De Moor B, De Causmaecker P, Moreau Y (2011) A guide to web tools to prioritize candidate genes. Brief Bioinform 12(1):22–32. doi:10.1093/bib/bbq007
64. Perez-Iratxeta C, Bork P, Andrade MA (2002) Association of genes to genetically inherited diseases using data mining. Nat Genet 31(3): 316–319
65. Perez-Iratxeta C, Wjst M, Bork P, Andrade MA (2005) G2D: a tool for mining genes associated with disease. BMC Genet 6:45
66. van Driel MA, Cuelenaere K, Kemmeren PP, Leunissen JA, Brunner HG (2003) A new web-based data mining tool for the identification of candidate genes for human genetic disorders. Eur J Hum Genet 11(1):57–63
67. van Driel MA, Cuelenaere K, Kemmeren PP, Leunissen JA, Brunner HG, Vriend G (2005) GeneSeeker: extraction and integration of human disease-related information from web-based genetic databases. Nucleic Acids Res 33(Web Server issue):W758–W761
68. Masseroli M, Galati O, Pinciroli F (2005) GFINDer: genetic disease and phenotype location statistical analysis and mining of dynamically annotated gene lists. Nucleic Acids Res 33(Web Server issue):W717–W723
69. Masseroli M, Martucci D, Pinciroli F (2004) GFINDer: Genome Function INtegrated Discoverer through dynamic annotation, statistical analysis, and mining. Nucleic Acids Res 32(Web Server issue):W293–W300
70. Rossi S, Masotti D, Nardini C, Bonora E, Romeo G, Macii E, Benini L, Volinia S (2006) TOM: a web-based integrated approach for identification of candidate disease genes. Nucleic Acids Res 34(Web Server issue):W285–W292
71. van Driel MA, Bruggeman J, Vriend G, Brunner HG, Leunissen JA (2006) A text-mining analysis of the human phenome. Eur J Hum Genet 14(5):535–542
72. Franke L, Bakel H, Fokkens L, de Jong ED, Egmont-Petersen M, Wijmenga C (2006) Reconstruction of a functional human gene network, with an application for prioritizing positional candidate genes. Am J Hum Genet 78(6):1011–1025

Chapter 13

Role of Text Mining in Early Identification of Potential Drug Safety Issues

Mei Liu, Yong Hu, and Buzhou Tang

Abstract

Drugs are an important part of today's medicine, designed to treat, control, and prevent diseases; however, besides their therapeutic effects, drugs may also cause adverse effects that range from cosmetic to severe morbidity and mortality. To identify these potential drug safety issues early, surveillance must be conducted for each drug throughout its life cycle, from drug development to different phases of clinical trials, and continued after market approval. A major aim of pharmacovigilance is to identify the potential drug–event associations that may be novel in nature, severity, and/or frequency. Currently, the state-of-the-art approach for signal detection is through automated procedures by analyzing vast quantities of data for clinical knowledge. There exists a variety of resources for the task, and many of them are textual data that require text analytics and natural language processing to derive high-quality information. This chapter focuses on the utilization of text mining techniques in identifying potential safety issues of drugs from textual sources such as biomedical literature, consumer posts in social media, and narrative electronic medical records.

Key words Text mining, Data mining, Natural language processing, Biomedical literature mining, Pharmacovigilance, Drug safety surveillance, Adverse drug reaction, Drug–drug interactions

1 Introduction

1.1 The Need for PhV

Clinical benefits of drugs depend on not only their efficacy in treating diseases but also their safety and tolerability in patients. Thus, use of drugs in medicine must balance between expected benefits and possible risks of adverse effects. Each drug carries its own risks for causing adverse drug reactions (ADRs), which are defined as those unintended and undesired responses to drugs beyond their anticipated therapeutic effects during clinical use at normal dosage [1, 2]. This excludes the adverse events arising from prescription faults and errors. Some ADRs may be minor, but others can have dire consequences leading to hospitalization and even death. As the population use of prescription drugs increases, ADRs have evolved to become a major public health problem accounting for up to 5 % of hospital admissions [2], 28 % of emergency department visits

Vinod D. Kumar and Hannah Jane Tipney (eds.), *Biomedical Literature Mining*, Methods in Molecular Biology, vol. 1159, DOI 10.1007/978-1-4939-0709-0_13, © Springer Science+Business Media New York 2014

[3], and 5 % of hospital deaths [4]. Over the past decade, both numbers of reported ADRs and related deaths have increased by ~2.6 times [5]. ADR is estimated to account for more than two million hospitalization incidents per year and more than 100,000 fatalities in the United States alone [6, 7].

Unfortunately, it is a challenge to identify ADRs before their widespread dissemination due to the short and biased nature of the pre-marketing drug trials [8]. In addition, many drugs have "off-label" usages; sometimes they are taken at a higher dose than recommended, for untested indications and populations. In short, severe ADRs may occur whenever prescribing patterns diverge from those used in pre-marketing trials. Therefore, for a good reason, it is crucial to monitor the drugs for their effectiveness and safety throughout their life cycles.

1.2 An Overview of PhV

Pharmacovigilance (PhV) has been defined by the World Health Organization (WHO) as "the science and activities relating to the detection, assessment, understanding and prevention of adverse effects or any other possible drug-related problems [9]." The principal aims of PhV programs are:

- To improve patient care and safety regarding the use of medicines and all medical and paramedical interventions.
- To improve public health and safety regarding the use of medicines.
- To assess the benefit, harm, effectiveness, and risk of medicines.
- To promote the awareness of PhV and its communication to healthcare professionals.

PhV research attempts to address the ADR problem from two perspectives: (1) pre-marketing surveillance—utilization of information collected from preclinical screening and phases I–III clinical trials to predict potential ADRs [10–22] and (2) post-marketing surveillance—utilization of data accumulated in the post-approval stage and throughout a drug's market life to detect previously unrecognized ADRs [23–48]. Existing PhV effort is heavily invested in the identification of unrecognized ADRs and drug–drug interactions (DDIs). DDI is a particularly important type of ADR that involves a combination of drugs causing consequences from diminished therapeutic effect to excessive response or toxicity as a result of pharmacokinetic, pharmacodynamic, or a combination of the mechanisms [49]. In other words, actions of one drug can be inhibited by another, resulting in reduced response to one or both drugs. On the other hand, one drug may alter the absorption, distribution, metabolism, or elimination of another drug, which may result in higher serum concentration of the agents and lead to excessive response or toxicity.

1.3 The Need for Text Mining

Recent progress in the medical research and the advancement in healthcare technologies (e.g., emergence of EMR systems) have generated enormous amount of medical data that exists in many forms including images from diagnostic procedures (e.g., X-ray), textual information described in research articles, or clinical observations from patient's medical records. Among these various forms of medical data, free-text descriptions constitute a substantial portion of it because the most common and natural way to express complex assumptions, interpretations of findings, and hypotheses is through natural language. Especially in the medical domain, a large portion of the patient's observations are documented as narrative text in clinical notes such as radiology reports, operative notes, and discharge summaries. According to Hale [50], only a small proportion of the information is available in the structured form, whereas approximately 80 % of the information is unstructured and expressed in natural language, which includes patient health records, EMRs, medical case reports, research articles, patents, blogs, forums, and news reports [51, 52]. Currently, there are over 19 million biomedicine-related articles indexed in MEDLINE and the amount of citations added to MEDLINE each fiscal year is rapidly increasing (from 392,354 new citations in 1995 to 760,903 new citations in 2012). Clearly, this is far too much for a researcher to keep up with manually. Thus, text mining is the critical approach to enable the identification of novel facts, hypotheses, and new associations from the vast quantities of free-text data. Since many of the complex relationships between drugs and adverse events are embedded in this mountain of free-text data, text mining techniques can be utilized to identify drug–adverse event or drug–drug interaction associations.

2 Materials

Before addressing the specific methodological issues, we provide an overview of the available data resources for text mining to support drug safety surveillance (Subheading 2.1), standard biomedical vocabularies (Subheading 2.2), common text processing techniques (Subheading 2.3), and information extraction tools to enable the identification of drug named entities and adverse events in free text (Subheading 2.4).

2.1 Sources of Textural Data for Drug Safety Surveillance

In both text mining and data mining, source of data is the key for a reliable result. Each data source may carry inherent biases because of the purpose for which they are created. The problem of identifying drug safety issues like ADRs and DDIs can be addressed using each of the following free-text sources.

- *Biomedical Literature*—There are now over 19 million journal articles with a concentration on biomedicine indexed in MEDLINE, a bibliographic database by the US National

Library of Medicine (NLM). It mainly consists of articles published from 1946 to the present and citations from approximately 5,600 worldwide journals in 39 languages. MEDLINE covers the subject scope of biomedicine and health, broadly defined as encompassing the areas of life sciences, behavioral sciences, chemical sciences, and bioengineering needed by health professionals and those engaged in basic research and clinical care, public health, health policy development, or related educational activities. MEDLINE is the primary component of PubMed (http://pubmed.gov) through which articles can be searched and free full text is often available for download.

- *EMRs*—EMR systems are mainly implemented to assist physicians in their daily practices by collecting health information about individual patients. EMRs usually include demographics, medical history, medication and allergies, immunization status, laboratory test results, radiology images, vital signs, personal statistics like age and weight, and billing information. When aggregated, EMR is a prominent resource for analyzing therapeutic outcomes across patient populations. Clinical data in EMR are highly heterogeneous, and their heterogeneity includes variation in both structured (e.g., laboratory test) and unstructured data (e.g., clinical notes). Much relevant clinical information such as drug exposure and responses is embedded in the narrative clinical text and requires NLP techniques to extract information for analysis.
- *Social Media*—Social network has become a popular means for interactions among people to create and share information. Recent changes in the healthcare system along with the rapid development of the Internet have resulted in many online health communities that provide a platform for patients to seek healthcare information and offer support to others in similar circumstances. Within these online health communities, users often discuss about their medical conditions, medications they take, and any side effects they may experience and user interactions can take a variety of forms including blogs, microblogs, and question/answer discussion forums. There are also sites specifically designed to collect drug side effect information and medicine ratings from consumers (e.g., Ask a Patient), while others can provide more options to patients. In short, the social media has created a rich textual data source for mining potential ADRs. Some of the popular health social media sites include PatientsLikeMe, MedHelp, DailyStrength, Ask a Patient, Yahoo Health and Wellness group, and iMedx.
- *Other Free-Text Documents*—ADR information may also exist in free-text documents such as approved drug labels regulated by the Food and Drug Administration (FDA) and pharmacological text in DrugBank. The side effect information

in drug labels or package inserts is typically obtained from clinical trials and post-marketing surveillance, so it is perfect for studying drug toxicity. DrugBank is an annotated database with more than 6,700 drug entries, and each entry contains more than 100 data fields that contain detailed chemical and pharmacological information (e.g., type, category, brand names, drug interactions) [53, 54]. Studies have recently explored the plain text under the field "interactions" in the DrugBank database to identify DDIs [55, 56].

2.2 Drug and Clinical Terminologies

It is a challenge to extract information from biomedical free text because people use different vocabularies and expressions to describe the same health concepts. Many standard medical terminologies are developed to address this issue. However one standard medical terminology may not be applicable to all forms of text. For instance, the standard medical lexicon United Medical Language System (UMLS) consists of ontologies primarily designed for medical professionals, and it may not be the best lexicon for social media data because consumers express problems in a completely different way than the professionals. Table 1 summarizes the various terminologies to support concept mapping for text mining in drug safety surveillance.

Table 1
Standard terminologies for text mining in drug safety surveillance

Terminology	Description	Corpus
ATC	Organizes drugs based on their pharmacotherapeutic properties	Drug
RxNorm	Normalized naming system for generic and branded drugs	Drug
NDF-RT	Organizes drugs into a formal representation	Drug
COSTART	Subset of UMLS for coding symbols for a thesaurus of adverse reaction terms	ADR
CHV	Health expressions derived from consumer comments linked to professional concepts. More appropriate for social media	ADR
ICD-9	Classification of diseases used to code diagnoses for billing	ADR
MedDRA	ADR classification dictionary used by regulatory authorities	ADR
MeSH	Controlled vocabulary thesaurus for indexing articles for PubMed	ADR
NCBO BioPortal	Open repository of over 250 biomedical ontologies	ADR
NCI Thesaurus	Covers vocabulary for clinical care, translational and basic research, and public information and administrative activities	ADR
SNOMED-CT	Core general terminology for EMR encoding	ADR
UMLS	Collection of biomedical vocabularies to enable interoperability between computer systems	ADR

2.3 Text Processing with NLP

In order for a computer to handle textual data, we need to first transform the unstructured data into some structures, which requires text processing with NLP that may involve the following frequently applied techniques (Fig. 1).

- *Sentence splitting* is to decompose a document into constitutive sentences. Sentence is the smallest unit of natural language to express a complete thought. However, detection of sentence boundaries is not trivial because the punctuation "." does not always occur at the end of sentences. Especially in biomedical text, they are full of entities and abbreviations with punctuations as part of their standard nomenclature (e.g., *E. coli*). Thus, accurate detection of sentence boundaries is an essential component of information extraction systems.
- *Tokenization* is to segment a stream of text into words or terms. It can be performed over sentences or full documents. The tokenization process normally depends on simple heuristics such as separation of tokens on whitespaces and punctuations [57].

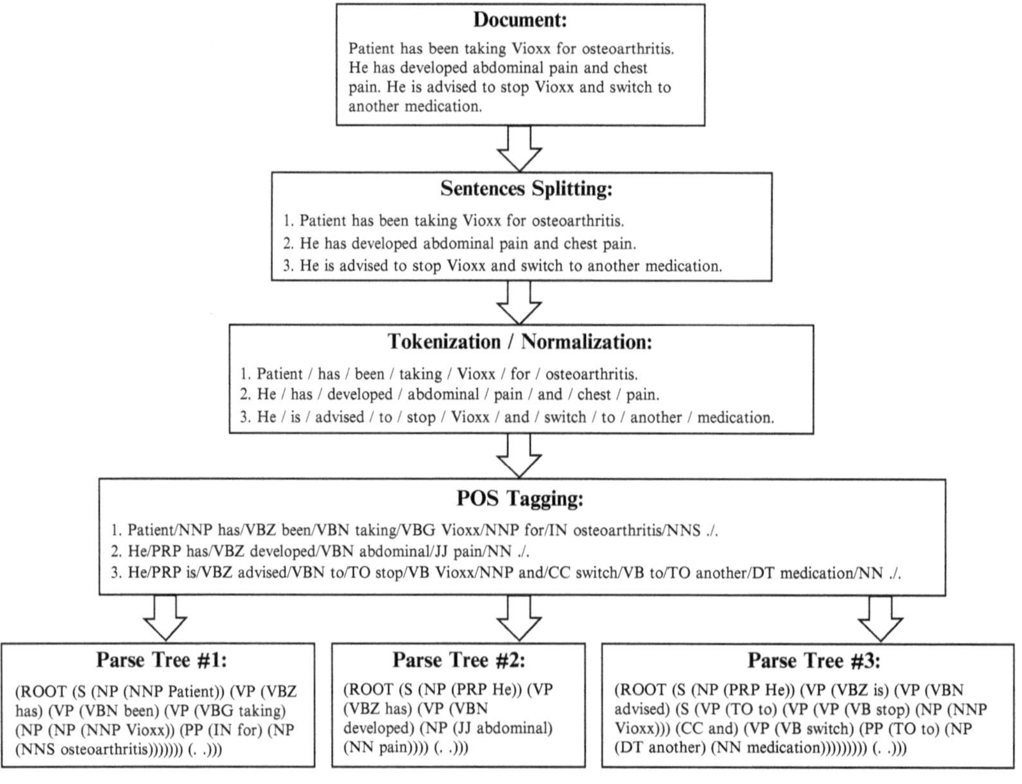

Fig. 1 Overview of the text processing techniques: (1) Sentence splitting, (2) tokenization/normalization, (3) POS tagging, and (4) parsing. The parse trees were generated using the Stanford Parser

- *Word normalization* is to reduce inflected words to their base forms, which can be performed through stemming or lemmatization. Stemming is a process that reduces words into their stems, for example, reduce "increasing" to "increase." Popular programs that perform stemming include Porter stemmer [58] and Snowball stemmer [59]. Lemmatization, on the other hand, is to reduce words into their lemmas, for example, reduce "increasing" to "increase." Lemmatization requires the understanding of the context the word appears in a sentence. Examples of lemmatization programs for English include MorphAdorner [60] and Dilemma-2 [61].
- *Part-of-speech (POS) tagging* is to assign words in a sentence with respective POS according to the grammatical context of the word in the sentence (i.e., nouns, verbs, adverbs). A word may have multiple POS depending on the context of its appearance. For example, the word "book" appears as a verb in the phrase "book a ticket" whereas it appears as a noun in "read a book." POS tagging is very important for the analysis of relations between words. Some specialized taggers optimized for the biomedical domain include the MedPost [62], the dTagger [63], and the Genia tagger [64].
- *Parsing* is a process to understand the grammatical structure of a sentence. It is most often performed over sentences rather than whole documents. A parser typically tokenizes a sentence into words first, then performs POS tagging, and finally generates a tree structure with words as nodes and edges connecting the interrelated words. Several parsers are available, and some have been successfully applied to biomedical text such as the Stanford parser [65], the McCloskey parser [66], and the Carnegie-Mellon Link Grammar parser [67].

2.4 Biomedical Information Extraction Tools

After identifying an appropriate source of data to address the research question, the next step to perform text mining is information extraction. Information extraction from narrative text is not a straightforward task because words often have more than one meaning (homonyms) and more than one word can be used to express the same meaning (synonyms). Also, natural language is very flexible and evolves rapidly. Especially, patient observations in EMRs are often expressed as fragmented, unstructured, and ungrammatical text, which makes the task more challenging. The informatics community has thus invested a lot of effort in developing methods that can abstract relevant information from the free text using NLP techniques. Following are some information extraction tools developed specifically for the biomedical domain that can be used to identify entities required for drug safety surveillance (i.e., drug entities and clinical event concepts).

2.4.1 Clinical Concept Extraction System

- *c*linical *T*ext *A*nalysis and *K*nowledge *E*xtraction *S*ystem (*cTAKES*) is an open-source NLP platform for information extraction from clinical text developed by Mayo Clinic as a part of Open Health Natural Language Processing (OHNLP) consortium [68, 69]. It builds on the IBM's Unstructured Information Management Architecture (UIMA) framework [70]. cTAKES contains several components: sentence splitter, tokenizer, POS tagger, shallow parser, and named entity recognizer. Each of the cTAKES components is specifically trained to handle medical text. Their named entity recognizer is based on dictionary lookup from UMLS, and it can handle negations and status of named entities. The negation recognizer is implemented through the NegEx algorithm [71], and the status annotator uses regular expressions to determine whether the named entity occurs as a history, current, or a family event.
- *KnowledgeMap* is another tool implemented to identify UMLS concepts in biomedical documents, and it employs rigorous NLP techniques and document- and context-based disambiguation methods [72, 73]. KnowledgeMap was originally created to provide curriculum management tools for students and faculty [74], which has been successfully exported to six other institutions [75]. KnowledgeMap was also successfully applied to map UMLS concepts from electrocardiogram (ECG) impressions [76] and used to identify QT prolongation, an important risk factor for sudden cardiac death [77].
- *Med*ical *L*anguage *E*xtraction and *E*ncoding (*MedLEE*) system is a comprehensive clinical NLP system to extract, structure, and encode clinical information in free-text patient reports [78]. MedLEE has been shown to be as accurate as physicians at extracting clinical concepts from chest radiology reports [79, 80]. It was later extended to many other types of clinical documents, such as mammography [81] and discharge summaries [82], and showed effective performance as well.
- *MetaMap* is an open-source and highly configurable program developed at the NLM to discover UMLS concepts referenced in biomedical text [83]. Given a text, MetaMap first parses the text into simple noun phrases using a shallow parser called SPECIALIST and then generates all variants of each phrase (i.e., synonyms, abbreviations, derivational variants, inflections, spelling variants) using a lexicon and databases of synonyms. Finally, UMLS concepts containing at least one of the variants are evaluated against the original input text by calculating the mapping strength using a function based on linguistic principles.
- *Sem*antic *Rep*resentation (*SemRep*) is a tool developed by NLM to identify UMLS concepts and relationships in any arbitrary text. It utilizes MetaMap to find the semantic concepts first

and then uses a rule-based approach to determine the relationships between concepts occurring within a sentence. SemRep can handle any free-text data including abstracts, full texts, and medical records.

- *SNO*MED *cat*egorizer (*SNOcat*) is a tool to identify SNOMED-CT concepts in biomedical texts [84]. It allows online submission of textual query such as abstracts, full-text articles, or medical reports and returns a ranked list of possible matches to the SNOMED concepts within the documents. SNOcat achieves the task through regular expression-based pattern matching of terms, vector space indexing and retrieval engine, and *tf-idf* weighting schema with cosine normalization.

2.4.2 Medication Entity Extraction System

- *MedEx* is a medication extraction system that can identify medication names and corresponding signature information such as dosage, route, administration frequency, and duration from clinical narratives [85]. It consists of a semantic tagger and a context-free grammar parser that parses medication sentences using a semantic grammar. MedEx starts extraction by preprocessing records into sentences, then tagging each sentence to identify drug names with the RxNorm lexicon, and finally linking the drug names to their respective signatures with a rule-based parser.
- *M*edication *E*xtraction and *R*econciliation *K*nowledge *I*nstrument (*MERKI*) is another open-source parser, similar to MedEx, for the extraction of medication information from medical text that can recognize drug names and signature information like dose, frequency, strength, and duration [86]. It was developed on discharge summaries and relies on parsing rules as a set of regular expressions.
- *ProMiner* is a rule-based extraction tool developed to identify protein and gene entities in biomedical text and is able to associate database identifiers to the extracted terms [87]. It can recognize biological, medical, and chemical named entities and their synonyms and spelling variants in text. ProMiner has shown competitive results for the extraction of gene and protein names during the 2004 and 2006 BioCreative open-assessment challenges. It has also been successfully applied to patents, medical reports, and various forms of free-text literature.
- *Textractor* is a hybrid medication information extraction system developed for the "2009 Center of Informatics for Integrating Biology and the Bedside (i2b2) medication extraction challenge" [88]. Textractor is based on the UMIA framework, and some of its modules employ machine learning algorithms while others use regular expressions, rules, and dictionaries. The system was among the ten best performing systems in the challenge.

3 Methods

The primary goal of text mining is to mine knowledge hidden in unstructured text, which generally involves five major activities described below (Fig. 2):

- Data acquisition to gather relevant text.
- Data extraction to extract specific information from the text of interest.
- Data selection to eliminate irrelevant ones to the problem of interest.
- Data analysis to find associations among information extracted from text.
- Evaluation to assess the system in identifying accurate associations.

Subheadings 3.1–3.4 provide details on data acquisition, extraction, selection, and analysis in the context of drug safety surveillance. Subheading 3.5 describes some commonly used datasets and performance measures for quantitative evaluations.

3.1 Data Acquisition

As described above in Subheading 2, various sources of text are available for drug safety surveillance. Here we discuss briefly on where and how each source of data can be gathered.

- *Biomedical Literature*—Most references to biomedical publications are indexed by MEDLINE. The MEDLINE database can be searched using Boolean expressions through the popular biomedical search engine PubMed (http://pubmed.gov).

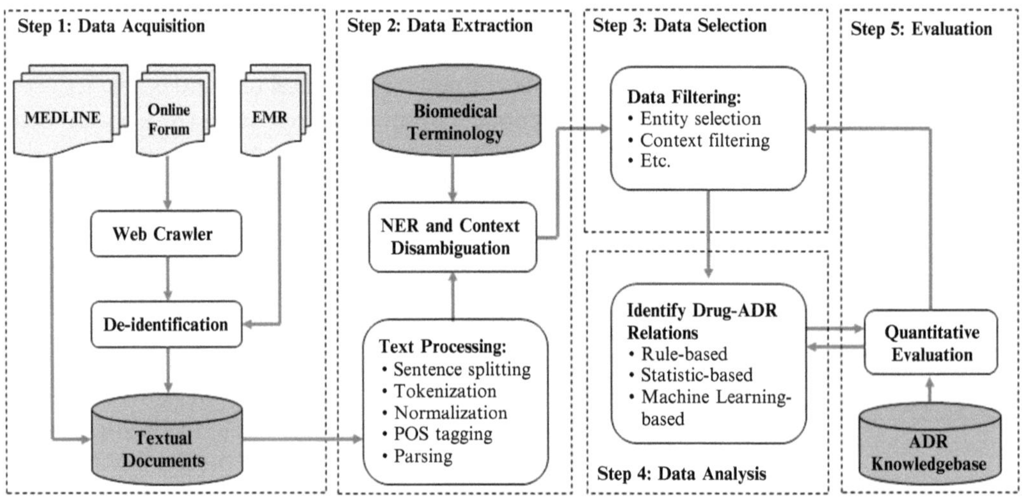

Fig. 2 Overview of the text mining process for the identification of ADRs: (1) Data acquisition, (2) data extraction, (3) data selection, (4) data analysis, and (5) evaluation

A researcher can obtain large clusters of abstracts containing specified keywords.

- *EMRs*—Patient medical records are usually collected and maintained by hospitals, clinics, or healthcare organizations; thus, its accessibility is mostly limited to people who are affiliated or collaborating with those organizations. Under all circumstances, the use of EMR data for research requires de-identification of the patients and must comply with the HIPAA privacy and security rule. Any study that uses EMR data is also subject to Institutional Review Board (IRB) approval.
- *Social Media Posts*—In common online communities, people often discuss with each other in the form of "threads" in which each thread is composed of an original post followed by a series of comments focusing on the same topic. Some of the popular health-related social network sites include PatientsLikeMe, MedHelp, DailyStrength, Ask a Patient, Yahoo Health and Wellness group, and iMedx. In order to gather posts and comments from these online communities, one may implement a web crawler to download pages. Since each site may be structured differently, the crawler oftentimes has to be customized. After downloading the pages, it is necessary to clean the data because a large portion of the text on a webpage is not related to the users' posts such as header, footer, navigation bar, and Javascript for the page. Finally, in order to protect patient privacy, it is a good practice to remove identifiers such as e-mail addresses, phone numbers, URLs, social and security numbers. To note, all data gathered must be in accordance with the site's Privacy Policy and Terms of Service and data generally should not be made publicly available without permission.
- *Other Text*—Here it mainly refers to the drug product labels and plain text descriptions in the DrugBank database. Most FDA-approved prescription drug labels can be obtained from DailyMed (http://dailymed.nlm.nih.gov/dailymed/). Plain text documents in the DrugBank database can be automatically downloaded using a web crawler.

3.2 Data Extraction

3.2.1 Identification of Biomedical Named Entities

To mine associations between drugs and adverse events from free text, it is crucial to find the named entities first such as drugs, diagnoses, and diseases. This task is commonly called Named Entity Recognition (NER). Text mining for drug safety surveillance typically starts with NER followed by normalization of the recognized entities and then mapping them to concepts in a biomedical ontology (e.g., UMLS, SNOMED). NER in biomedicine is not trivial because many named entities are ambiguous, synonyms (different terms describing similar condition), homonyms (same spelling but different meaning), abbreviations, and many more. Current approaches for

NER in biomedical text can be divided into dictionary based, rule based, and machine learning based.

- *Dictionary-based NER* relies on the existing dictionaries to identify named entities in free text. The dictionaries are typically derived from standard terminological resources. Dictionary lookup is usually based on string matching or string similarity algorithms. Performance of dictionary-based NER depends on how comprehensive the underlying terminological resource is. There are several successful information extraction systems that perform dictionary-based NER. For instance, both MedLEE and cTAKES perform dictionary-based NER and can accurately identify and encode patient-related information such as diagnoses, diseases, and treatments.
- *Rule-based NER* utilizes manually curated rules to identify named entities in free text [89]. The rules generally describe term formation patterns using grammatical (e.g., POS), syntactic (e.g., word precedence), lexical, morphological, and orthographic features (e.g., capitalization). The rule-based NER systems often rely on combinations of regular expressions, heuristics, and rules designed by domain experts. Because of the dependence on expert knowledge, this type of NER system lacks scalability and adaptability to other domains.
- *Machine learning-based NER* applies different learning algorithms to train models to recognize and extract named entities in text. Although machine learning-based approach can easily adapt to new domains, it requires manually annotated corpora to train the models. The power of machine learning models depends on the discriminative power of the textual features utilized as well as the algorithm [90]. Conditional Random Fields (CRF) is one of the most commonly applied machine learning techniques for biomedical NER and has demonstrated good results in medical entity recognition [91–93].

3.2.2 Context Disambiguation

Sometimes accurate identification of named entities in free text requires additional processing such as disambiguation. Disambiguation of medical entities can be in the form of entity disambiguation to solve the homonym problems or assertion classification to determine the opinion made over the entity.

- *Entity disambiguation* is also known as word sense disambiguation. It is a process of determining the appropriate sense of a named entity when it is associated with multiple meanings (i.e., homonym). For example, "cold" may occur in a patient's medical history as a disease denoting "the common cold" or a symptom "patient feels cold." Many approaches have been developed to address the word sense disambiguation problem [94–96].

- *Assertion classification* is important in the medical domain because simple occurrence of a medical entity is not sufficient to conclude that the patient has the medical condition. Physicians often mention medical problems in clinical notes that are present, absent, or hypothetical. For example, in the sentence "patient does not have colon cancer," the medical entity "colon cancer" is absent, whereas in the sentence "patient's problem is likely to be due to hypertension," "hypertension" is present. Assertion classification was part of the 2010 i2b2 NLP challenge, and 21 systems were developed for the task [97]. Results showed that machine learning algorithms can be combined with rule-based systems to accurately determine assertions.

3.3 Data Selection

Accurate information extraction is the key to optimize the performance of the next step, data analysis. Sometimes data filtering is necessary to eliminate irrelevant information based on knowledge in the application domain. For instance, one may want to limit the extracted clinical entities to certain semantic classes to maximize the likelihood of retrieving correct concepts. Using drug safety surveillance as an example, to ensure that extracted entities are possible ADRs, it may be necessary to select entities from the following UMLS semantic classes: finding, disease or symptom, mental or behavioral dysfunction, sign or symptom, and neoplastic process. Filtering is also needed to eliminate entities that are negative, past, or family history events. For example, when extracting patient clinical information from EMRs, it is desirable to exclude all medical concepts from the "past history" section of the notes as it indicates past rather than present conditions. Similarly, medication mentions may also be filtered to eliminate medications not given at the present time. Effective filtering techniques have been demonstrated to yield better results for mining drug–ADR associations from narrative EMRs [34, 37, 98]. Since EMRs vary across institutions, expert knowledge is essential to determine the most appropriate filtering criteria.

3.4 Data Analysis

After extracting the named entities of interest (i.e., drugs and clinical events) from free text, the next step is data analysis to discover the relationships or the associations between the entities. There are many organized challenges to investigate relationship extraction from free text. For instance, the 2010 BioCreative challenge focused on the identification of biological relationships such as protein–protein interactions [99]. For medical relationship extraction, the 2010 i2b2 NLP challenge asked participants to extract relations between three classes of medical concepts (i.e., problem, treatment, and test) [97]. More recently, the 2011 DDIExtraction challenge focused on the extraction of DDIs from free text [100].

In general, for drug safety surveillance using free text, the relationships between drugs and clinical events can be either derived based on semantic patterns or mined using statistical and machine learning methods.

3.4.1 Rule-Based Drug–ADR Detection

Many drug–ADR relationship detection methods are based on semantic patterns, more specifically, using patterns or rules to match the drug name and ADR mention in text. For instance, Leaman et al. [36] implemented a filter to exclude irrelevant ADR events extracted from the user posts on DailyStrength based on the closest verb to the left of the ADR mention, achieving a result of 0.78 precision, 0.70 recall, and 0.74 *F*-score. Sohn et al. [42] manually developed rules by examining the keywords and expression patterns of ADRs in clinical notes to identify drug–ADR relationships. Their rule-based system produced *F*-score of 0.80. Karimi et al. [101] proposed to find patterns of ADR reporting using both heuristics and automatically extracted rules from online posts. Haerian et al. [98] identified ADR-related events from EMRs by combining NLP and expert-generated knowledge source. Furthermore, Segura-Bedmar et al. [55, 56] used a set of linguistic rules to extract DDIs from pharmacological documents with a reasonable 0.67 precision and a low 0.14 recall. Tari et al. [102] identified DDIs by representing the general knowledge of drug metabolism and interactions in the form of logic rules.

3.4.2 Statistic-Based Drug–ADR Detection

The ADR problem can also be treated as finding associations between a drug and an adverse event based on co-occurrence of the two in text documents. In this case, statistic-based methods are widely used to detect associations between the drug and ADR entities.

One of the most extensively investigated statistic-based approaches for ADR detection is disproportionality analysis that involves frequency analyses of contingency tables to quantify the degree to which a drug and ADR co-occur “disproportionally” compared with what would be expected if there were no association. For instance, Shetty and Dalal [41] compared the observed number of drug–ADR literature citations with the expected count under the null hypothesis that the drug and ADR are independent. LePendu et al. [103] developed an annotation tool for clinical text and examined the resulting annotations by computing the risks of myocardial infarction for patients with rheumatoid arthritis that take Vioxx using odds ratio. Vilar et al. [104] calculated the odds ratio to measure statistical associations between drugs and the event pancreatitis in EMRs. LePendu et al. [105] proposed a novel framework for annotating the unstructured clinical notes and used the free-text derived features to detect drug–ADR associations and DDIs by calculating odds ratio. Moreover, Leeper et al. [106] analyzed EMRs to identify drug–ADR associations using odds ratio.

Duke et al. [107] used literature to screen for potential DDIs based on mechanistic properties followed by EMR-based validation to identify clinically important DDIs that synergistically increase the risk of myopathy by calculating their relative risks. Gurulingappa et al. [108] applied Multi-item Gamma Poisson Shrinkage (MGPS) method for drug–ADR signal detection from text and open-source data and used the detected signals to predict drug label changes. In three separate studies, Wang et al. used MedLEE to identify medication and ADR entities from narrative discharge summaries in EMRs and calculated χ^2 statistic [34, 37] and mutual information [33] to detect associations between the entities. Benton et al. [109] identified drug–ADR pairs from medical message boards by calculating the one-tailed Fisher's exact value. Last but not least, Henriksson et al. [110] calculated distributional similarities for drug–symptom pairs based on co-occurrence information in EMRs.

Other studies have utilized classic data mining algorithms such as association rule mining and regression to identify drug–ADR pairs. Yang et al. [111] applied association mining and proportional reporting ratios to identify the associations between drug and ADR from user posts in social media. Nikfarjam and Gonzalez [112] applied association rule mining to extract the underlying expression patterns about ADRs from user reviews on drugs, which achieved 0.70 precision, 0.66 recall, and 0.68 *F*-score. Wang et al. [113] proposed to determine the relationship between a drug and ADR event based on PubMed citations using logistic regression. Liu et al. [114] analyzed the statistical enrichment of the drug–ADR co-occurrences extracted from clinical notes with local regression models.

3.4.3 Machine Learning-Based Drug–ADR Detection

Drug–ADR relationship extraction can also be formulated as a learning task. More specifically, given a sentence with mentions of a drug and clinical event, we want to determine whether association exists between the drug and event based on the feature vectors generated from the surrounding words or the sentence. All classic machine learning and data mining algorithms are applicable here.

Chee et al. [38] explored ensemble learning, a common machine learning technique utilizing multiple classifiers, to identify drug–ADR pairs from online health forums. Bisgin et al. [115] applied the latent Dirichlet allocation (LDA) algorithm to the FDA drug labels to discover topics that group drugs together with similar safety concerns. Yang et al. [116] compared different machine learning algorithms, Naïve Bayes, Decision Tree, and support vector machine (SVM), for ADR detection using letters to the editor. Gurulingappa et al. [117, 118] applied the maximum entropy classifier to identify drug-related ADR events from medical case reports, and this resulted in 0.70 *F*-score.

Bjorne et al. [119] applied the SVM and regularized least-squares (RLS) classifiers to extract interaction relations for drug mention pairs found in biomedical texts. Their system achieved a performance of 0.63 *F*-score. Thomas et al. [120] built an ensemble of contrasting machine learning classifiers to extract DDIs through majority voting. The best single classifier achieved 0.57 *F*-score, and the best ensemble achieved 0.61 *F*-score. Segura-Bedmar et al. [121] proposed a shallow linguistic kernel for DDI extraction and achieved a precision of 0.51, a recall of 0.73, and an *F*-score of 0.60. Zhang et al. [122] proposed a single kernel-based approach to extract DDIs from biomedical literature. Percha et al. [123] inferred DDIs by training a random forest classifier to score potential DDIs based on the assertions extracted from literature that relate two drugs to a gene product. Kolchinsky et al. [124] evaluated six classifiers (variable trigonometric threshold classifier, SVM, logistic regression, Naïve Bayes, linear discriminant analysis, diagonal version of LDA) for the identification of pharmacokinetic DDIs using PubMed abstracts. Boyce et al. [125] also evaluated different machine learning algorithms for their ability to identify pharmacokinetic DDIs but from package inserts rather than literature. In their experiment, SVM performed the best with an *F*-score of 0.86. He et al. [126] combined feature-based, graph, and tree kernels to extract DDIs from literature and achieved an *F*-score of 0.69.

3.5 Quantitative Evaluation

3.5.1 Annotated Dataset for Extraction System Evaluation

Over the years, organized NLP extraction challenges have made available valuable annotated datasets for training and testing extraction systems, which include the following.

- *DDIExtraction Challenge* aims to stimulate research in the automatic extraction of DDIs from free text. The challenge has created the DrugDDI, a benchmark corpus to study the phenomenon of interactions among drugs. DrugDDI contains 3,160 annotated DDIs at the sentence level by a researcher with pharmaceutical background. Only sentences with two or more drug mentions were annotated in the corpus. The corpus contains 5,806 sentences with average 10.3 sentences per document, and 2,044 sentences contain at least one interaction.
- *i2b2 NLP Challenge* aims to enhance the ability of NLP tools to extract more fine-rained information from clinical records by providing sets of fully de-identified notes from the Research Patient Data Repository at Partners HealthCare. Approximately 1,500 notes have been released from the first four i2b2 NLP challenges. To access the notes, researchers will need to register and an expedited review of the request will be conducted; if accepted, the requester will need to sign a standard Data Use Agreement before accessing the annotated notes.

3.5.2 Knowledgebase for Drug–ADR Association Evaluation

There also exists public knowledgebase of known drug–ADR associations, often referred to as reference standard or gold standard, to evaluate the correctness of the identified associations between drugs and ADRs. Following are some of the publicly available sources.

- *DrugBank* (http://www.drugbank.ca/interax/drug_lookup) also contains drug interaction information and can be searched by the Interax Interaction Search engine on its website. In a DDI prediction study based on chemical structures, Vilar et al. [127] constructed a dataset of 9,454 unique DDIs among 928 drugs using the Interax engine from DrugBank, and they have made the dataset publicly available.
- *Epocrates Online* (http://www.epocrates.com) aims to provide information about drugs to doctors and other healthcare professionals. Among its various functions, one can check for drug adverse effects and interactions. It is widely used by physicians (~40 %) in the United States.
- *Micromedex* (http://www.micromedex.com) is a well-respected, evidence-based database that contains referenced information about drugs, toxicology, diseases, acute care, and alternative medicine for healthcare professionals.
- *SIDER* (http://www.sideeffects.embl.de) is a side effect resource that contains information on marketed medicines and their recorded ADRs extracted from public documents and package inserts. Available information includes side effect frequency, drug and side effect classifications, as well as links to further information. The current version of SIDER (released on October 17, 2012) contains 996 drugs, 4,192 ADRs, and 99,423 drug–ADR pairs [128].
- *STITCH* (http://stitch.embl.de) is a searchable database for interaction of chemicals that integrates information on interactions from metabolic pathways, crystal structures, binding experiments, and drug–target relationships [129]. The relationships between chemicals are predicted using text mining and chemical structure similarity. STITCH contains interaction information for over 68,000 different chemicals, including 2,200 drugs.

3.5.3 Evaluation Metrics

There are various evaluation metrics to assess the performance of a system. Following are the most commonly used evaluation metrics in text mining.

- *F-score* measures the overall completeness and correctness of a system. It is calculated by comparing the system's output against a reference standard. Elements identified by the system are also present in the reference standard and are called "true positives." On the other hand, elements identified by the

Table 2
Table of basic truth measures

		Reference standard	
		Positive	**Negative**
System output	Positive	True positive (TP)	False positive (FP)
	Negative	False negative (FN)	True negative (TN)

system but not present in the reference standard are referred to as "false positives." Moreover, elements present in the reference standard but was not identified by the system are referred to as "false negatives." Table 2 illustrates the basic truth measures. The truth measures are used to determine the system's precision and recall which are then combined to calculate the *F*-score. Precision is the portion of true positives against all predicted positive results, which is calculated as follows:

$$\text{Precision} = \frac{\text{TP}}{\text{TP} + \text{FP}}$$

- Recall is the fraction of true positives among all positives and is calculated as follows:

$$\text{Recall} = \frac{\text{TP}}{\text{TP} + \text{FN}}$$

- Finally, *F*-score is a harmonic mean of the precision and recall and can be calculated as follows:

$$F\text{-score} = \frac{2 \times \text{Precision} \times \text{recall}}{\text{Precision} + \text{recall}}$$

- For all the above metrics, a result of 1 indicates the best performance and 0 indicates the worst.

- *Mean average precision* (*MAP*) score is a performance measure commonly used to evaluate systems that produce a ranked list of results. This measure considers the order of the results presented. The MAP score is similar to the area under the precision–recall curve that penalizes both types of misclassifications: false positives and false negatives. MAP can be calculated as follows where Q is the set of queries (Q= 1 for the general identification of relations between drugs and ADRs) and R_{jk} is the set of ranked relations until you get K true relations where $[r_1, \ldots, r_{mj}]$ is the set of true relation in R_{jk}:

$$\text{MAP}(Q) = \frac{1}{|Q|} \sum_{j=1}^{Q} \frac{1}{m_j} \sum_{k=1}^{m_j} \text{Precision}(R_{jk})$$

4 Discussion

In this chapter, we have discussed various components of text mining to support the identification of drug–ADR associations from free-text documents including biomedical literature, narrative EMR notes, drug labels, and user posts on social media. Text mining in this domain can be generally divided into five tasks: data acquisition, data extraction, data selection, data analysis, and evaluation. In the previous section, we provided detailed information regarding the tasks and the common approaches involved in each task.

Although the existing PhV research based on text mining has produced promising results, the identified associations between drugs and ADRs remain to be the potentially important signals not meant to be proven and that must be further validated through appropriate medical and epidemiological evaluations. Such validation analysis requires sophisticated clinical and laboratory evaluations that exclude biases and confounding variables. The primary goal of using computational methods for PhV is to efficiently search for a list of significant drug–ADR pairs to be considered for further validation analysis.

While working with different textual sources, one has to keep in mind the intrinsic characteristics and limitations of individual sources. For instance, several studies have demonstrated that mining EMRs can generate earlier and important ADR signals than the conventional spontaneous reporting data. However, EMR data is not available to everyone, and even when it is available, most researchers only have one dataset from a single affiliating or collaborating health unit which may have a limited population. In addition, patients may visit different hospitals and different doctors for treatment, thus resulting in missing data in EMRs. This problem may be solved by the integration of EMRs from multiple resources, but it is still currently a technical and policy challenge.

Social media data, on the other hand, may be publicly available and include patients from a wide range of geographical regions around the world, but the user posts are often much noisier than data collected from health professionals. Since drug names in social networks are specified by the user when ADR comments are submitted, the problem of relationship determination is eliminated but the main challenge becomes the determination of what the consumer-posted ADR terms are and what they really mean. In addition, the main difference between text mining from social media and other text is the language and length of the text. Compared to the text in biomedical literature, consumer posts are usually informal, general, and short. This is the main reason for the inadequate performance of current NLP systems in extracting and analyzing online health forums. Another important reason is that most existing medical terminologies are developed for professionals and thus may not cover the vocabulary of the consumers.

In conclusion, our key messages are the following: (1) consider all aspects related to data quality when choosing your data source for ADR signal detection with text mining; (2) tailor your algorithm selection to each case, keeping in mind the biases and confounders that may influence the algorithm performance; (3) there is still a large room for improvement in this area; and (4) the detected ADR signals should always be guided by appropriate clinical evaluation to support actual patients' needs.

Acknowledgment

Mei Liu is supported by funds from the New Jersey Institute of Technology. Yong Hu is supported by the National Science Foundation of China (71271061, 70801020); Science and Technology Planning Project of Guangdong Province, China (2010B010600034, 2012B091100192); and Business Intelligence Key Team of Guangdong University of Foreign Studies (TD1202). Buzhou Tang is supported by the China Postdoctoral Science Foundation (2011 M500669).

References

1. World Health Organization (1966) International drug monitoring: the role of the hospital. In: Technical report series no. 425. World Health Organization, Geneva
2. Pirmohamed M, Breckenridge AM, Kitteringham NR et al (1998) Adverse drug reactions. BMJ 316:1295–1298
3. Patel P, Zed PJ (2002) Drug-related visits to the emergency department: how big is the problem? Pharmacotherapy 22:915–923
4. Juntti-Patinen L, Neuvonen PJ (2002) Drug-related deaths in a university central hospital. Eur J Clin Pharmacol 58:479–482
5. Moore TJ, Cohen MR, Furberg CD (2007) Serious adverse drug events reported to the Food and Drug Administration, 1998–2005. Arch Intern Med 167:1752–1759
6. Lazarou J, Pomeranz BH, Corey PN (1998) Incidence of adverse drug reactions in hospitalized patients: a meta-analysis of prospective studies. JAMA 279:1200–1205
7. Jha AK, Kuperman GJ, Rittenberg E et al (2001) Identifying hospital admissions due to adverse drug events using a computer-based monitor. Pharmacoepidemiol Drug Saf 10: 113–119
8. Griffin MR, Stein CM, Ray WA (2004) Postmarketing surveillance for drug safety: surely we can do better. Clin Pharmacol Ther 75:491–494
9. Edwards IR, Aronson JK (2000) Adverse drug reactions: definitions, diagnosis, and management. Lancet 356:1255–1259
10. Fliri AF, Loging WT, Thadeio PF et al (2005) Analysis of drug-induced effect patterns to link structure and side effects of medicines. Nat Chem Biol 1:389–397
11. Bender A, Scheiber J, Glick M et al (2007) Analysis of pharmacology data and the prediction of adverse drug reactions and off-target effects from chemical structure. ChemMedChem 2:861–873
12. Campillos M, Kuhn M, Gavin AC et al (2008) Drug target identification using side-effect similarity. Science 321:263–266
13. Fuzuzaki M, Seki M, Kashima H et al (2009) Side effect prediction using cooperative pathways. In: IEEE international conference on bioinformatics and biomedicine. Washington, DC, pp 142–147
14. Scheiber J, Chen B, Milik M et al (2009) Gaining insight into off-target mediated effects of drug candidates with a comprehensive systems chemical biology analysis. J Chem Inf Model 49:308–317
15. Scheiber J, Jenkins JL, Sukuru SC et al (2009) Mapping adverse drug reactions in chemical space. J Med Chem 52:3103–3107
16. Xie L, Li J, Bourne PE (2009) Drug discovery using chemical systems biology: identification

of the protein-ligand binding network to explain the side effects of CETP inhibitors. PLoS Comput Biol 5:e1000387

17. Hammann F, Gutmann H, Vogt N et al (2010) Prediction of adverse drug reactions using decision tree modeling. Clin Pharmacol Ther 88:52–59
18. Yamanishi Y, Kotera M, Kanehisa M et al (2010) Drug-target interaction prediction from chemical, genomic and pharmacological data in an integrated framework. Bioinformatics 26:i246–i254
19. Brouwers L, Iskar M, Zeller G et al (2011) Network neighbors of drug targets contribute to drug side-effect similarity. PLoS One 6:e22187
20. Pauwels E, Stoven V, Yamanishi Y (2011) Predicting drug side-effect profiles: a chemical fragment-based approach. BMC Bioinformatics 12:169
21. Pouliot Y, Chiang AP, Butte AJ (2011) Predicting adverse drug reactions using publicly available PubChem BioAssay data. Clin Pharmacol Ther 90:90–99
22. Lounkine E, Keiser MJ, Whitebread S et al (2012) Large-scale prediction and testing of drug activity on side-effect targets. Nature 486:361–367
23. Lindquist M, Edwards IR (2001) The WHO Programme for International Drug Monitoring, its database, and the technical support of the Uppsala Monitoring Center. J Rheumatol 28:1180–1187
24. Szarfman A, Machado SG, O'Neill RT (2002) Use of screening algorithms and computer systems to efficiently signal higher-than-expected combinations of drugs and events in the US FDA's spontaneous reports database. Drug Saf 25:381–392
25. Hauben M, Reich L, Chung S (2004) Postmarketing surveillance of potentially fatal reactions to oncology drugs: potential utility of two signal-detection algorithms. Eur J Clin Pharmacol 60:747–750
26. Chan KA, Hauben M (2005) Signal detection in pharmacovigilance: empirical evaluation of data mining tools. Pharmacoepidemiol Drug Saf 14:597–599
27. Berlowitz DR, Miller DR, Oliveria SA et al (2006) Differential associations of beta-blockers with hemorrhagic events for chronic heart failure patients on warfarin. Pharmacoepidemiol Drug Saf 15:799–807
28. Bjornsson E, Olsson R (2006) Suspected drug-induced liver fatalities reported to the WHO database. Dig Liver Dis 38:33–38
29. Hauben M, Reich L, Gerrits CM (2006) Reports of hyperkalemia after publication of RALES: a pharmacovigilance study. Pharmacoepidemiol Drug Saf 15:775–783
30. Brown JS, Kulldorff M, Chan KA et al (2007) Early detection of adverse drug events within population-based health networks: application of sequential testing methods. Pharmacoepidemiol Drug Saf 16:1275–1284
31. Jin HD, Chen J, He HX et al (2008) Mining unexpected temporal associations: applications in detecting adverse drug reactions. IEEE Trans Inf Technol Biomed 12:488–500
32. Matthews EJ, Kruhlak NL, Benz RD et al (2009) Identification of structure-activity relationships for adverse effects of pharmaceuticals in humans: Part C: use of QSAR and an expert system for the estimation of the mechanism of action of drug-induced hepatobiliary and urinary tract toxicities. Regul Toxicol Pharmacol 54:43–65
33. Wang X, Hripcsak G, Friedman C (2009) Characterizing environmental and phenotypic associations using information theory and electronic health records. BMC Bioinformatics 10(Suppl 9):S13
34. Wang X, Hripcsak G, Markatou M et al (2009) Active computerized pharmacovigilance using natural language processing, statistics, and electronic health records: a feasibility study. J Am Med Inform Assoc 16:328–337
35. Harpaz R, Chase HS, Friedman C (2010) Mining multi-item drug adverse effect associations in spontaneous reporting systems. BMC Bioinformatics 11(Suppl 9):S7
36. Leaman R, Wojtulewicz L, Sullivan R et al (2010) Towards internet-age pharmacovigilance: extracting adverse drug reactions from user posts to health-related social networks. In: Workshop on biomedical natural language processing, pp 117–125
37. Wang X, Chase H, Markatou M et al (2010) Selecting information in electronic health records for knowledge acquisition. J Biomed Inform 43:595–601
38. Chee BW, Berlin R, Schatz B (2011) Predicting adverse drug events from personal health messages. AMIA Annu Symp Proc 2011:217–226, Washington, DC
39. Harpaz R, Perez H, Chase HS et al (2011) Biclustering of adverse drug events in the FDA's spontaneous reporting system. Clin Pharmacol Ther 89:243–250
40. Ji Y, Ying H, Dews P et al (2011) A potential causal association mining algorithm for screening adverse drug reactions in postmarketing surveillance. IEEE Trans Inf Technol Biomed 15:428–437
41. Shetty KD, Dalal SR (2011) Using information mining of the medical literature to

improve drug safety. J Am Med Inform Assoc 18:668–674
42. Sohn S, Kocher JP, Chute CG et al (2011) Drug side effect extraction from clinical narratives of psychiatry and psychology patients. J Am Med Inform Assoc 18(Suppl 1):i144–i149
43. Zorych I, Madigan D, Ryan P et al (2011) Disproportionality methods for pharmacovigilance in longitudinal observational databases. Stat Methods Med Res 22:39–56
44. Harpaz R, Vilar S, Dumouchel W et al (2012) Combing signals from spontaneous reports and electronic health records for detection of adverse drug reactions. J Am Med Inform Assoc 20:413–419
45. Liu M, McPeek Hinz ER, Matheny ME et al (2012) Comparative analysis of pharmacovigilance methods in the detection of adverse drug reactions using electronic medical records. J Am Med Inform Assoc 20:420–426
46. Liu M, Wu Y, Chen Y et al (2012) Large-scale prediction of adverse drug reactions using chemical, biological, and phenotypic properties of drugs. J Am Med Inform Assoc 19:e28–e35
47. Warrer P, Hansen EH, Juhl-Jensen L et al (2012) Using text-mining techniques in electronic patient records to identify ADRs from medicine use. Br J Clin Pharmacol 73:674–684
48. Yoon D, Park MY, Choi NK et al (2012) Detection of adverse drug reaction signals using an electronic health records database: comparison of the Laboratory Extreme Abnormality Ratio (CLEAR) algorithm. Clin Pharmacol Ther 91:467–474
49. Scripture CD, Figg WD (2006) Drug interactions in cancer therapy. Nat Rev Cancer 6:546–558
50. Hale R (2005) Text mining: getting more value from literature resources. Drug Discov Today 10:377–379
51. Cohen AM, Hersh WR (2005) A survey of current work in biomedical text mining. Brief Bioinform 6:57–71
52. Van De Belt TH, Engelen LJ, Berben SA et al (2010) Definition of Health 2.0 and Medicine 2.0: a systematic review. J Med Internet Res 12:e18
53. Wishart DS, Knox C, Guo AC et al (2008) DrugBank: a knowledgebase for drugs, drug actions and drug targets. Nucleic Acids Res 36:D901–D906
54. Knox C, Law V, Jewison T et al (2010) DrugBank 3.0: a comprehensive resource for 'omics' research on drugs. Nucleic Acids Res 39:D1035–D1041
55. Segura-Bedmar I, Crespo M, de Pablo-Sanchez C et al (2010) Resolving anaphoras for the extraction of drug–drug interactions in pharmacological documents. BMC Bioinformatics 11(Suppl 2):S1
56. Segura-Bedmar I, Martinez P, de Pablo-Sanchez C (2011) A linguistic rule-based approach to extract drug–drug interactions from pharmacological documents. BMC Bioinformatics 12(Suppl 2):S1
57. Jiang J, Zhai C (2007) An empirical study of tokenization strategies for biomedical information retrieval. Inf Retr 10:341–363
58. Porter M (1997) An algorithm for suffix stripping. In: Sparck Jones K, Willett P (eds) Readings in information retrieval. Morgan Kaufmann Publishers Inc., San Francisco, CA, pp 313–6
59. Porter M (2001) Snowball: a language for stemming algorithms. Available from: http://snowball.tartarus.org/texts/introduction.html
60. Burns PR (2013) MorphAdorner v2: a Java Library for the morphological adornment of English language texts. Northwestern University, Evanston, IL
61. Paulussen H, Martin W (1992) DILEMMA-2: a lemmatizer-tagger for medical abstracts. In: Third conference on applied natural language processing
62. Smith L, Rindflesch T, Wilbur WJ (2004) MedPost: a part-of-speech tagger for bioMedical text. Bioinformatics 20:2320–2321
63. Divita G, Browne A, Loane R (2006) dTagger: a POS Tagger. AMIA Annu Symp Proc 2006:200–203
64. Tsuruoka Y, Tateishi Y, Kim J-D et al (2005) Developing a robust part-of-speech tagger for biomedical text. Lect Notes Comput Sci 3746:382–392
65. Klein D, Manning C (2003) Accurate unlexicalized parsing. In: Proceedings of the 41st meeting of the Association for Computational Linguistics, vol 2003, pp 423–430
66. McClosky D (2006) Effective self-training for parsing. In: Proceedings of North American chapter of the Association for Computational Linguistics, vol 2006, pp 152–159
67. Grinberg D, Lafferty J, Sleator D (1995) A robust parsing algorithm for link grammars. In: Proceedings of the fourth international workshop on parsing technologies, vol 1995
68. Open Health Natural Language Processing (OHNLP) Consortium
69. Savova GK, Masanz JJ, Ogren PV et al (2010) Mayo clinical Text Analysis and Knowledge Extraction System (cTAKES): architecture, component evaluation and applications. J Am Med Inform Assoc 17:507–513

70. Unstructured Information Management Architecture (UIMA). Available from: http://uima-framework.sourceforge.net
71. Mitchell KJ, Becich MJ, Berman JJ et al (2004) Implementation and evaluation of a negation tagger in a pipeline-based system for information extract from pathology reports. Stud Health Technol Inform 107:663–667
72. Denny JC, Smithers JD, Miller RA et al (2003) "Understanding" medical school curriculum content using KnowledgeMap. J Am Med Inform Assoc 10:351–362
73. Denny J, Peterson J (2007) Identifying QT prolongation from ECG impressions using natural language processing and negation detection. Stud Health Technol Inform 129: 1283–1288
74. Denny JC, Pr I, Wehbe FH et al (2003) The KnowledgeMap project: development of a concept-based medical school curriculum database. AMIA Annu Symp Proc 2003:195–199
75. Denny JC, Soriano RP, Stein G et al (2009) POGOe: a national repository of geriatric education materials. Proc AMIA Annu Fall Symp 2009:1–192
76. Denny JC, Spickard A, Miller RA et al (2005) Identifying UMLS concepts from ECG impressions using KnowledgeMap. AMIA Annu Symp Proc 2005:196–200
77. Denny JC, Miller RA, Waitman LR et al (2009) Identifying QT prolongation from ECG impressions using a general-purpose Natural Language Processor. Int J Med Inform 78(Suppl 1):S34–S42
78. Friedman C, Alderson PO, Austin JH et al (1994) A general natural-language text processor for clinical radiology. J Am Med Inform Assoc 1:161–174
79. Hripcsak G, Friedman C, Alderson PO et al (1995) Unlocking clinical data from narrative reports: a study of natural language processing. Ann Intern Med 122:681–688
80. Hripcsak G, Austin JH, Alderson PO et al (2002) Use of natural language processing to translate clinical information from a database of 889,921 chest radiographic reports. Radiology 224:157–163
81. Friedman C (1997) Towards a comprehensive medical language processing system: methods and issues. Proc AMIA Annu Fall Symp 1997:595–599
82. Friedman C, Knirsch C, Shagina L et al (1999) Automating a severity score guideline for community-acquired pneumonia employing medical language processing of discharge summaries. Proc AMIA Symp 1999:256–260
83. Aronson AR, Lang FM (2010) An overview of MetaMap: historical perspective and recent advances. J Am Med Inform Assoc 17: 229–236
84. Ruch P, Gobeill J, Lovis C et al (2008) Automatic medical encoding with SNOMED categories. BMC Med Inform Decis Mak 8(Suppl 1):S6
85. Xu H, Stenner SP, Doan S et al (2010) MedEx: a medication information extraction system for clinical narratives. J Am Med Inform Assoc 17:19–24
86. Gold S, Elhadad N, Zhu X et al (2008) Extracting structured medication event information from discharge summaries. AMIA Annu Symp Proc 2008:237–241
87. Hanisch D, Fundel K, Mevissen HT et al (2005) ProMiner: rule-based protein and gene entity recognition. BMC Bioinformatics 6(Suppl 1):S14
88. Meystre SM, Thibault J, Shen S et al (2010) Textractor: a hybrid system for medications and reason for their prescription extraction from clinical text documents. J Am Med Inform Assoc 17:559–562
89. Cohen B, Hunter L (2004) Natural language processing and systems biology. In: Pereira F, Dubitzky W (eds) Artificial intelligence methods and tools for systems biology. Springer, Netherlands
90. Krauthammer M, Nenadic G (2004) Term identification in the biomedical literature. J Biomed Inform 37:512–526
91. Leaman R, Gonzalez G (2008) BANNER: an executable survey of advances in biomedical named entity recognition. Pac Symp Biocomput 2008:652–663
92. Mahbub Chowdhury F, Lavelli A (2010) Disease mention recognition with specific features. In: Proceedings of the 2010 workshop on biomedical natural language processing (BioNLP), vol 2010, p 91 98
93. Jiang M, Chen Y, Liu M et al (2011) A study of machine-learning-based approaches to extract clinical entities and their assertions from discharge summaries. J Am Med Inform Assoc 18:601–606
94. Savova GK, Coden AR, Sominsky IL et al (2008) Word sense disambiguation across two domains: biomedical literature and clinical notes. J Biomed Inform 41: 1088–1100
95. Jimeno-Yepes AJ, McInnes BT, Aronson AR (2011) Exploiting MeSH indexing in MEDLINE to generate a data set for word sense disambiguation. BMC Bioinformatics 12:223

96. Stevenson M, Agirre E, Soroa A (2012) Exploiting domain information for Word Sense Disambiguation of medical documents. J Am Med Inform Assoc 19:235–240
97. Uzuner O, South BR, Shen S et al (2011) 2010 i2b2/VA challenge on concepts, assertions, and relations in clinical text. J Am Med Inform Assoc 18:552–556
98. Haerian K, Varn D, Vaidya S et al (2012) Detection of pharmacovigilance-related adverse events using electronic health records and automated methods. Clin Pharmacol Ther 92:228–234
99. Leitner F, Mardis SA, Krallinger M et al (2010) An overview of BioCreative II.5. IEEE/ACM Trans Comput Biol Bioinform 7:385–399
100. Segura-Bedmar I, Martinez P, Sanchez-Cisneros D (2011) The 1st DDIExtraction-2011 challenge task: extraction of drug–drug interactions from biomedical texts. In: Proceedings of workshop on first challenge task: drug–drug interaction extraction, vol 2011, p 1–9
101. Karimi S, Kim SN, Cavedon L (2011) Drug side-effects: what do patients forums reveal? In: The second international workshop on Web science and information exchange in the medical Web. ACM, Glasgow
102. Tari L, Anwar S, Liang S et al (2010) Discovering drug–drug interactions: a text-mining and reasoning approach based on properties of drug metabolism. Bioinformatics 26:i547–i553
103. Lependu P, Iyer SV, Fairon C et al (2012) Annotation analysis for testing drug safety signals using unstructured clinical notes. J Biomed Semantics 3(Suppl 1):S5
104. Vilar S, Harpaz R, Santana L et al (2012) Enhancing adverse drug event detection in electronic health records using molecular structure similarity: application to pancreatitis. PLoS One 7:e41471
105. LePendu P, Iyer SV, Bauer-Mehren A et al (2013) Pharmacovigilance using clinical notes. Clin Pharmacol Ther 93:547–555
106. Leeper NJ, Bauer-Mehren A, Iyer SV et al (2013) Practice-based evidence: profiling the safety of cilostazol by text-mining of clinical notes. PLoS One 8:e63499
107. Duke JD, Han X, Wang Z et al (2012) Literature based drug interaction prediction with clinical assessment using electronic medical records: novel myopathy associated drug interactions. PLoS Comput Biol 8:e1002614
108. Gurulingappa H, Toldo L, Rajput AM et al (2013) Automatic detection of adverse events to predict drug label changes using text and data mining techniques. Pharmacoepidemiol Drug Saf 22:1189–1194
109. Benton A, Ungar L, Hill S et al (2011) Identifying potential adverse effects using the web: a new approach to medical hypothesis generation. J Biomed Inform 44:989–996
110. Henriksson A, Kvist M, Hassel M et al (2012) Exploration of adverse drug reactions in semantic vector space models of clinical text. In: Proceedings of the 29th international conference on machine learning, vol 2012. Edinburgh, Scotland
111. Yang CC, Yang H, Jiang L et al (2012) Social media mining for drug safety signal detection. In: SHB '12 Proceedings of the 2012 international workshop on smart health and wellbeing, vol 2012. ACM, pp 33–40
112. Nikfarjam A, Gonzalez GH (2011) Pattern mining for extraction of mentions of adverse drug reactions from user comments. AMIA Annu Symp Proc 2011:1019–1026
113. Wang W, Haerian K, Salmasian H et al (2011) A drug-adverse event extraction algorithm to support pharmacovigilance knowledge mining from PubMed citations. AMIA Annu Symp Proc 2011:1464–1470
114. Liu Y, Lependu P, Iyer S et al (2012) Using temporal patterns in medical records to discern adverse drug events from indications. AMIA Summits Transl Sci Proc 2012:47–56
115. Bisgin H, Liu Z, Fang H et al (2011) Mining FDA drug labels using an unsupervised learning technique: topic modeling. BMC Bioinformatics 12(Suppl 10):S11
116. Yang C, Srinivasan P, Polgreen PM (2012) Automatic adverse drug events detection using letters to the editor. AMIA Annu Symp Proc 2012:1030–1039
117. Gurulingappa H, Mateen-Rajput A, Toldo L (2012) Extraction of potential adverse drug events from medical case reports. J Biomed Semantics 3:15
118. Gurulingappa H, Rajput AM, Roberts A et al (2012) Development of a benchmark corpus to support the automatic extraction of drug-related adverse effects from medical case reports. J Biomed Inform 45:885–892
119. Bjorne J, Airola A, Pahikkala T et al (2011) Drug–drug interaction extraction from biomedical texts with SVM and RLS classifiers. In: Proceedings of the 1st challenge task on drug–drug interaction extraction (DDI Extraction 2011), September 2011, Huelva, Spain, pp 35–42
120. Thomas P, Neves M, Solt I et al (2011) Relation extraction for drug–drug interactions using ensemble learning. In: Proceedings of the 1st challenge task on drug–drug interaction

extraction (DDI Extraction 2011), September 2011. Huelva, Spain

121. Segura-Bedmar I, Martinez P, de Pablo-Sanchez C (2011) Using a shallow linguistic kernel for drug–drug interaction extraction. J Biomed Inform 44:789–804
122. Zhang Y, Lin H, Yang Z et al (2012) A single kernel-based approach to extract drug–drug interactions from biomedical literature. PLoS One 7:e48901
123. Percha B, Garten Y, Altman RB (2012) Discovery and explanation of drug–drug interactions via text mining. Pac Symp Biocomput 2012:410–421
124. Kolchinsky A, Lourenco A, Li L et al (2013) Evaluation of linear classifiers on articles containing pharmacokinetic evidence of drug–drug interactions. Pac Symp Biocomput 2013:409–420
125. Boyce R, Gardner G, Harkema H (2012) Using natural language processing to identify pharmacokinetic drug–drug interactions described in drug package inserts. In: Proceedings of the 2012 workshop on biomedical natural language processing (BioNLP 2012), June 8, 2012, Association for Computational Linguistics, Montreal, Canada, p 206–213
126. He L, Yang Z, Zhao Z et al (2013) Extracting drug–drug interaction from the biomedical literature using a stacked generalization-based approach. PLoS One 8:e65814
127. Vilar S, Harpaz R, Uriarte E et al (2012) Drug–drug interaction through molecular structure similarity analysis. J Am Med Inform Assoc 19:1066–1074
128. Kuhn M, Campillos M, Letunic I et al (2010) A side effect resource to capture phenotypic effects of drugs. Mol Syst Biol 6:343
129. Kuhn M, von Mering C, Campillos M et al (2008) STITCH: interaction networks of chemicals and proteins. Nucleic Acids Res 36:D684–D688

Chapter 14

Systematic Drug Repurposing Through Text Mining

Luis B. Tari and Jagruti H. Patel

Abstract

Drug development remains a time-consuming and highly expensive process with high attrition rates at each stage. Given the safety hurdles drugs must pass due to increased regulatory scrutiny, it is essential for pharmaceutical companies to maximize their return on investment by effectively extending drug life cycles. There have been many effective techniques, such as phenotypic screening and compound profiling, which identify new indications for existing drugs, often referred to as drug repurposing or drug repositioning. This chapter explores the use of text mining leveraging several publicly available knowledge resources and mechanism of action representations to link existing drugs to new diseases from biomedical abstracts in an attempt to generate biologically meaningful alternative drug indications.

Key words Drug repurposing, Alternative drug indications, Drug repositioning

1 Introduction

The current drug discovery and development model is perceived as a costly and time-consuming process [1]. To reduce cost and shorten the duration for drug development, drug repurposing, also known as drug repositioning, has become an attractive alternative to traditional drug development. Drug repurposing is the process of finding a new indication for existing drug compounds. In other words, it is a research process on how an existing drug can be used for disease treatment other than its original indication. Drug re-profiling is advantageous because it bypasses many expensive drug development steps, such as in vitro and in vivo screening, chemical optimization, toxicology studies, and formulation development. Consequently, financial and development risks are reduced, and the typical 10–17-year drug development process can be shortened to 3–12 years [2]. The most cited success story for drug repositioning is sildenafil, an angina treatment developed by Pfizer. During clinical trials, it was noted that patients suffering from erectile dysfunction had improvement in their conditions. Sildenafil went on to become the blockbuster drug more commonly known as Viagra®.

Vinod D. Kumar and Hannah Jane Tipney (eds.), *Biomedical Literature Mining*, Methods in Molecular Biology, vol. 1159, DOI 10.1007/978-1-4939-0709-0_14, © Springer Science+Business Media New York 2014

Further studies showed yet another therapeutic indication in treating pulmonary arterial hypertension, whereby sildenafil was marketed as Revatio®. Mechanistically, the additional indications could be explained. Sildenafil is an inhibitor of phosphodiesterase-5 (PDE-5), which is known to be expressed in pulmonary hypertensive lungs and plays a role in regulating blood flow to the penis [3].

The main concept behind drug repurposing is that novel drug indications can be identified based on the principle that the primary drug target can be associated with diseases other than its original drug indication. In addition, as drugs can act on multiple targets, secondary targets can be utilized for novel drug indications as well. Several systematic approaches for finding new uses for old drugs have been proposed. One method with much literature support leverages chemical compound similarity [4]. Since similar drug compounds have comparable target profiles, novel targets can be identified for a compound by analyzing similar compound activity. Another approach to identify alternative drug indications includes finding drugs that share a significant number of side effects [5, 6]. Drugs with similar effects may have similar actions, linking the side effects to disease. Drug *D* is proposed to be a candidate for the treatment of disease *Dise* if *D* shares side effects that are induced by a drug class currently used for *Dise* treatment [5]. Finally, gene expression signatures have been used to reposition drugs whereby a drug signature opposite to a disease signature is proposed to be a potential treatment for the disease [7]. Readers can refer to [8] for a comprehensive computational drug repurposing method review.

With the vast pharmacological and biological knowledge available in literature, finding novel drug indications using *in silico* approaches has become increasingly feasible. Literature-based discovery methods go a step further by identifying relevant knowledge through text mining so that new knowledge can be inferred from existing knowledge [9]. Swanson's ABC model [10] is a popular literature-based discovery methodology that links two concepts through a commonly shared concept. A notable finding identified from the Swanson's ABC model was the proposed use of fish oil to treat Raynaud's syndrome, which was later clinically validated [10, 11]. Scientific concepts *A* and *C* form a relationship when concept *A* co-occurs with concept *B* in one publication while concepts *B* and *C* co-occur in another publication. Variations of Swanson's ABC model have been described in the literature for indirect relationship identification [12, 13]. However, approaches based on concept co-occurrence within abstracts tend to generate too many hypotheses. Another direction for network-based approaches aims to uncover knowledge through biological networks. DrugMap Central [14] is a network-based approach that utilizes information on chemical structures, drug targets, and signaling pathways for users to visualize and identify alternative drug indications. However, these co-occurrence and network-based approaches also generate many hypotheses, and identifying new drug indications from large networks can be time consuming.

2 Materials

A critical step in performing systematic drug repurposing is choosing appropriate knowledge sources. While there is an extensive list of publicly available databases that capture assorted biological knowledge (*see* http://www.pathguide.org/ for a list of interaction databases), most manually curated databases have poor literature coverage due to the resource-demanding curation process. More importantly, it is common for interaction databases not to capture the interaction details required for inference. For example, interaction-type descriptions are typically not included in these interaction databases. Medline is an ideal resource to obtain such detailed information on biologic interactions. For drug repurposing, Medline abstracts are utilized for obtaining gene–disease relationships that describe associations between gene expression regulation and diseases as well as protein–protein interaction relationships that capture the induction or the inhibition between protein pairs.

For the remainder of this section, we describe resources that complement the Medline knowledge source. Specifically, these resources are targeted for acquiring knowledge on cancer-related genes and drug–target interactions.

2.1 Gene Ontology

The Gene Ontology [15] is a hierarchical controlled vocabulary that includes three independent ontologies for biological processes, molecular functions, and cellular components. Standardized terms in the Gene Ontology describe gene roles and gene products in any organism. The Gene Ontology itself does not contain organism gene products. Rather, gene product biological roles are kept in Gene Ontology annotation form. For example, the Gene Ontology annotation is a useful resource in identifying genes that are associated with cancer-related biological processes. In particular, the following Gene Ontology terms and the corresponding descendants are selected as antitumor biological processes: negative regulation of cell proliferation (GO:0008285), positive regulation of apoptosis (GO:0043065), and negative regulation of angiogenesis (GO:0016525). On the other hand, tumor-promoting biological processes include these Gene Ontology terms and corresponding descendants: positive regulation of cell proliferation (GO:0008284), negative regulation of apoptosis (GO:0043066), and positive regulation of angiogenesis (GO:0045766).

2.2 UniProt

The UniProt Knowledge Base (UniProtKB) (http://www.uniprot.org/) is the largest protein sequence repository. In addition to protein sequence information, UniProtKB also includes manual annotation on proteins, and it is an ideal resource to obtain a cancer gene list. In particular, the keywords "oncogene" and "tumor suppressor" are used as search criteria with the results limited to human genes only.

2.3 NCBI Gene

The NCBI Gene (http://www.ncbi.nlm.nih.gov/gene) is a knowledge base that contains about 12 million curated gene records. Similar to UniProt, NCBI Gene is leveraged to identify cancer genes using the keywords "oncogene" and "tumor suppressor" with results restricted to human genes.

2.4 CancerQuest

CancerQuest (http://www.cancerquest.org) is a resource with information on cancer biology and treatment. The CancerQuest tool maintains a list of oncogenes (http://www.cancerquest.org/oncogene-table) and tumor suppressors (http:// www.cancerquest.org/tumor-suppressors-table).

2.5 DrugBank

Drug–target interactions are also essential for systematic drug repurposing. DrugBank [16] is a comprehensive knowledge base for drugs, drug actions, and drug targets. The drug–target interactions obtained from DrugBank include modulation definitions such as antagonist or agonist.

3 Methods

An important component behind our approach in performing systematic drug repurposing is in acquiring information relevant to drug mechanism. The semantics of the acquired biological interactions is leveraged to infer novel drug indications. By utilizing semantics and automated reasoning, we aim to produce novel drug indications that are accompanied by the biological mechanism behind the hypotheses as explanation.

Our approach, as shown in Fig. 1, can be divided into three main components: (1) the *knowledge representation component*, (2) the *knowledge acquisition component*, and (3) the *reasoning component*. In order to automatically propose alternative drug indications, it is necessary to first represent the drug mechanism in logic rule form. The knowledge acquisition component includes publicly available curated sources as well as relevant facts for drug indication identification acquired using text mining. With the facts gathered by the knowledge acquisition component and the logic rules defined in the knowledge representation component, the reasoning engine utilizes the logic rules to find interactions that link drugs with corresponding drug indications.

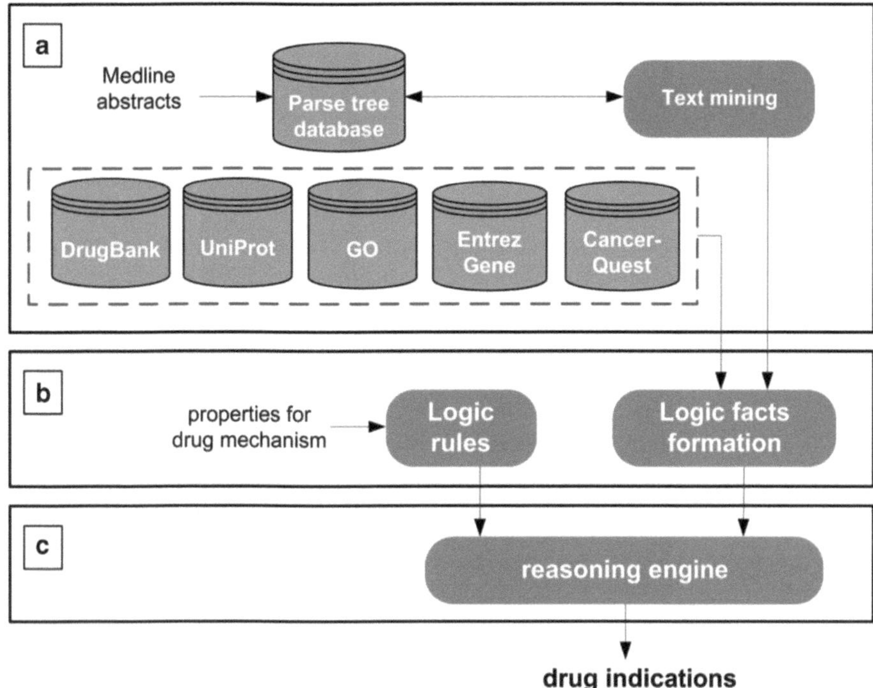

Fig. 1 An overview of the components involved in identifying alternative drug indications through text mining. These components include (**a**) the knowledge acquisition component, (**b**) the knowledge representation component, and (**c**) the reasoning component

3.1 Knowledge Representation

Basic drug mechanisms include the modulation, either activation or inhibition, of a protein target that is responsible for disease. These drug–target interactions then translate into clinical effects. For example, erlotinib is an epidermal growth factor receptor (EGFR) antagonist which alters the oncogenic EGFR signal transduction pathway. The key to identifying alterative drug indications is based on the principle that drug targets can be involved in diseases other than the original drug indication. Although compounds may interact with multiple targets, the primary target usually determines the first indication for development. Novel drug indications can be hypothesized through identifying alternative relations between primary targets and diseases as well as examining the secondary targets and their corresponding roles in disease.

Drug action representation involves initially identifying antagonists as triggering drug target inhibition and agonists as initiating drug target activation. With rich knowledge about cancer and its mechanisms, we applied our approach in identifying drugs that can be used as cancer treatments. A drug is identified as a treatment for cancer in one of the following scenarios:

- When the drug inhibits a protein that is known to be an oncogene.
- When the drug induces a protein that is known to be a tumor suppressor.

- When the drug inhibits a protein that is involved in a tumor-promoting biological process.
- When the drug induces a protein that is involved in a tumor-suppressing biological process.

Alternatively, a drug can also be identified as a cancer treatment when the drug activates protein function leading to an increase in tumor-suppressing protein expression or activates protein function leading to a decrease in tumor-promoting protein expression.

3.2 Relationship Extraction

While databases such as PharmGKB [17] and IntAct [18] are great resources for gene–disease relations and protein–protein interactions, they are limited in literature coverage due to the time-intensive process involved in manual curation. More importantly, it is commonly the case that interaction types are not captured in these databases. The information deficiency becomes an obstacle when the interactions from these databases are used in new knowledge discovery. As an example, let us assume that we know that protein *A* interacts with oncoprotein *B*. The interaction consequence (e.g., whether the function of *B* is ultimately activated or suppressed by *A*) is an important factor when the interaction is considered as a cancer drug treatment mechanism. To capture the interactions, we utilize text mining so that appropriate interactions can be identified efficiently from the literature.

Our text mining approach relies on grammatical structures and keywords to capture the directionality and the interaction types during gene–disease relation and protein–protein interaction extractions. The *parse tree query language (PTQL)* [19] is a suitable language that allows extraction patterns to be defined over keywords and grammatical structures. PTQL is designed for information extraction over a database of text known as the *parse tree database (PTDB)*. A parse tree is composed of a constituent tree and a linkage. A constituent tree is a syntactic sentence tree with the nodes represented by parts-of-speech tags and words in the sentence. A linkage represents the syntactic dependencies (or links) between word pairs in a sentence. The Stanford parser [20] is utilized to create parse sentence trees. BANNER [21] is used for gene name recognition from text, and the recognized gene names are then mapped to official gene symbols using GNAT [22]. The syntactic and semantic information is stored in our parse tree database, and extraction is performed by database queries. By storing the syntactic and semantic information, document collection reprocessing for every extraction is avoided. On-the-fly extraction is suitable for mining various interaction types as needed for identifying alternative drug indications.

A PTQL query is composed of four components: (1) tree patterns, (2) link conditions, (3) proximity conditions, and (4) return expression. The components in a PTQL query are separated by the symbol ":". Here we describe the PTQL query syntax by the following query:

```
//S{/NP{//?[Value='high']=>//?[Value='lev
els']=>
//?[Tag='GENE'](kw1)}=>/VP{//?[Tag='DISE']
(kw2)}} :::
distinct kw1.value, kw2.value
```

The above tree query pattern specifies that within a noun phrase (denoted as `NP`), a gene name (denoted as variable `kw1`) has to be preceded by keywords "high" and "levels" through the operator =>. This gene name also needs to be followed by a verb phrase (denoted as `VP`), which contains a disease name mention (denoted as `kw2`). With the PTQL query, we obtained the relation that ADA overexpression is associated with acute lymphoblastic leukemia from the following sentence:

High levels of *adenosine deaminase (ADA)* activity have been associated with normal T cell differentiation and T cell disease, such as *acute lymphoblastic leukemia* (PMID: 6981287).

Readers can refer to [19] for more detailed information on PTQL and its implementation. By defining the keywords and extraction patterns over parse trees in the form of PTQL queries, it becomes possible to extract not only the interactions but also the directionality and interaction types. Specifically, the following interaction types are extracted:

1. Association between overexpressed or underexpressed genes and diseases.
2. Protein stimulation or inhibition by other proteins.

Sample interactions are listed in Table 1.

3.3 Logic Forms

In order to identify drug indications through automated reasoning, it is important to have proper drug mechanism knowledge representation. *Logic facts* are formed based on the knowledge acquired from the various sources as described in the previous subsection. In addition, *logic rules* are used to represent drug mechanism properties. We adopted a popular knowledge representation language called answer set programming (ASP) [23, 24] for logic fact and rule representation.

ASP is a declarative language that is useful for reasoning, including reasoning with incomplete information. An advantage for using a declarative language is that we define what the program should achieve and not how it should be achieved. Here we give a brief introduction to ASP syntax.

An *ASP rule* is in the form

$$l \leftarrow l_0, \ldots, l_m, \mathbf{not} l_{m+1}, \ldots, \mathbf{not} l_n$$

where l_is are literals and **not** represents *default negation*. The intuitive meaning of the above rule is that if it is known that literals l_0, …, l_m

Table 1
Sample extracted gene–disease relationships and protein–protein interactions with their support evidences

Evidences	Extracted relationships
The results of our study demonstrate that *AMACR* expression is *upregulated* in *gastric cancer* (PMID: 18787636)	<overexpressed AMACR, associated with, gastric cancer>
Therefore, *inactivation* of *Rb protein* by HPV 18 E7 protein may be associated with carcinogenesis of *small-cell carcinoma* (PMID: 14506638)	<underexpressed RB1, associated with, small cell carcinoma>
Moreover, *HER-2* expression was *stimulated* by *EGF* addition in young cells (PMID: 8028398)	<EGF, induces, ERBB2>
Inhibition of *PPARgamma* activity by *TNF-alpha* is involved in pathogenesis of insulin resistance (PMID: 18655773)	<TNF, inhibits, PPARG>

are to be true and if l_{m+1}, …, l_n are assumed to be false, then l must be true. A literal is defined as either an atom or an atom preceded by the symbol ¬ that indicates *classical negation.* If there is no literal l in the rule *head*, then the rule is referred to as a *constraint.* On the other hand, if there are no literals in the rule *body*, then the rule is referred to as a *fact*, and its representation fact short hand is simply the head literal itself. A set of ASP rules composes an answer set program, and an answer set program interpretation is called an answer set. Readers can refer to [25] for more details on ASP syntax and semantics.

Two basic logic fact types are used to represent the drug mechanisms: (1) concepts such as proteins and drugs and (2) interactions such as gene–disease relationships. The concept protein is represented in the *protein(Prot)* form, where *Prot* is a concept variable. For example, *protein(tp53)* indicates that tp53 is a protein concept instance. A complete concept and logic forms list is shown in Table 2. Interactions are represented with the predicate *interaction* for drug–target and protein–protein interactions and *relation* for gene–disease and gene-biological process relations. For instance, the logic form *relation(overexpressed(amacr), associated_with, gastric_cancer)* translates to overexpressed AMACR is associated with gastric cancer, and the logic form *interaction(egf, induces, erbb2)* represents that EGF induces ERBB2 activity. Table 3 shows a complete list of interaction types and their corresponding logic forms.

3.4 Automated Reasoning

With the knowledge denoted in logic fact form, drug mechanisms now need to be represented using ASP rules. The idea is to characterize and encode the mechanisms as pre- and post-interaction conditions, in which the precondition is represented in the body of

Table 2
Logic forms for the classes and entities involved in the drug mechanism domain

Facts	Logic forms	Examples
Prot is a protein, e.g., P53	*protein(Prot)*	*protein(tp53)*
Prot is an oncogene, e.g., EGFR	*oncogene(Prot)*	*oncogene(egfr)*
Prot is a tumor suppressor, e.g., P53	*suppressor(Prot)*	*suppressor(tp53)*
Dr is a drug, e.g., moclobemide	*drug(Dr)*	*drug(moclobemide)*
Dise is a disease, e.g., depression	*disease(Dise)*	*disease(depression)*
Bp is a cancer-promoting biological process, e.g., positive regulation of cell proliferation	*cancer_promoting_bioprocess(Bp)*	*cancer_promoting_bioprocess (pos_reg_cell_proliferation)*
Bp is a cancer-resisting biological process, e.g., positive regulation of apoptosis	*cancer_resisting_bioprocess(Bp)*	*cancer_resisting_bioprocess (pos_reg_apoptosis)*

Table 3
Logic forms for the interactions involved in the drug mechanism domain

Relations	Logic forms
Drug *Dr* induces the activity of protein *Prot*	*interaction(Dr, induces, Prot)*
Drug *Dr* inhibits the activity of protein *Prot*	*interaction(Dr, inhibits, Prot)*
Protein *Prot1* induces the activity of protein *Prot2*	*interaction(Prot1, induces, Prot2)*
Protein *Prot1* inhibits the activity of protein *Prot2*	*interaction(Prot1, inhibits, Prot2)*
Overexpressed protein *Prot* is associated with disease *Dise*	*relation(overexpressed(Prot), associated_with, Dise)*
Underexpressed protein *Prot* is associated with disease *Dise*	*relation(underexpressed(Prot), associated_with, Dise)*
Protein *Prot* plays a role in biological process *Bp*	*relation(Prot, is_associated, Bp)*

an ASP rule while the head represents the post-condition. Three rule sets are needed to perform inference for alterative drug indications: *initial triggers*, *inference rules*, and *constraints*. The initial triggers specify the criteria to initiate an inference. For drug mechanisms, the initial triggers correspond to drug target activation or inactivation by agonists or antagonists, respectively. The triggers are captured by the following rules:

- Initial trigger 1: Drug *Dr* activates drug target *Prot* function when *Dr* induces *Prot* expression:

  trigger(Dr, activates, Prot, 1) ← *interaction(Dr, induces, Prot), protein(Prot), drug(Dr).*

- Initial trigger 2: Drug *Dr* inactivates drug target *Prot* function when *Dr* inhibits *Prot* expression:

  trigger(Dr, inactivates, Prot, 1) ← *interaction(Dr, inhibits, Prot), protein(Prot), drug(Dr).*

The following rules are used to represent other types of direct inference:

- Inference rule 1: Cancer is identified as an indication for drug *Dr* in step *S*+1 when tumor-suppressor *Prot* has been activated in the previous step *S*:

  trigger(Dr, treats, cancer, S+1) ← *trigger(Dr, activates, Prot, S), suppressor(Prot), drug(Dr), step(S).*

- Inference rule 2: Cancer is identified as an indication for drug *Dr* in step *S*+1 when oncogene *Prot* has been inhibited in the previous step *S*:

  trigger(Dr, treats, cancer, S+1) ← *trigger(Dr, inactivates, Prot, S), oncogene(Prot), drug(Dr), step(S).*

- Inference rule 3: Cancer is identified as an indication for drug *Dr* in step *S*+1 when protein *Prot*, which is involved in a cancer-promoting biological process *Bp*, has been inhibited in the previous step *S*:

  trigger(Dr, treats, cancer, S+1) ← *relation(Prot, is_associated, Bp), trigger(Dr, inactivates, Prot, S), protein(Prot), drug(Dr), cancer_promoting_bioprocess(Bp), step(S).*

- Inference rule 4: Cancer is identified as an indication for drug *Dr* in step *S*+1 when protein *Prot*, which is involved in a tumor-suppressing biological process *Bp*, has been activated in the previous step *S*:

  trigger(Dr, treats, cancer, S+1) ← *relation(Prot, is_associated, Bp), trigger(Dr, activates, Prot, S), protein(Prot), drug(Dr), cancer_resisting_bioprocess(Bp), step(S).*

- Inference rule 5: Cancer is identified as an indication for drug *Dr* in step *S*+1 when protein *Prot* has been inhibited in the previous step *S* and overexpressed *Prot* is known to be associated with cancer:

  trigger(Dr, treats, cancer, S+1) ← *trigger(Dr, inactivates, Prot, S), relation(overexpressed(Prot), associated_with, cancer), drug(Dr), protein(Prot), step(S).*

- Inference rule 6: Cancer is identified as an indication for drug *Dr* in step *S*+1 when protein *Prot* has been induced in the previous step *S* and underexpressed *Prot* is known to be associated with cancer:

  trigger(Dr, treats, cancer, S+1) ← *trigger(Dr, activates, Prot, S), relation(underexpressed(Prot), associated_with, cancer), drug(Dr), protein(Prot), step(S).*

In the above rules, the variable S is used to indicate a time stamp. Such time stamps represent interaction sequence, indicating that the different interactions that must occur prior to inferring a drug can be a cancer therapy. Such scenarios are considered *direct inferences* for cancer treatment. Furthermore, cancer therapies that are derived through drug-activated protein–protein interactions are considered *indirect inferences*, and they are represented with the ASP rules below:

- Inference rule 7: Drug *Dr* triggers protein *Prot2* functional activation in step $S+1$ when protein *Prot1* has been activated in the previous step S and activated *Prot1* increases *Prot2* expression:

  trigger(Dr, activates, Prot2, S+1) ← *trigger(Dr, activates, Prot1, S), interaction(Prot1, induces, Prot2), drug(Dr), protein(Prot1), protein(Prot2), step(S).*

- Inference rule 8: Drug *Dr* triggers protein *Prot2* functional inactivation in step $S+1$ when protein *Prot1* has been activated in the previous step S and activated *Prot1* decreases *Prot2* expression:

  trigger(Dr, inactivates, Prot2, S+1) ← *trigger(Dr, activates, Prot1, S), interaction(Prot1, inhibits, Prot2), drug(Dr), protein(Prot1), protein(Prot2), step(S).*

With the initial triggers and inference rules in place, constraints are used to define the valid inference criteria as follows:

- Constraint 1: An inference is valid only if goal becomes true, e.g., the series of steps must include the inference for a drug to be used as a cancer treatment, which is indicated by *trigger(Dr, treats, cancer, S)*:

  goal ← *trigger(Dr, treats, cancer, S), drug(Dr), step(S).*

  ← **not** *goal.*

- Constraint 2: No other interactions should follow *trigger(Dr, treats, cancer, S)* in a valid inference, ensuring that *trigger(Dr, treats, cancer, S)* is the last valid inference step:

  ← *trigger(Dr, activates, Prot, S), trigger(Dr, treats, cancer, S1), protein(Prot), drug(Dr), step(S), step(S1),* $S \geq S1$.

  ← *trigger(Dr, inactivates, Prot, S), trigger(Dr, treats, cancer, S1), protein(Prot), drug(Dr), step(S), step(S1),* $S \geq S1$.

To compute the answer sets that infer drug indications, an ASP solver called clingo [26] is utilized to compute direct and indirect inferences based on the rules and the acquired logic facts.

3.5 Dipyridamole as a Treatment for Cancer

Here we use the drug dipyridamole as an example to illustrate the direct inference of drug indications. Dipyridamole is prescribed to reduce blood clots through ADA inhibition [source: PubMed Health]. To find alternative indications for dipyridamole, we first acquire the necessary knowledge such as drug–target interactions and gene–disease relations. In this case, the following fact is acquired:

- Dipyridamole acts as an antagonist for ADA [source: DrugBank]:

  interaction(dipyridamole, inhibits, ada).

  This interaction acts as the precondition for ADA functional inhibition through initial trigger 2, which results in *trigger(dipyridamole, inactivates, ada, 1)* being true. Mining biomedical abstracts reveals that ADA overexpression is associated with cancers like acute lymphoblastic leukemia.

- *High levels* of *ADA* activity have been *associated with* normal T cell differentiation and T cell disease, such as *acute lymphoblastic leukemia* [source: PMID: 6981287]:

  relation(overexpressed(ada), associated_with, cancer).

  With the above interactions, inference rule 5 turns *trigger(dipyridamole, treats, cancer, 2)* to be true, indicating that dipyridamole is proposed as a potential treatment for cancer as ADA can be inhibited by dipyridamole and ADA overexpression is associated with cancer. This hypothesis together with the drug mechanisms is illustrated in Fig. 2.

3.6 Tazarotene as a Cancer Therapy

Here we use the drug tazarotene as an illustration for drug indication indirect inference. Tazarotene is approved for psoriasis and acne treatment. The facts below are acquired from different sources to identify an alternative indication for tazarotene.

- Tazarotene acts as an agonist for retinoic acid receptor alpha (RARA) [source: DrugBank]:

  interaction(tazarotene, induces, rara).

  The above interaction results in *trigger(tazarotene, activates, RARA, 1)* to be true through initial trigger 1. By mining biomedical abstracts, it is discovered that RARA is known to inhibit EGFR oncogenic activity.

- These results suggest that RAR ligand-associated downregulation of EGFR activity reduces cell proliferation by reducing the magnitude and duration of EGF-dependent ERK1/2 activation [source: PMID: 11788593]:

  interaction(rara, inhibits, egfr).

- EGFR is a known oncogene [source: CancerQuest]:

  oncogene(efgr).

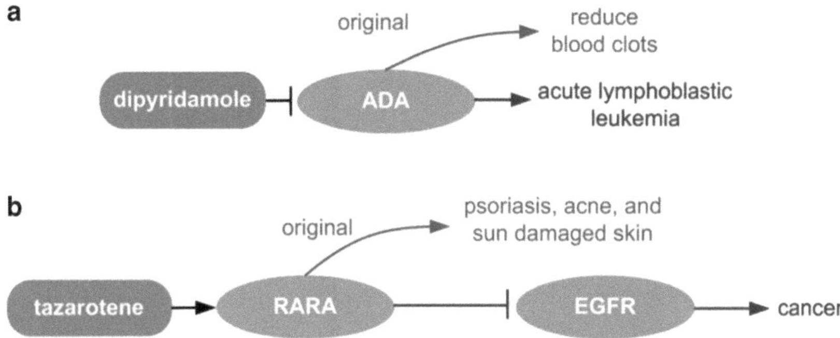

Fig. 2 A diagrammatic view of (**a**) direct and (**b**) indirect inferences for dipyridamole and tazarotene novel cancer indications

With the acquired facts and inference rule 8, *trigger(tazarotene, inactivates, EGFR, 2)* becomes true and subsequently turns *trigger(tazarotene, treats, cancer, 3)* to be true based on inference rule 2. The hypothesis generated through indirect inference indicates that the agonist tazarotene activates RARA which in turn inhibits EGFR, indicating the potential use of tazarotene as an oncology therapy. This hypothesis together with the drug mechanisms is illustrated in Fig. 2.

4 Conclusions

Drug repurposing plays an increasingly important role for pharmaceutical companies to minimize the time spent in the drug development process while maximizing previous investments. We described an approach that acquires knowledge from publicly available resources including Medline abstracts through text mining and generates alternative drug indication hypotheses through automated reasoning based on the acquired knowledge and drug mechanism logic representations. Using an evaluation set of 943 drugs obtained from DrugBank, 81 drugs are currently used for cancer treatments, while 289 drugs not having cancer as an original indication are currently being investigated as cancer therapies according to clinicaltrials.gov. Our method suggested 507 drugs that have the potential to be used for cancer treatments with a subset of 211 confirmed to be cancer related. Further analysis revealed that our approach was able to make 67 suggestions for cancer therapies among the 81 known cancer drugs (a recall of 82.7 %), and the remaining 144 suggestions are non-oncology drugs that are currently being tested in cancer clinical trials (a recall of 49.8 %). A more detailed result analysis can be found in [27].

It is important to note that there are a few important features that distinguish our approach from other literature-based approaches. These include (1) interaction-type extraction and utilization, (2) the mining and application of directional interactions, and (3) the drug mechanism representation. Typical literature-based approaches that adopt the Swanson's ABC model usually produce large biological networks based on co-occurrence. Then researchers have to engage in the time-consuming process of using network visualization tools to sift through the networks and manually identify novel drug indications. The distinguishing features adopted in our approach reduce search space size so that it not only becomes computationally feasible, but, more importantly, the hypotheses generated reflect the drug mechanism of action as well as the key cancer mechanisms. These features lead to deriving potential alternative drug indications with scientific evidences explaining the mechanism behind the hypotheses.

The ability to identify alternative drug indications illustrates that combining text mining and automated reasoning is a powerful technique that enables knowledge inference in the biomedical domain.

References

1. DiMasi JA, Hansen RW, Grabowski HG (2003) The price of innovation: new estimates of drug development costs. J Health Econ 22:151–185. doi:10.1016/S0167-6296(02)00126-1
2. Ashburn TT, Thor KB (2004) Drug repositioning: identifying and developing new uses for existing drugs. Nat Rev Drug Discov 3:673–683. doi:10.1038/nrd1468
3. Ghofrani HA, Osterloh IH, Grimminger F (2006) Sildenafil: from angina to erectile dysfunction to pulmonary hypertension and beyond. Nat Rev Drug Discov 5:689–702. doi:10.1038/nrd2030
4. Dubus E, Ijjaali I, Barberan O, Petitet F (2009) Drug repositioning using in silico compound profiling. Future Med Chem 1:1723–1736. doi:10.4155/fmc.09.123
5. Yang L, Agarwal P (2011) Systematic drug repositioning based on clinical side-effects. PLoS One 6(12)
6. Duran-Frigola M, Aloy P (2012) Recycling side-effects into clinical markers for drug repositioning. Genome Med 4(1):3
7. Dudley JT, Sirota M, Shenoy M, Pai RK, Roedder S, Chiang AP, Morgan AA, Sarwal MM, Pasricha PJ, Butte AJ (2011) Computational repositioning of the anticonvulsant topiramate for inflammatory bowel disease. Sci Transl Med 3:96ra76
8. Hurle MR, Yang L, Xie Q, Rajpal DK, Sanseau P, Agarwal P (2013) Computational drug repositioning: from data to therapeutics. Clin Pharmacol Ther 93(4):335–341
9. Deftereos SN, Andronis C, Friedla EJ, Persidis A, Persidis A (2011) Drug repurposing and adverse event prediction using high-throughput literature analysis. Wiley Interdiscip Rev Syst Biol Med 3:323–334. doi:10.1002/wsbm.147
10. Swanson DR (1990) Medical literature as a potential source of new knowledge. Bull Med Libr Assoc 78:29–37
11. Swanson DR (1986) Fish oil, Raynaud's syndrome, and undiscovered public knowledge. Perspect Biol Med 30:7–18
12. Weeber M, Vos R, Klein H, De Jong-Van Den Berg LTW, Aronson AR et al (2003) Generating hypotheses by discovering implicit associations in the literature: a case report of a search for new potential therapeutic uses for thalidomide. J Am Med Inform Assoc 10:252–259. doi:10.1197/jamia.M1158
13. Yetisgen-Yildiz M, Pratt W (2006) Using statistical and knowledge-based approaches for literature-based discovery. J Biomed Inform 39:600–611. doi:10.1016/j.jbi.2005.11.010
14. Fu C, Jin G, Gao J, Zhu R, Ballesteros-Villagrana E, Wong ST (2013) DrugMap Central: an on-line query and visualization tool to facilitate drug repositioning studies. Bioinformatics 29(14):1834–1836. doi:10.1093/bioinformatics/btt279

15. The Gene Ontology Consortium (2000) Gene ontology: tool for the unification of biology. Nat Genet 25(1):25–29
16. Knox C, Law V, Jewison T et al (2011) DrugBank 3.0: a comprehensive resource for 'omics' research on drugs. Nucleic Acids Res 39(Database issue):D1035–D1041
17. Klein TE, Chang JT, Cho MK, Easton KL, Fergerson R et al (2001) Integrating genotype and phenotype information: an overview of the PharmGKB project. Pharmacogenetics Research Network and Knowledge Base. Pharmacogenomics J 1:167–170
18. Aranda B, Achuthan P, Alam-Faruque Y, Armean I, Bridge A et al (2010) The IntAct molecular interaction database in 2010. Nucleic Acids Res 38:D525–D531. doi:10.1093/nar/gkp878
19. Tari L, Tu PH, Hakenberg J, Chen Y, Son TC et al (2010) Incremental information extraction using relational databases. IEEE Trans Knowledge Data Eng 24:86–99. doi:10.1109/TKDE.2010.214
20. Klein D, Manning CD (2003) Accurate unlexicalized parsing. Proceedings of the 41st Annual meeting on association for computational linguistics (ACL'03), Vol 1, pp 423–430. doi:10.3115/1075096.1075150
21. Leaman R, Gonzalez G (2008) BANNER: an executable survey of advances in biomedical named entity recognition. Pac Symp Biocomput. pp 652–663
22. Hakenberg J, Plake C, Leaman R, Schroeder M, Gonzalez G (2008) Inter-species normalization of gene mentions with GNAT. Bioinformatics 24:i126–i132. doi:10.1093/bioinformatics/btn299
23. Gelfond M, Lifschitz V (1988) The stable model semantics for logic programming. In International symposium on logic programming, pp 1070–1080
24. Gelfond M, Lifschitz V (1991) Classical negation in logic programs and disjunctive databases. New Generation Computing 9:365–387
25. Baral C (2003) Knowledge representation, reasoning and declarative problem solving. Cambridge University Press, New York
26. Gebser M, Ostrowski M, Schaub T (2009) Constraint answer set solving. In Proceedings of the 25th International conference on logic programming (ICLP'09), Vol 5649, pp 235–249. doi:10.1007/978-3-642-02846-5
27. Tari L, Vo N, Liang S, Patel J, Baral C, Cai J (2012) Identifying novel drug indications through automated reasoning. PLoS One 7(7):e40946

Chapter 15

Mining the Electronic Health Record for Disease Knowledge

Elizabeth S. Chen and Indra Neil Sarkar

Abstract

The growing amount and availability of electronic health record (EHR) data present enhanced opportunities for discovering new knowledge about diseases. In the past decade, there has been an increasing number of data and text mining studies focused on the identification of disease associations (e.g., disease–disease, disease–drug, and disease–gene) in structured and unstructured EHR data. This chapter presents a knowledge discovery framework for mining the EHR for disease knowledge and describes each step for data selection, preprocessing, transformation, data mining, and interpretation/validation. Topics including natural language processing, standards, and data privacy and security are also discussed in the context of this framework.

Key words Electronic health record, Knowledge discovery in databases, Data mining, Text mining, Natural language processing, Data warehouse, Data privacy and security, Standards

1 Introduction

The increased adoption of electronic health record (EHR) systems has the potential for enhanced collection and access to a wide range of information about an individual's lifetime health status and health care to support a range of "secondary uses" including genomic, clinical, and public health research [1–6]. The use of knowledge discovery and data mining approaches for transforming health data into actionable knowledge has been an active area of research and will continue to advance in the era of "big data" [7–10]. These approaches have been applied to EHR data for studying patterns and relationships between diseases (comorbidity analysis) [11–13], drugs and adverse events (pharmacovigilance) [14–16], as well as drugs and genes (pharmacogenomics) [17]. Potential uses of this knowledge range from hypothesis generation to clinical decision support.

The knowledge discovery in databases (KDD) process is defined as "the nontrivial process of identifying valid, novel,

Vinod D. Kumar and Hannah Jane Tipney (eds.), *Biomedical Literature Mining*, Methods in Molecular Biology, vol. 1159, DOI 10.1007/978-1-4939-0709-0_15,

potentially useful, and ultimately understandable patterns in data" [18, 19]. The major steps in this interactive and iterative process are data selection, preprocessing, transformation, data mining, and interpretation/validation. While *data mining* is traditionally applied to collections of "structured" data from databases, *text mining* or *text data mining* is the application of data mining techniques to collections of "unstructured" or "semi-structured" textual data [20]. The text mining process typically involves the use of natural language processing (NLP) techniques to extract structured data from unstructured narrative for subsequent use [21].

This chapter presents a general framework for mining the EHR for disease knowledge that is adapted from the KDD and text mining processes. The sets of techniques associated with each step in this "disease knowledge discovery" (DKD) process are described along with a discussion of key issues ranging from data privacy and security to standardization. Given the breadth of methods and applications, this protocol is focused on describing an approach for identifying pairwise disease associations (e.g., disease–disease, disease–drug, and disease–gene) where the driving example is a study focused on identifying comorbidities for type 2 diabetes mellitus.

2 Materials

2.1 Electronic Health Record

An EHR system includes a "longitudinal collection of electronic health information for and about persons, where health information is defined as information pertaining to the health of an individual or health care provided to an individual" [22] (*see* **Note 1**). Typically, an EHR system interfaces with and integrates data from ancillary systems such as registration, billing, laboratory, radiology, and pharmacy [23]. Data within the EHR fall into two major categories: *structured*—discrete data elements (e.g., demographics, billing diagnoses, problems, procedures, medications, and allergies) that may be associated with codes and are "computer understandable" and *unstructured* (or *semi-structured*)—free-text narrative (e.g., clinical notes and reports) that may include some structure and can be converted into a structured form using methods such as NLP (*see* **Note 2**) [24]. The underlying real-time database for EHR systems is often referred to as a Clinical Data Repository (CDR) [25].

2.2 Data Warehouse

For reporting and data analysis purposes, a separate database or data warehouse is typically available that is populated with data through an extract, transform, and load (ETL) process [25]. Depending on the intended use (e.g., for quality or research) and contents, this is referred to as a Clinical Data Warehouse, Research Data Warehouse, Enterprise Data Warehouse, Integrated Data Warehouse, or Integrated Data Repository (*see* **Note 3**) [26]; for the remainder of this chapter, the term Enterprise Data Warehouse (EDW) will be used. In addition to the EHR and ancillary systems, the EDW may also incorporate data from other systems such as disease registries, biobanks, payer claims databases, and literature databases. The standard process for accessing and requesting data from an EDW for research purposes involves Institutional Review Board (IRB) approval and training in the protection of human subjects in research (*see* **Note 4**).

2.3 Other Resources

In addition to the aforementioned institutional resources, there are several publicly accessible repositories of de-identified EHR data that are each associated with an approval process (e.g., requiring IRB approval or a Data Use Agreement). These include the following:

- Multiparameter Intelligent Monitoring in Intensive Care II (MIMIC-II) research database [27, 28]—intensive care unit data from Beth Israel Deaconess Medical Center.
- i2b2 NLP Research Datasets [29]—discharge summaries from Partners HealthCare that have been used in a series of NLP challenges (e.g., for de-identification [30], smoking status [31], and medications [32]).
- Integrating data for analysis, anonymization, and sharing (iDASH) Data Repository [33, 34]—open, community-serving, crowd-sourced resource for high-quality collections of medical data, including images and text.

3 Methods

The DKD process aligns with the DIKW hierarchy that represents the relationship between *d*ata, *i*nformation, *k*nowledge, and *w*isdom [35]. As depicted in Fig. 1, the DKD process involves the following: *Data selection*—identifying and extracting data from the EHR (through the EDW) and potentially other sources for a particular application or study; *preprocessing*—de-identifying, cleaning, and enriching the dataset, which may include the use of NLP techniques for unstructured data and applying standards for integrating structured data; *transformation*—reducing and converting the dataset in preparation for the data mining step; *data mining*—choosing and implementing the algorithms for generating patterns; and *interpretation/evaluation*—visualizing and validating the patterns, which may lead to revising and reiterating through the previous steps. The result of this process is the validation of known knowledge and perhaps more significantly the potential discovery of new disease knowledge. Key issues underlying these steps include *data privacy and security* (*see* **Note 4**) and *standards* (*see* **Note 5**).

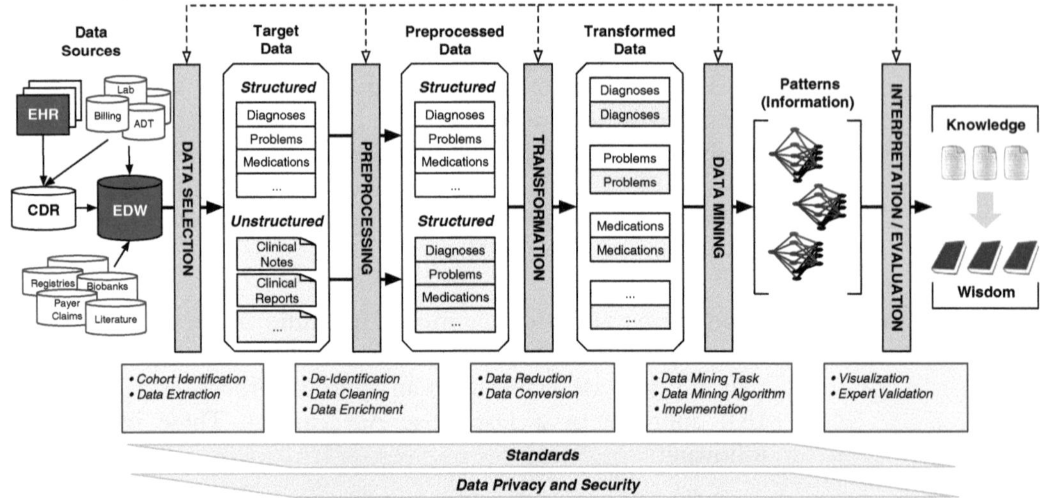

Fig. 1 Overview of disease knowledge discovery process

3.1 Example Studies

1. The driving example in this section will be a study focused on identifying comorbidities associated with type 2 diabetes mellitus (henceforth referred to as the "diabetes comorbidity study").
2. For additional examples, there are several studies involving the use of NLP to extract information from clinical notes (e.g., discharge summaries) and biomedical literature (e.g., PubMed/MEDLINE) to acquire disease–disease [11, 12], disease–finding [36, 37], disease–drug [38, 39], and disease–symptom [40, 41] relationships.
3. Other example studies have focused on structured data such as billing diagnoses, problems, procedures, medications, laboratory results, and vital signs to generate disease co-occurrences (e.g., disease–disease [13], disease–drug [42, 43], disease–lab [42, 43], disease–procedure [44], and disease–gene [45]).
4. While the aforementioned examples are focused on pairwise associations that do not take into consideration time or the sequence of events, there have been studies focused on the discovery of larger associations (e.g., triplet combinations such as disease–disease–disease [46]) and temporal associations [47], which are out of scope for this protocol.

3.2 Defining the Study

Prior to starting the DKD process, there should be a good understanding of the application of interest in order to guide each of the steps. For example, high-level questions include the following:

- How to access, extract, and protect the EHR data?
- What type(s) of EHR data to use (e.g., structured diagnoses or unstructured discharge summaries)?
- Are NLP techniques needed to process unstructured data?

- What standards have been used or what standards can be applied to facilitate data integration?
- How to select and identify the patient cohort (e.g., focus on particular disease(s) or specific time frame)?
- Which data mining techniques to select and implement?
- How to evaluate (e.g., involve clinical experts or compare results to medical knowledge resources)?
- What are anticipated uses of the discovered knowledge?

3.3 Data Selection

The goal of data selection is to create a target dataset by first identifying a cohort based on a set of inclusion/exclusion criteria and then extracting the requisite data. Depending on the institution, there may be resources and tools available to facilitate cohort identification and data extraction (*see* **Note 3**).

1. Criteria that can be used for *cohort identification* include patient demographics (e.g., age, sex, race/ethnicity, and zip code), encounters, diagnoses, procedures, medications, and laboratory results. These latter four types of data may be associated standardized codes such as ICD, CPT, RxNorm, and Logical Observation Identifiers Names and Codes (LOINC), respectively (*see* **Note 5**). In addition, keyword searching, regular expression matching, or NLP could enable the use of clinical notes and reports for meeting specified criteria (*see* **Note 2**).
2. The cohort identification process is often challenging due to heterogeneous data sources, lack of standards, and wealth of unstructured text. Recent efforts in "EHR-based phenotyping" have been focused on the development and validation of standardized phenotyping algorithms [48, 49]. For example, Phenotype KnowledgeBase (PheKB) [50], developed as part of the Electronic Medical Records and Genomics (eMERGE) Network [51], includes algorithms for conditions such as cataracts, dementia, and type 2 diabetes mellitus. This latter algorithm defines a set of data elements (e.g., lists of ICD-9-CM, RxNorm, and LOINC codes associated with both type 1 and type 2 diabetes mellitus), definitions (e.g., abnormal lab values such as a hemoglobin A1c ≥6.0 %), and logic for identifying cases and controls. Depending on the study, further criteria could be applied such as limiting to patients with encounters during a specified time period (e.g., January 1, 2012, to December 31, 2012) or a specific age group (e.g., children [<18] or adults [≥18]).
3. Once the cohort has been identified, the next step is *data extraction*. This process involves determining what data are needed for those patients, what is available in the EDW, and how to obtain and securely store the resulting dataset.

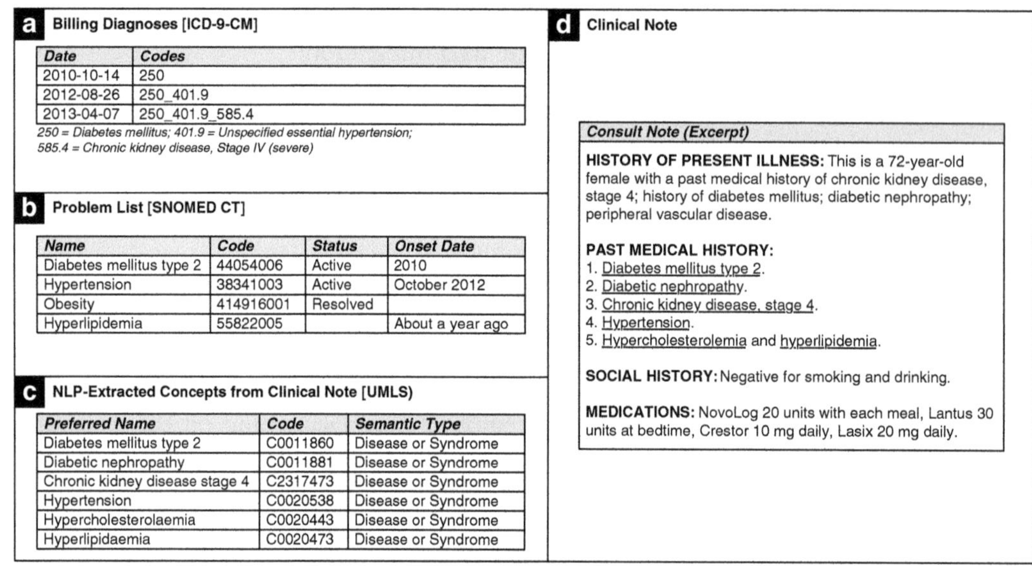

a Billing Diagnoses [ICD-9-CM]

Date	Codes
2010-10-14	250
2012-08-26	250_401.9
2013-04-07	250_401.9_585.4

250 = Diabetes mellitus; 401.9 = Unspecified essential hypertension; 585.4 = Chronic kidney disease, Stage IV (severe)

b Problem List [SNOMED CT]

Name	Code	Status	Onset Date
Diabetes mellitus type 2	44054006	Active	2010
Hypertension	38341003	Active	October 2012
Obesity	414916001	Resolved	
Hyperlipidemia	55822005		About a year ago

c NLP-Extracted Concepts from Clinical Note [UMLS]

Preferred Name	Code	Semantic Type
Diabetes mellitus type 2	C0011860	Disease or Syndrome
Diabetic nephropathy	C0011881	Disease or Syndrome
Chronic kidney disease stage 4	C2317473	Disease or Syndrome
Hypertension	C0020538	Disease or Syndrome
Hypercholesterolaemia	C0020443	Disease or Syndrome
Hyperlipidaemia	C0020473	Disease or Syndrome

d Clinical Note

Consult Note (Excerpt)

HISTORY OF PRESENT ILLNESS: This is a 72-year-old female with a past medical history of chronic kidney disease, stage 4; history of diabetes mellitus; diabetic nephropathy; peripheral vascular disease.

PAST MEDICAL HISTORY:
1. Diabetes mellitus type 2.
2. Diabetic nephropathy.
3. Chronic kidney disease, stage 4.
4. Hypertension.
5. Hypercholesterolemia and hyperlipidemia.

SOCIAL HISTORY: Negative for smoking and drinking.

MEDICATIONS: NovoLog 20 units with each meal, Lantus 30 units at bedtime, Crestor 10 mg daily, Lasix 20 mg daily.

Fig. 2 Example EHR data—billing diagnoses (**a**), problem list (**b**), and NLP-extracted concepts (**c**) from clinical note (**d**; content adapted from [52])

For the diabetes comorbidity study, potential data types include *billing diagnoses* where a single patient encounter may be associated with multiple ICD codes (Fig. 2a); *problem lists* where an entry may be associated with a name and code (from ICD or other coding systems such as Systematized Nomenclature of Medicine—Clinical Terms, SNOMED CT), status (e.g., active, inactive, or resolved), onset date, and other information (Fig. 2b); and *clinical notes* that typically include information such as note type (e.g., progress note, consult note, or discharge summary) and author type (e.g., attending or resident) as part of the metadata (Fig. 2d; adapted from [52]). Each of these data types offers a unique perspective (e.g., administrative vs. clinical and structured vs. unstructured) and could be analyzed individually for comparison or collectively.

4. As part of the formal EDW data request, the request type (e.g., aggregate, de-identified, or identified dataset), data types, specific data elements, and any additional restrictions should be provided for creating the data extract (*see* **Notes 3** and **4**). These additional restrictions could include retrieving billing diagnoses for a specific time period (e.g., last 5 years), limiting to active problem list entries, and obtaining only the most recent consult note. Depending on the request type, the resulting dataset should be stored in the appropriate environment for further analysis (*see* **Note 4**).

3.4 Preprocessing

Once the dataset is obtained, major preprocessing tasks include de-identification, data cleaning, and data enrichment.

1. In some cases, the dataset may already be de-identified; however, in other cases, there may be a need to perform *de-identification* if the dataset includes any of the 18 Protected Health Information (PHI) elements as defined by the Health Insurance Portability and Accountability Act (HIPAA) in the United States (*see* **Note 4**).
2. *Data cleaning* involves a set of tasks for handling noisy, incomplete, and inconsistent data in order to improve data quality (e.g., completeness, correctness, concordance, plausibility, and currency [53]) [54]. While EHR systems include error checking and other functionalities for improving data entry in some areas, there is still the possibility for erroneous and missing values. For example, in a dataset containing ages, there may be invalid values (e.g., "1BC" or "1.5.67"), outliers (e.g., "–100.0" or "1000000") or missing or null values. Depending on the intended use, options include keeping these cases (e.g., if the study goal is to identify comorbidities for the cohort as a whole) or removing these cases (e.g., if the goal is to compare comorbidities based on particular age groups).
3. In cases where data originate from different sources, there may be variation in content and format where standards could play a key role in facilitating data integration (*see* **Note 5**). For example, use of different timestamp formats (e.g., January 1, 2013, vs. 01/01/2013) where a standard like ISO 8601 for the representation of dates and time could be used (e.g., YYYY-MM-DD or YYYYMMDD) [55]. Another example of reformatting is separating data that may be concatenated within a single field such as in Fig. 2a where each diagnosis (represented by an ICD-9-CM code) is delimited by the "_" character.
4. *Data enrichment* is focused on enhancing the dataset, which may involve NLP, standards, and other external sources such as the Unified Medical Language System (UMLS) [56]. For datasets including unstructured data, NLP techniques can be used to extract, structure, and encode disease and other information from clinical notes (either the entire note or particular sections within the note) with UMLS concepts (*see* **Notes 2** and **5**). Figure 2c depicts the UMLS concepts (represented by Concept Unique Identifiers [CUI]) and semantic types identified from the Past Medical History section of the clinical note in Fig. 2d.
5. As described earlier, billing diagnoses, problem lists, and clinical notes can each provide disease information; however, there may be challenges integrating these sources due to use of

different coding systems (e.g., local terminology vs. ICD-9-CM vs. SNOMED CT) and different levels of granularity (e.g., "Diabetes mellitus" vs. "Diabetes mellitus type 2"). Resources such as the UMLS, which contains over 150 source vocabularies and provides linkages between them (e.g., mappings between ICD-9-CM and SNOMED CT), can be leveraged to facilitate integration and potentially handle granularity issues through the use of hierarchical relationships. For example, for hypertension, ICD-9-CM code 401.9 and SNOMED CT code 38341003 are linked through UMLS CUI C0020443 (Fig. 2a–c).

3.5 Transformation

After completing the preprocessing tasks, the next step is to reduce and convert the dataset in preparation for the data mining step.

1. Among the *data reduction* strategies are aggregation, generalization, and dimensionality reduction. In the context of the diabetes comorbidity study, a particular disease may be mentioned multiple times in the billing diagnoses across encounters, represented at different levels of granularity in the problem list, and mentioned throughout different clinical notes. If the interest is only if the disease is present rather than when the disease was present (e.g., patient had disease x at time y), then these multiple occurrences could be aggregated into a single occurrence (considering time and temporality is the focus of temporal data mining [57], which is out of scope for this chapter).
2. Generalization techniques can also be applied that involve transforming specific values to more general ones using domain concept hierarchies [58]. For diseases, this could involve using customized disease classes defined by clinical experts such as the PheWAS code groups for ICD-9-CM codes [45, 59], existing classification schemes such as the Clinical Classifications Software (CCS) for grouping ICD-9-CM codes [60], or leveraging the hierarchies in terminological systems such as ICD, SNOMED CT, and UMLS (*see* **Note 5**). For example, the ICD-9-CM hierarchy can be used to generalize specific codes to 250.0 (as depicted in Fig. 3) whereas CCS categorizes 250.00 and 250.01 (along with 10 other ICD-9-CM codes) into "Diabetes mellitus without complications" and 250.02 and 250.03 (along with 55 other ICD-9-CM codes) into "Diabetes mellitus with complications."
3. Dimensionality reduction can be used to limit the number of dimensions (or features or attributes) by only selecting or transforming those that are essential to the analysis [61]. For example, this could involve excluding attributes such as the status associated with a problem or combining the status value with the problem name (e.g., "Hypertension-*Active*" or "Hypertension-*Resolved*").

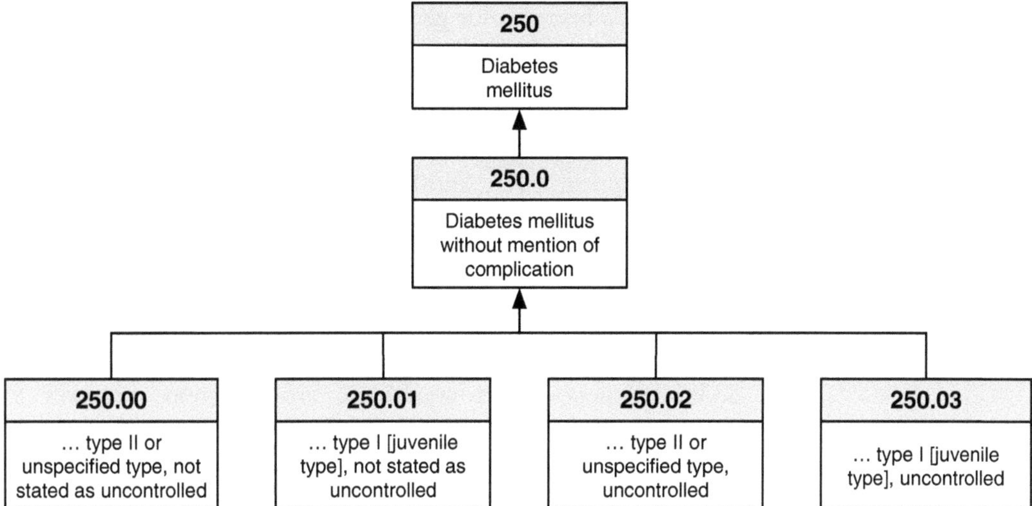

Fig. 3 Portion of ICD-9-CM hierarchy for diabetes mellitus

4. Depending on the data mining implementation, there may be specific formatting requirements for the dataset. To accommodate these and other implementation-specific requirements, *data conversion* needs to be performed. For example, the dataset may include one row per disease per patient whereas a specific data mining tool requires a single row per patient that includes the list of diseases separated by a particular delimiter (e.g., comma or tab).

3.6 Data Mining

Two primary goals of data mining are: (1) *prediction* for identifying patterns for predicting future behavior based on past behavior (e.g., as reflected in historical data) and (2) *description* for identifying patterns and relationships in data for further exploration [18, 62]. Predictive tasks include classification, regression, and time series analysis, while descriptive tasks include clustering, association rules, and sequence discovery. Given the broad array of methods and algorithms associated with each of these tasks, this section focuses on *association rules.* For comprehensive reviews of data mining methods that have been used in biomedicine and health care, *see* refs. 7, 9, 63.

1. Several open-source tools are available for performing data mining tasks and visualizing the results, including Weka and R that can be used for association rule mining [64, 65].
2. Association rule mining (or association rule learning or association rule generation) is a commonly used approach to discover interesting relationships between items in large datasets [62, 66]. An important consideration is that association rules convey co-occurrence relationships rather than causality but may

serve as a first step for generating hypotheses relating to causal relationships. Typically, association rule mining is performed on a dataset that includes a set of "transactions" (e.g., patients) where each transaction contains a subset of "items" (e.g., diseases) referred to as an "itemset." A number of algorithms exist (e.g., Apriori, Eclat, and FP-Growth [67, 68]) where the general algorithm is to first generate frequent itemsets (e.g., combinations of diseases that satisfy a specified threshold) and then generate rules in the form of $X \Rightarrow Y$ where X and Y are sets of items (e.g., {Diabetes mellitus type 2} ⇒ {Hypertension} or {Diabetes mellitus type 2, Hypertension} ⇒ {Obesity}).

3. While support and confidence are common measures for conveying the strength of each rule, there are a number of other established "interestingness" measures that can be used [69, 70]. For example, in a comparison of five common measures (support, confidence, chi-square [χ^2], interest [or lift], and conviction), χ^2 appeared to produce more accurate relationships [42] and has been the primary statistic used in several studies related to discovering disease associations in the EHR [36, 38, 41]. Other studies have used measures such as odds ratio [12, 13], relative risk [71], and relative reporting ratio [72].
4. A known challenge of association rule mining is the potentially intractable search space due to the generation of large numbers of rules that may be redundant or irrelevant [68]. Numerous algorithms and techniques exist for addressing these issues that involve filtering or pruning rules to produce a more limited and valid set of rules for further review. For example, as was described in Subheading 3.5, generalization techniques can be used to generate rules at different levels of granularity using concept hierarchies.
5. Another important consideration is ensuring that potentially important rules are not missed due to setting thresholds that are too high (e.g., for support and confidence values). This is known as the "rare item problem" where a variety of techniques have been proposed such as using multiple minimum support values [73].

3.7 Interpretation/Evaluation

1. Different *visualization techniques* can be used to facilitate exploration of the resulting patterns. For example, the ability to review and interact with patterns represented in a graphical network can provide the "big picture" as well as allow users to focus on particular parts of the network. There are a number of open-source visualization tools (e.g., GraphViz [74] and Cytoscape [75]) as well as data mining tools that include visualization (e.g., Weka [76] or Orange [77]).
2. For assessing the validity of patterns, one or more *clinical experts* are often involved who manually review the patterns

(all or a subset) to determine if each one is "known" or "unknown" (and thus potentially novel) as well as provide any interpretations or explanations. For example, in a study focused on identifying disease-finding associations in discharge summaries, the evaluation involved having two experts categorize each association as a "direct association," "indirect association," or "non-association" [36, 37].

3. Other validation approaches involve the use of established *medical knowledge sources* such as biomedical literature (e.g., PubMed/MEDLINE and CINAHL) and medical references (e.g., Micromedex and UpToDate). In one published study, searches for supporting literature in PubMed/MEDLINE were conducted for both well-known and unknown (or poorly known) disease associations identified from free-text problem list entries [13]. Another pair of studies involved the use of the Lexi-Comp drug reference database, Mosby's Diagnostic and Laboratory Test Reference, Harrison's Principles of Internal Medicine, eMedicine, and UpToDate to validate disease–drug and disease–laboratory associations generated from structured EHR data [42, 43]. Other studies have involved comparing disease–disease and disease–drug associations generated from EHR data with those obtained from biomedical literature using text mining approaches [12, 38, 39].
4. For the diabetes comorbidity study, a search for "Diabetes Mellitus, Type 2"[mh] AND "Hypertension"[mh] returns almost 6,000 articles (as of September 18, 2013), thus suggesting support for this association in the literature. Manual review of a subset of these articles would then be required to fully validate an association between these two diseases.

4 Notes

1. *EHR Systems*: As described in the 2003 Institute of Medicine report on "Key Capabilities of an Electronic Health Record System," criteria for these systems include the following: (1) improve patient safety, (2) support the delivery of effective patient care, (3) facilitate management of chronic conditions, (4) improve efficiency, and (5) feasibility of implementation [22]. To address these criteria, eight categories of core functionality for an EHR system are defined: (1) health information and data, (2) result management, (3) order entry/management, (4) decision support, (5) electronic communication and connectivity, (6) patient support, (7) administrative process, and (8) reporting and population health management [22]. Options for EHR implementation range from homegrown (e.g., StarChart/StarPanel at Vanderbilt University

Medical Center [78]) to commercial (e.g., Cerner [79] and Epic [80]) to open-source (e.g., OpenVista [81]) systems. While organizations such as Health Level 7 (HL7) [82] have been actively involved with defining and developing EHR standards, there is often variation in features and functions as EHR products evolve where particular systems may be more advanced in some areas than others [25]. In addition, an institution may have multiple EHR systems (e.g., one vendor for inpatient and another for outpatient) and institution-specific customizations may lead to variations in how and where data are collected. In considering secondary uses such as research, the heterogeneous nature of EHR systems needs to be considered and accounted for.

2. *NLP*: Automated methods such as NLP enable a new level of functionality by providing a means to extract, encode, and structure relevant information from free text in a timely fashion [83]. With respect to healthcare data, the development and use of NLP tools can facilitate the capture and exchange of essential information for a variety of subsequent uses ranging from patient care to quality reporting to research. A variety of NLP algorithms and systems have emerged over the last few decades to support the tasks of extracting information captured within clinical (e.g., discharge summaries and progress notes) and biomedical (e.g., literature) text for EHR enrichment, decision support, surveillance, terminology management, and text mining [84]. Many of these systems are focused on extracting named entities (e.g., diseases, medications, or procedures), identifying contextual information such as negation and temporality (e.g., "*no* hypertension" and "myocardial infarction *five years ago*"), and may also provide mappings to an appropriate standard such as ICD-9-CM, SNOMED CT, and UMLS [85, 86]. This standardized encoding of information provides domain knowledge and is essential for promoting interoperability and facilitating subsequent analyses such as data mining (*see* **Note 5**). Example systems include Medical Language Extraction and Encoding (MedLEE) system developed at Columbia University [87–89], MetaMap from the National Library of Medicine [90], clinical Text Analysis and Knowledge Extraction System (cTAKES) released by the Open Health Natural Language Processing Consortium [91], and Health Information Text Extraction (HITEx) that is part of the i2b2 framework [92] (*see* **Note 3**).

3. *Data Warehouses*: In recent years there has been increased attention to the development and use of data warehouses that integrate administrative, clinical, biomedical, and research data from EHR and other systems. Depending on the institution, different names may be used such as Clinical Data

Warehouse, Research Data Warehouse, EDW, Integrated Data Warehouse, or Integrated Data Repository [26]. One prominent example is the i2b2 (Informatics for Integrating Biology and the Bedside) framework that has been adopted by over 60 academic health centers nationally and internationally [93]. The open-source i2b2 platform can be used for de-identified cohort discovery and hypothesis testing by researchers as well as for creation of identified datasets with the requisite approvals (*see* **Note 4**). To enable federated queries between i2b2 instances, Shared Health Research Information Network (SHRINE) implementations can be created for data sharing and cohort identification across institutions [94]. In addition to i2b2, other examples include the Synthetic Derivative at Vanderbilt University [95], Stanford Translational Research Integrated Database Environment (STRIDE) at Stanford University [96], the Enterprise Data Trust at Mayo Clinic [97], Biomedical Translational Research Information System (BTRIS) at the National Institutes of Health [98], and Translational Research Informatics and Data Management Grid (TRIAD) at Ohio State University [99]. Many of these systems provide an interface that allows researchers to first search for cohorts based on specified criteria (e.g., demographics, ICD diagnosis codes, and CPT procedure codes). After a cohort has been identified, there is typically a formal request and approval process for obtaining a data extract containing de-identified or identified data (*see* **Note 4**).

4. *Data Privacy and Security*: With the increased sharing and use of electronic health data for research purposes, attention to data privacy and security issues is essential [100–103]. In the United States, the Health Insurance Portability and Accountability Act (HIPAA) is aimed at protecting the privacy of an individual's health information [104]. HIPAA includes two major rules: (1) Privacy Rule that governs the use and disclosure of PHI and (2) Security Rule that specifies a series of administrative, physical, and technical safeguards. Following the HIPAA Privacy Rule, a "de-identified" dataset is one where the 18 defined PHI elements (e.g., names, addresses, dates, ages, and unique identifying numbers) are removed using either the safe harbor method or expert determination while a "limited" dataset involves removal of 16 of these elements and maintains dates and geographic information [105, 106]. For datasets including unstructured data (e.g., clinical notes), text de-identification approaches are needed to remove or mask these PHI elements [107]. While a de-identified or a limited dataset typically include codes or "pseudonyms" for enabling re-linkages to individuals if needed, an "anonymized" dataset does not maintain any linkages [108]. However, despite

the use of de-identification and anonymization techniques, there may still be risks of misuse and re-identification [109]. Thus, researchers have the ethical and legal obligation as well as the responsibility to adhere to both local and federal regulations in the protection of human subjects and responsible conduct of research, including receiving the requisite training and conducting research in a HIPAA-compliant environment. For institutional data warehouses (*see* **Note 3**), there are typically several categories of requests that align with different levels of privacy: aggregate counts, limited dataset, de-identified dataset, and identified dataset [110]. Each of these request types may be associated with a set of required approvals (e.g., IRB) and forms (e.g., Data Use Agreement [DUA]).

5. *Standards*: To support semantic interoperability, there is a need for common representations and use of vocabulary standards for specifying the syntax and semantics of health information [111, 112]. Vocabulary standards include International Classification of Diseases (ICD) for diagnoses where the transition from ICD-9-CM (Ninth Revision, Clinical Modification) to ICD-10 is under way [113]; Current Procedural Terminology (CPT) for procedures [114]; LOINC for laboratory and other clinical observations [115, 116]; RxNorm for clinical drugs and drug delivery devices [117, 118]; SNOMED CT for comprehensive coding of health information [119]; and Medical Subject Headings (MeSH) for indexing biomedical literature [120]. The aforementioned standards are among the 150 source vocabularies integrated within the UMLS Metathesaurus, developed at the National Library of Medicine, that includes millions of concepts and their relationships, thus providing linkages between the source vocabularies [56, 121]. Complementary resources such as the BioPortal maintained at the National Center for Biomedical Ontology (NCBO) offer additional concepts from those vocabularies that may not be part of the UMLS [122, 123].

Acknowledgment

The example clinical note in Fig. 2d was obtained with permission from MTSamples (http://www.mtsamples.com). This work was supported in part by the National Library of Medicine of the National Institutes of Health under award number R01LM011364. The content is solely the responsibility of the authors and does not necessarily represent the official views of the National Institutes of Health.

References

1. Institute of Medicine (U.S.), Committee on Improving the Patient Record (eds), Dick RS, Steen EB, Detmer DE (1997) The computer-based patient record: an essential technology for health care. Revised edition. National Academy Press, Washington, DC
2. Stewart WF, Shah NR, Selna MJ, Paulus RA, Walker JM (2007) Bridging the inferential gap: the electronic health record and clinical evidence. Health Aff (Millwood) 26: w181–w191
3. Kohane IS (2011) Using electronic health records to drive discovery in disease genomics. Nat Rev Genet 12:417–428
4. Coorevits P, Sundgren M, Klein GO, Bahr A, Claerhout B, Daniel C et al (2013) Electronic health records: new opportunities for clinical research. J Intern Med 274(6):547–560
5. Kukafka R, Ancker JS, Chan C, Chelico J, Khan S, Mortoti S et al (2007) Redesigning electronic health record systems to support public health. J Biomed Inform 40:398–409
6. Denny JC (2012) Chapter 13: Mining electronic health records in the genomics era. PLoS Comput Biol 8:e1002823
7. Bath P (2004) Data mining in health and medical information. Annu Rev Inform Sci Technol 38:331–369
8. van Bemmel JH, van Mulligen EM, Mons B, van Wijk M, Kors JA, van der Lei J (2006) Databases for knowledge discovery. Examples from biomedicine and health care. Int J Med Inform 75:257–267
9. Iavindrasana J, Cohen G, Depeursinge A, Muller H, Meyer R, Geissbuhler A (2009) Clinical data mining: a review. Yearb Med Inform:121–133
10. Murdoch TB, Detsky AS (2013) The inevitable application of big data to health care. JAMA 309:1351–1352
11. Roque FS, Jensen PB, Schmock H, Dalgaard M, Andreatta M, Hansen T et al (2011) Using electronic patient records to discover disease correlations and stratify patient cohorts. PLoS Comput Biol 7:e1002141
12. Holmes AB, Hawson A, Liu F, Friedman C, Khiabanian H, Rabadan R (2011) Discovering disease associations by integrating electronic clinical data and medical literature. PLoS One 6:e21132
13. Hanauer DA, Rhodes DR, Chinnaiyan AM (2009) Exploring clinical associations using '-omics' based enrichment analyses. PLoS One 4:e5203
14. Wilson AM, Thabane L, Holbrook A (2004) Application of data mining techniques in pharmacovigilance. Br J Clin Pharmacol 57: 127–134
15. Wang X, Hripcsak G, Markatou M, Friedman C (2009) Active computerized pharmacovigilance using natural language processing, statistics, and electronic health records: a feasibility study. J Am Med Inform Assoc 16:328–337
16. Harpaz R, Perez H, Chase HS, Rabadan R, Hripcsak G, Friedman C (2011) Biclustering of adverse drug events in the FDA's spontaneous reporting system. Clin Pharmacol Ther 89:243–250
17. Wilke RA, Xu H, Denny JC, Roden DM, Krauss RM, McCarty CA et al (2011) The emerging role of electronic medical records in pharmacogenomics. Clin Pharmacol Ther 89:379–386
18. Fayyad U, Piatetsky-Shapiro G, Smyth P (1996) From data mining to knowledge discovery in databases. AI Mag 17:37–54
19. Fayyad U, Piatetsky-Shapiro G, Smyth P (1996) The KDD process for extracting useful knowledge from volumes of data. Commun ACM 39:27–34
20. Hearst M (1999) Untangling text data mining. Proceedings of the 37th annual meeting of the Association for Computational Linguistics on computational linguistics, pp 3–10
21. Zweigenbaum P, Demner-Fushman D, Yu H, Cohen KB (2007) Frontiers of biomedical text mining: current progress. Brief Bioinform 8:358–375
22. Institute of Medicine (2003) Key capabilities of an electronic health record system. National Academies Press, Washington, DC
23. National Institutes of Health National Center for Research Resources and MITRE Corporation (2006) Electronic health records overview. http://www.himss.org/files/HIMSSorg/content/files/Code%20180%20MITRE%20Key%20Components%20of%20an%20EHR.pdf
24. ASTM Standard E1384 (2013) Standard guide for content and structure of the Electronic Health Record (EHR). ASTM International, West Conshohocken, PA
25. Carter J (2008) Electronic health records for clinicians and administrators: infrastructure and supporting technologies. In: Carter J (ed) Electronic health records, 2nd edn. American College of Physicians, Philadelphia, PA
26. MacKenzie SL, Wyatt MC, Schuff R, Tenenbaum JD, Anderson N (2012) Practices and perspectives on building integrated data repositories: results from a 2010 CTSA survey. J Am Med Inform Assoc 19: e119–e124
27. http://mimic.physionet.org/
28. Scott DJ, Lee J, Silva I, Park S, Moody GB, Celi LA et al (2013) Accessing the public

MIMIC-II intensive care relational database for clinical research. BMC Med Inform Dec Mak 13:9
29. https://i2b2.org/NLP/DataSets/
30. Uzuner O, Luo Y, Szolovits P (2007) Evaluating the state-of-the-art in automatic de-identification. J Am Med Inform Assoc 14:550–563
31. Uzuner O, Goldstein I, Luo Y, Kohane I (2008) Identifying patient smoking status from medical discharge records. J Am Med Inform Assoc 15:14–24
32. Uzuner O, Solti I, Cadag E (2010) Extracting medication information from clinical text. J Am Med Inform Assoc 17:514–518
33. Ohno-Machado L, Bafna V, Boxwala AA, Chapman BE, Chapman WW, Chaudhuri K et al (2012) iDASH: integrating data for analysis, anonymization, and sharing. J Am Med Inform Assoc 19:196–201
34. http://idash.ucsd.edu/data-repository-0
35. Ackoff R (1989) From data to wisdom. J Appl Syst Anal 16:3–9
36. Cao H, Markatou M, Melton GB, Chiang MF, Hripcsak G (2005) Mining a clinical data warehouse to discover disease-finding associations using co-occurrence statistics. AMIA Annu Symp Proc:106–110
37. Cao H, Hripcsak G, Markatou M (2007) A statistical methodology for analyzing co-occurrence data from a large sample. J Biomed Inform 40:343–352
38. Chen ES, Hripcsak G, Xu H, Markatou M, Friedman C (2008) Automated acquisition of disease drug knowledge from biomedical and clinical documents: an initial study. J Am Med Inform Assoc 15:87–98
39. Chen ES, Stetson PD, Lussier YA, Markatou M, Hripcsak G, Friedman C (2007) Detection of practice pattern trends through Natural Language Processing of clinical narratives and biomedical literature. AMIA Annu Symp Proc:120–124
40. Wang X, Hripcsak G, Friedman C (2009) Characterizing environmental and phenotypic associations using information theory and electronic health records. BMC Bioinforma 10(Suppl 9):S13
41. Wang X, Chase H, Markatou M, Hripcsak G, Friedman C (2010) Selecting information in electronic health records for knowledge acquisition. J Biomed Inform 43:595–601
42. Wright A, Chen ES, Maloney FL (2010) An automated technique for identifying associations between medications, laboratory results and problems. J Biomed Inform 43: 891–901
43. Wright A, Pang J, Feblowitz JC, Maloney FL, Wilcox AR, Ramelson HZ et al (2011) A method and knowledge base for automated inference of patient problems from structured data in an electronic medical record. J Am Med Inform Assoc 18:859–867
44. Doddi S, Marathe A, Ravi SS, Torney DC (2001) Discovery of association rules in medical data. Med Inform Internet Med 26: 25–33
45. Denny JC, Ritchie MD, Basford MA, Pulley JM, Bastarache L, Brown-Gentry K et al (2010) PheWAS: demonstrating the feasibility of a phenome-wide scan to discover gene-disease associations. Bioinformatics 26:1205–1210
46. Mullins IM, Siadaty MS, Lyman J, Scully K, Garrett CT, Miller WG et al (2006) Data mining and clinical data repositories: insights from a 667,000 patient data set. Comput Biol Med 36:1351–1377
47. Concaro S, Sacchi L, Cerra C, Fratino P, Bellazzi R (2011) Mining health care administrative data with temporal association rules on hybrid events. Methods Inf Med 50: 166–179
48. Newton KM, Peissig PL, Kho AN, Bielinski SJ, Berg RL, Choudhary V et al (2013) Validation of electronic medical record-based phenotyping algorithms: results and lessons learned from the eMERGE network. J Am Med Inform Assoc 20:e147–e154
49. Richesson RL, Hammond WE, Nahm M, Wixted D, Simon GE, Robinson JG et al (2013) Electronic health records based phenotyping in next-generation clinical trials: a perspective from the NIH Health Care Systems Collaboratory. J Am Med Inform Assoc 20(e2):e226–e231
50. http://www.phekb.org/
51. Gottesman O, Kuivaniemi H, Tromp G, Faucett WA, Li R, Manolio TA et al (2013) The Electronic Medical Records and Genomics (eMERGE) Network: past, present, and future. Genet Med 15(10):761–771
52. http://www.mtsamples.com/site/pages/sample.asp?type=97-Consult%20-%20History%20and%20Phy.&sample=2063-Gen%20Med%20Consult%20-%2049
53. Weiskopf NG, Weng C (2013) Methods and dimensions of electronic health record data quality assessment: enabling reuse for clinical research. J Am Med Inform Assoc 20: 144–151
54. Rahm E, Do H (2000) Data cleaning: problems and current approaches. IEEE Data Eng Bull 23:3–13
55. http://www.w3.org/TR/NOTE-datetime
56. Bodenreider O (2004) The Unified Medical Language System (UMLS): integrating biomedical terminology. Nucleic Acids Res 32:D267–D270
57. Post AR, Harrison JH Jr (2008) Temporal data mining. Clin Lab Med 28:83–100, vii

58. Carter C, Hamilton H (1995) A fast, on-line generalization algorithm for knowledge discovery. Appl Math Lett 8:5–11
59. http://knowledgemap.mc.vanderbilt.edu/research/content/phewas
60. http://www.hcup-us.ahrq.gov/toolssoftware/ccs/ccs.jsp
61. Liu H, Motoda H (1998) Feature extraction, construction and selection: a data mining perspective. Kluwer, Boston
62. Dunham MH (2003) Data mining introductory and advanced topics. Prentice Hall, Upper Saddle River, NJ
63. Sarkar IN (2013) Methods in biomedical informatics: a pragmatic approach, 1st edn. Academic, New York
64. Zupan B, Demsar J (2008) Open-source tools for data mining. Clin Lab Med 28:37–54, vi
65. http://www.kdnuggets.com/software/index.html
66. Tan P-N, Steinbach M, Kumar V (2006) Introduction to data mining, 1st edn. Pearson Addison Wesley, Boston
67. Agrawal R, Srikant R (1994) Fast algorithms for mining association rules in large databases. Proceedings of the 20th International conference on very large data bases, pp 487–499
68. Hipp J, Guntzer U, Nakhaeizadeh G (2000) Algorithms for association rule mining—a general survey and comparison. ACM SIGKDD Explor Newslett 2:58–64
69. Tan P, Kumar V, Srivastava J (2002) Selecting the right interestingness measure for association patterns. Proceedings of the 8th ACM SIGKDD International conference on knowledge discovery and data mining, pp 32–41
70. Ohsaki M, Abe H, Tsumoto S, Yokoi H, Yamaguchi T (2007) Evaluation of rule interestingness measures in medical knowledge discovery in databases. Artif Intell Med 41:177–196
71. Hidalgo CA, Blumm N, Barabasi AL, Christakis NA (2009) A dynamic network approach for the study of human phenotypes. PLoS Comput Biol 5:e1000353
72. Harpaz R, Chase HS, Friedman C (2010) Mining multi-item drug adverse effect associations in spontaneous reporting systems. BMC Bioinforma 11(Suppl 9):S7
73. Liu B, Hsu W, Ma Y (1999) Mining association rules with multiple minimum supports. KDD '99 Proceedings of the 5th ACM SIGKDD International conference on knowledge discovery and data mining, pp 337–341
74. http://www.graphviz.org/
75. http://www.cytoscape.org/
76. http://www.cs.waikato.ac.nz/ml/weka/
77. http://orange.biolab.si/
78. http://informatics.mc.vanderbilt.edu/archives/starchart
79. http://cerner.com/
80. http://www.epic.com/
81. http://medsphere.com/vista-to-openvista
82. http://www.hl7.org
83. Friedman C, Johnson S (2006) Natural language and text processing in biomedicine. In: Shortliffe E, Cimino JJ (eds) Biomedical informatics computer applications in health care and biomedicine, 3rd edn. Springer, New York
84. Meystre SM, Savova GK, Kipper-Schuler KC, Hurdle JF (2008) Extracting information from textual documents in the electronic health record: a review of recent research. Yearb Med Inform:128–144
85. Cimino JJ (1996) Review paper: coding systems in health care. Methods Inf Med 35: 273–284
86. Cimino JJ, Zhu X (2006) The practical impact of ontologies on biomedical informatics. Yearb Med Inform:124–135
87. Friedman C (2000) A broad-coverage natural language processing system. Proc AMIA Symp:270–274
88. Friedman C, Hripcsak G, Shagina L, Liu H (1999) Representing information in patient reports using natural language processing and the extensible markup language. J Am Med Inform Assoc 6:76–87
89. Friedman C, Shagina L, Lussier Y, Hripcsak G (2004) Automated encoding of clinical documents based on natural language processing. J Am Med Inform Assoc 11:392–402
90. Aronson AR, Lang FM (2010) An overview of MetaMap: historical perspective and recent advances. J Am Med Inform Assoc 17:229–236
91. Savova GK, Masanz JJ, Ogren PV, Zheng J, Sohn S, Kipper-Schuler KC et al (2010) Mayo clinical Text Analysis and Knowledge Extraction System (cTAKES): architecture, component evaluation and applications. J Am Med Inform Assoc 17:507–513
92. Zeng QT, Goryachev S, Weiss S, Sordo M, Murphy SN, Lazarus R (2006) Extracting principal diagnosis, co-morbidity and smoking status for asthma research: evaluation of a natural language processing system. BMC Med Inform Dec Mak 6:30
93. Kohane IS, Churchill SE, Murphy SN (2012) A translational engine at the national scale: informatics for integrating biology and the bedside. J Am Med Inform Assoc 19: 181–185
94. McMurry AJ, Murphy SN, MacFadden D, Weber G, Simons WW, Orechia J et al (2013) SHRINE: enabling nationally scalable multisite disease studies. PLoS One 8:e55811

95. Roden DM, Pulley JM, Basford MA, Bernard GR, Clayton EW, Balser JR et al (2008) Development of a large-scale de-identified DNA biobank to enable personalized medicine. Clin Pharmacol Ther 84:362–369
96. Lowe HJ, Ferris TA, Hernandez PM, Weber SC (2009) STRIDE—an integrated standards-based translational research informatics platform. AMIA Annu Symp Proc 2009:391–395
97. Chute CG, Beck SA, Fisk TB, Mohr DN (2010) The Enterprise Data Trust at Mayo Clinic: a semantically integrated warehouse of biomedical data. J Am Med Inform Assoc 17:131–135
98. Cimino JJ, Ayres EJ (2010) The clinical research data repository of the US National Institutes of Health. Stud Health Technol Inform 160:1299–1303
99. Payne P, Ervin D, Dhaval R, Borlawsky T, Lai A (2011) TRIAD: the Translational Research Informatics and Data Management Grid. Appl Clin Inform 2:331–344
100. Wylie JE, Mineau GP (2003) Biomedical databases: protecting privacy and promoting research. Trends Biotechnol 21:113–116
101. Malin B, Karp D, Scheuermann RH (2010) Technical and policy approaches to balancing patient privacy and data sharing in clinical and translational research. J Investig Med 58: 11–18
102. Krishna R, Kelleher K, Stahlberg E (2007) Patient confidentiality in the research use of clinical medical databases. Am J Public Health 97:654–658
103. Berman JJ (2002) Confidentiality issues for medical data miners. Artif Intell Med 26: 25–36
104. http://www.hhs.gov/ocr/privacy/index.html
105. Gunn PP, Fremont AM, Bottrell M, Shugarman LR, Galegher J, Bikson T (2004) The Health Insurance Portability and Accountability Act Privacy Rule: a practical guide for researchers. Med Care 42:321–327
106. Nosowsky R, Giordano TJ (2006) The Health Insurance Portability and Accountability Act of 1996 (HIPAA) privacy rule: implications for clinical research. Annu Rev Med 57:575–590
107. Meystre SM, Friedlin FJ, South BR, Shen S, Samore MH (2010) Automatic de-identification of textual documents in the electronic health record: a review of recent research. BMC Med Res Methodol 10:70
108. Kushida CA, Nichols DA, Jadrnicek R, Miller R, Walsh JK, Griffin K (2012) Strategies for de-identification and anonymization of electronic health record data for use in multicenter research studies. Med Care 50(Suppl): S82–S101
109. El Emam K, Jonker E, Arbuckle L, Malin B (2011) A systematic review of re-identification attacks on health data. PLoS One 6:e28071
110. Murphy SN, Gainer V, Mendis M, Churchill S, Kohane I (2011) Strategies for maintaining patient privacy in i2b2. J Am Med Inform Assoc 18(Suppl 1):i103–i108
111. Hammond WE (2005) The making and adoption of health data standards. Health Aff (Millwood) 24:1205–1213
112. Chen ES, Melton GB, Sarkar IN (2012) Translating standards into practice: experiences and lessons learned in biomedicine and health care. J Biomed Inform 45:609–612
113. http://www.who.int/classifications/icd/en/
114. http://www.ama-assn.org/go/cpt
115. http://loinc.org/
116. Vreeman DJ, McDonald CJ, Huff SM (2010) LOINC(R)—a universal catalog of individual clinical observations and uniform representation of enumerated collections. Int J Funct Inform Personal Med 3:273–291
117. http://www.nlm.nih.gov/research/umls/rxnorm/
118. Nelson SJ, Zeng K, Kilbourne J, Powell T, Moore R (2011) Normalized names for clinical drugs: RxNorm at 6 years. J Am Med Inform Assoc 18:441–448
119. http://www.ihtsdo.org/snomed-ct/
120. http://www.nlm.nih.gov/mesh/
121. http://www.nlm.nih.gov/research/umls/
122. http://bioportal.bioontology.org/
123. Noy NF, Shah NH, Whetzel PL, Dai B, Dorf M, Griffith N et al (2009) BioPortal: ontologies and integrated data resources at the click of a mouse. Nucleic Acids Res 37: W170–W173

Index

A

B

C

D

E

G

H

I

Vinod D. Kumar and Hannah Jane Tipney (eds.), *Biomedical Literature Mining*, Methods in Molecular Biology, vol. 1159, DOI 10.1007/978-1-4939-0709-0, © Springer Science+Business Media New York 2014

FSC
www.fsc.org
MIX
Papier aus verantwortungsvollen Quellen
Paper from responsible sources
FSC® C105338

If you have any concerns about our products,
you can contact us on
ProductSafety@springernature.com

In case Publisher is established outside the EU,
the EU authorized representative is:
Springer Nature Customer Service Center GmbH
Europaplatz 3, 69115 Heidelberg, Germany

Printed by Libri Plureos GmbH
in Hamburg, Germany